Information Technology

For Digital Library Management and Automation

Information Technology

For Digital Library Management and Automation

V.K. Jain

B.E., M.Tech., FIE, FIETE, MISHRAE

ATLANTIC

PUBLISHERS & DISTRIBUTORS (P) LTD

Published by
ATLANTIC
PUBLISHERS & DISTRIBUTORS (P) LTD
7/22, Ansari Road, Darya Ganj,
New Delhi-110002
Phones : 2327 3880, 23275880, 23280451
Fax : +91-11-23285873
Web : www.atlanticbooks.com
E-mail : info@atlanticbooks.com

Branch Office
5, Nallathambi Street, Wallajah Road,
Chennai-600 002
Phones : +91-44-64611085, 32413319
E-mail : chennai@atlanticbooks.com

ISBN:978-81-269-1014-4

Printed in India at Nice Printing Press, A-33/3A, Site-IV, Industrial Area, Sahibabad, Ghaziabad, U.P.

Preface

Library Science is an interdisciplinary subject incorporating the humanities, law and applied science to study topics related to libraries, the collection, organization, preservation and dissemination of information resources, and the political economy of information. Academic courses in Library Science typically include collection management, information systems and technology, cataloging and classification, preservation, reference, statistics and management.

Computer networks and electronic information systems are essential resources required for the progress of society in an age of Information Technology. These resources are valuable community assets to be used and managed responsibly to ensure their integrity, security, and availability for appropriate educational and business activities. All authorized users of these resources are required to use them in an effective, efficient, and responsible manner. This book has been composed keeping in view the above needs.

Computer science has the greatest drawback in its non-standardization of the stuff not only in terms of software but in courseware also. A vast amount of curriculum exists at different levels. The UGC has done a commendable job by accrediting the courses and thus prescribing the syllabi for different levels.

This book has been composed as per the syllabi for degree level (also MCA) courses prescribed by various universities and institutions including IITs and NIITs, etc.

The various facets of computer science are changing its attributes with such a fast speed that it is very difficult to keep track of the latest developments and be aware of the latest happenings in the field. Therefore, in presenting the state-of-the-art textware, immense help has been sought from various sources. I express my deep gratitude to the authors from whose works the information has been derived.

I acknowledge with great sense of gratitude the hard work put in by my wife Heera Jain. I also thank Jyotsna Agarwal for her hard work and cooperation in proofreading and preparing figures for the manuscript.

V.K. Jain

Contents

1
UNIT

Information Technology

1. 1 INFORMATION TECHNOLOGY

Information is the representation of data (or raw facts) to a receiver; it is the principal resource of a modern organization. Information is data in a usable form, processed in some way; it is data plus interpretation. Technology is the combination of skills, knowledge, abilities, materials, machines, and tools that people use to convert or change raw materials into valuable goods and services. Information Technology (IT) is technology that is used to store, communicate, and manipulate information.

Information technology includes computer hardware and software, operating systems, web-based information and applications, telephones and other telecommunications products, video equipment and multimedia products, information kiosks, and office products such as photocopiers and fax machines.

Informally, all of these devices are commonly referred to simply as "information technology", or "IT". However, from a legal standpoint, there was a need to expand upon an existing definition of information technology, while maintaining consistency with that early definition.

1.1.1 Definition of Information Technology

Any equipment or interconnected system or subsystem of equipment, that is used in the automatic acquisition, storage, manipulation, management, movement, control, display, switching, interchange, transmission, or reception of data or information. The term 'information technology' includes computers, ancillary equipment, software, firmware and similar procedures, services (including support services), and related resources.

According to Indian Information Technology Act, 2000. "Data" means a representation of information, knowledge, facts, concepts or instructions which are being prepared or have been prepared in a formalised manner, and is intended to be processed, is being processed or has been processed in a computer system or computer network, and may be in any form (including computer printouts magnetic or optical storage media, punched cards, punched tapes) or stored internally in the memory of the computer; "Information" includes data, text, images, sound, voice, codes, computer programmes, software and databases or micro film or computer generated micro fiche.

Electronic and information technology is a term used in the 1998 amendments to Section 508 of the Rehabilitation Act of USA. The term is used to define the scope of products covered under Section 508. Section 508 requires that electronic and information technology that is developed, procured, maintained, or used by the federal government be accessible.

1.2 NEED OF IT

Modern civilisation has become so complicated and sophisticated that to survive one has to be competitive. This compels the people to keep himself informed of all types of happenings in the society. With the advent of educational reforms in society, mankind is surrounded with a vast amount of data available. In USA alone about 1 trillion documents are being created every year and this rate is increasing by 70 per cent every year. Modern business management system has also rendered itself to bulk collection of data from various sources which needs to be rearranged in a fashion so that it can be utilized with minimum possible time. All this needs a high amount of filing either at data stage or at information stage. No office can be without files. If you go to any tax collection department or municipal office you will find a high amount of files stacked here and there. Modern rules, regulation and law requires every transaction to happen in a written form, may be an agreement, application, voucher, bill, letter, memo, order etc.

Paper files require a high amount of storage space and paper storage creates several other problems like fire risk, spoilage and deterioration by way of aging micro-organism and humidity etc. Computerization of documents and files has solved this problem to a great extent. Not only this, but it has led to a high amount of relaxation to human mind, as everything can be automated. Now the document can be created in a number of ways, styles and any number of documents can be created as and when required, without making any mistake.

Humanity is currently leading to office automation where every transaction in business is conducted through computers. Electronic Data Processing (EDP) systems makes and stores documents in magnetic media *i.e.* the soft copy, which is in very condensed form of storage of information. A floppy disk contains information worth 300 typed pages and CD ROM (laser disk) contains matter worth 45000 typed papers.

The urge for converting facts into useful information is not a phenomenon of modern life. Throughout history, and even pre-history, people have found it necessary to sort data into forms that were easier to understand. For example, the ancient Egyptians recorded the ebb and flow of the Nile River and used this information to predict yearly crop yields. Today, computers convert data about land and water into recommendations to farmers on crop planting. The great Mougal emperor Akbar appointed Raja Todarmal to keep record of property possessed by all farmers and citizens to keep track of personal and state's properties. That seems to be first ever database organized in written manner by Indians. In modern days information is needed to run man's own livelihood, to run a system or process or to command a business. Information is needed to:

- Gain knowledge about the surroundings, and whatever is happening in the society and universe.
- Keep the system up-to-date.
- Know about the rules and regulations and bye-laws of society, local government, provincial and central government, associations, clients etc., as ignorance is no bliss.

Based on above three, to arrive at a particular decision for planning current and prospective actions in process of forming, running and protecting a process or system.

Organizations use technology in general and IT in particular to become more efficient, more effective, and to innovate. In the stone age of IT, when the mainframe computer was the Center of the

Universe, there was the EDP Department. EDP stood for Electronic Data Processing and that is all it did: processing data using an electronic device called a computer. This had a purely supporting role, which was usually to assist the accounting department.

In the beginning of the 1980s, a silent revolution took place: IBM introduced its first Personal Computer (PC) and published the technical details of this machine, thus opening the possibility for other companies to make IBM compatible computers and cheap clones. These devices were first used by individuals in their home environment but quickly infiltrated organizations in the offices and on the production floors. At first, the EDP professionals did not know how to cope with this phenomenons. Some thought it was just hype that would disappear as fast as it came, others simply ignored it and some began a war against these intruders. After a while, it became clear that the PC was there to stay; it had conquered its place in the organization. First as a stand-alone workstation, but rapidly the need to act as a terminal for the mainframe arose.

A few years later, two other evolutions took place: the introduction of departmental systems and Local Area Networks (LANs). These technologies allowed people to share information and work together in small workgroups or company wide. These evolutions had another consequence: where the good old EDP department only had one "client" within the organization, the newly born IT department suddenly had to cope with the whole organization. Almost every white-collar or blue-collar employee had a PC connected to the network. Where the users of the traditional EDP systems were specialists in their branch and were consequently well trained to work with these systems, the newbies only had a very limited knowledge of computers. Specialized services like training classes and help desks had to be set up by the IT department. Because of the improved productivity of the employees and because of the new ways to communicate and collaborate, the traditional hierarchies within the organizations began to fall apart. Terms like flat organizations and horizontal departments became common. Also, and this is an important issue, the dependency of IT increased dramatically.

Almost every aspect of the business was supported by or based on Information Technology. As more and more aspects of the business were covered by IT, this technology was more and more used to get a better insight in the business that was supported. Management

Information Systems, Business Intelligence, Data Warehousing and Data Mining became part of the normal vocabulary of the business managers.

In the meantime, another technological evolution took place: networks and the digitalization of telecommunications. People and their computers started to collaborate over networks and the analog communication technology was digitized. Because of this technological convergence many organizations took the decision to put communications under the same person who was in charge of the data processing business.

Today, IT is very close to or even part of the core business in many organizations. This is the case in service organizations (banking, insurance, publishing, government...) whose non-material products cannot exist without IT. Think of a simple bank account: this is a purely imaginary object without any physical manifestation; it only exists on the disks and in the memories of the bank's computers. The modern stock exchange is another example; all the transactions are merely bits and bytes floating around in the networks of the stockbrokers and again no physical exchanges take place. In many organizations, this is reflected by the presence of a representative in the senior management team: the Chief Information Officer (CIO). The CIO is in charge of both technology managers and business managers and is the link between these two. In other organizations, where IT is merely a support function, the IT Manager plays his role as the line manager of a technologically oriented department.

IT-Enabled Services, or Remote Processing, has emerged as the next major driver of the technology services industry. With competitive telecommunication costs, well developed infrastructure and a huge pool of English-speaking and computer-literate graduate manpower, India rates higher than many other countries as a hub for IT-enabled services. Call centers and business process outsourcing is emerging as the next wave of growth in India. The McKinsey study indicated that India will earn $18 billion in revenues through these services and create additional employment of one million jobs. Already companies like GE Caps, British Airways, Swissair, American Express and British Telecom are using Indian companies for these services.

The IT-Enabled Services segment currently employs around 70,000 people and accounts for 10.6 per cent of the total IT software and services industry revenues.

Outsourcing software requirements depends mostly on quality of services, and quality has continued to remain a prime edge for Indian software companies. According to the NASSCOM survey, the Indian software industry continued to win recognition for its quality in software development over the last year. Out of the top 400 companies, more than 250 have already acquired ISO 9000 certifications. However, it is in the SEI CMM (Software Engineering Institution Capability Maturity Model) Level 5 segment that the real accolades have come Indias way. Out of the 54 companies in the world that acquired SEI CMM Level 5 certification, 27 companies are located in India.

1.3 SCOPE OF INFORMATION TECHNOLOGY

Information Technology includes information technology and any equipment or interconnected system or subsystem of equipment, that is used in the creation, conversion, or duplication of data or information. The term electronic and information technology includes, but is not limited to, telecommunications products (such as telephones), information kiosks and transaction machines, World Wide Web sites, multimedia, and office equipment such as copiers and fax machines. The term does not include any equipment that contains embedded information technology that is used as an integral part of the product, but the principal function of which is not the acquisition, storage, manipulation, management, movement, control, display, switching, interchange, transmission, or reception of data or information. For example, HVAC (heating, ventilation, and air conditioning) equipment such as thermostats or temperature control devices, and medical equipment where information technology is integral to its operation, are not information technology.

Changes in the hardware, software and telecommunication technologies play a major role in the way our society is evolving. During the last decade, the rate of change in information technology has increased. Indeed it is clear that we are now entering an era where explosive change in telecommunication technology combined with ever increasing computing power will lead to profound changes in information systems that support our organizations. These changes will affect the way our organizations function, will lead to new business opportunities and will create a need for new non-profit organizations. Governments and international organizations do and will have to scramble to create policies and laws for control of public goods and services such as airwaves and public networks. Educational institutions will continue to change the content of educational materials they deliver to include new knowledge and skills. In addition these institutions will change the delivery mechanisms for disseminating these materials.

An information system environment may be seen as the domain of two or more interacting human activity systems, *e.g.* teams, functional areas, departments, etc. These systems may be parts of a larger system or may be conceived as autonomous systems. In any case the boundaries of a particular information systems environment may be defined in terms of dependencies between the concerned human activity systems. The source of these dependencies may be:

- Functional, *i.e.* participating in some task or sharing some common resource.
- Morphological, *i.e.* sharing the same concepts, language, information, norms , values, etc.
- Structural, *i.e.* obeying the same rules, directives and authority.

Human activity systems are able to:

1. Acquire and process information
2. Adapt to their environments
3. Cope with conflicts
4. Create their own culture that results in decisions as well as being determined by them
5. Learn and fail to learn.

The dependencies between human activity systems may be supportive or conflicting. In any case these dependencies may be seen as the source of uncertainty. Uncertainty reflects a state of doubt or ambiguity. The more the conceived uncertainty the more the need for cooperation, coordination and information exchange between the involved human activity systems.

1.3.1 Human Activity Systems as Information Systems

Generally speaking, every human activity system forms a well delimited information system. Cognitive limits, interests, experiences, specialized knowledge, social status, responsibilities, etc. may be seen as the natural grounds for the infological closure of any human activity system. There are two parts that together form a human activity system. The first part called actors system is characterized of its ability to perform certain task in a correct way. The second part called information system is characterized of its ability to systematically collect, organize and process information in order to provide the correct answers to some stated issues.

Thus, the main distinction between an actors' system and a formal information system is derived from the concept of knowledge. In this sense every human activity system may be conceived as information systems or information environments, performing programmed and non-programmed informational tasks. There are many fruitful theoretical models which describe how human beings acquire and process information. There are also many interesting theoretical models which describe how to organize the enterprise of formal information systems.

1.4 THE OBJECTIVE OF INFORMATION TECHNOLOGY

The objective of IT is to promote and encourage interactions among professionals from practice and research and advancement of investigation of concepts, methods, techniques, tools, and issues related to information systems in organisations.

In addition to the general function of improving knowledge, information assists management in several ways including.

- **The reduction of uncertainty:** Uncertainty exists where there is less perfect knowledge. Rarely if ever is there perfect knowledge but relevant information helps to reduce the unknown. This is particularly relevant in planning and decision-making.
- **As an aid to monitoring and control:** By providing information about performance and the extent of deviations from planned levels of performance, management are better able to control operation.
- **As a means of communication:** Managers need to know about developments, plans, forecasts, impending changes and so on.
- **As a memory supplement:** By having historical information about performance, transactions, results of past actions and decisions available for reference, personal memories are supplemented.
- As an aid to simplification by reducing uncertainty and enhancing understanding, problems and situations are simplified and become more manageable.

2.1 EVOLUTION OF COMPUTER

Counting calculating and computing devices has been an essential component of the process of calculation and data storage since it became useful for numerical values to be processed and shared. The earliest counting hardware was probably some form of tally stick; later recording devices include the Phoenician clay shapes which represented counts of items, probably livestock or grains, in containers. These seem to have been used by the merchants, accountants, and government officials of the time.

The history of calculation starts from about 2000 years ago, at the birth of the abacus. Chinese and others frustrated with counting on their fingers invented the abacus. Humanity has used devices to aid in computation for millennia. One example is a device for establishing equality by weight: the classic scales, later used to symbolize equality in justice. Another is simple enumeration: the checkered cloths of the counting houses served as simple data structures for enumerating stacks of coins, by weight. A more arithmetic-oriented machine is the abacus. One of the earliest machines of this type was the Chinese abacus.

Abacus is a wooden rack holding two horizontal wires with beads strung on them. When these beads are moved around, according to programming rules memorized by the user, all regular arithmetic problems can be done.

The suanpan of the Chinese is similar to the Roman abacus in principle, though has a different construction.

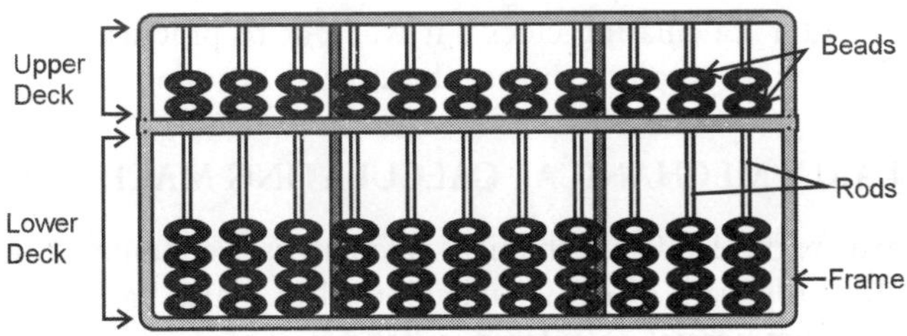

Fig. 1.1: Abacus

Chinese abacus can be used for functions other than counting. Unlike the simple counting board used in elementary schools, very efficient suanpan techniques have been developed to do multiplication, division, addition, subtraction, square root and cube root operations at high speed.

Bead arithmetic is the calculating technique used with various types of abaci, in particular the Chinese abacus. The beads in the upper deck has a value of 5 and that in the lower deck of unit value.

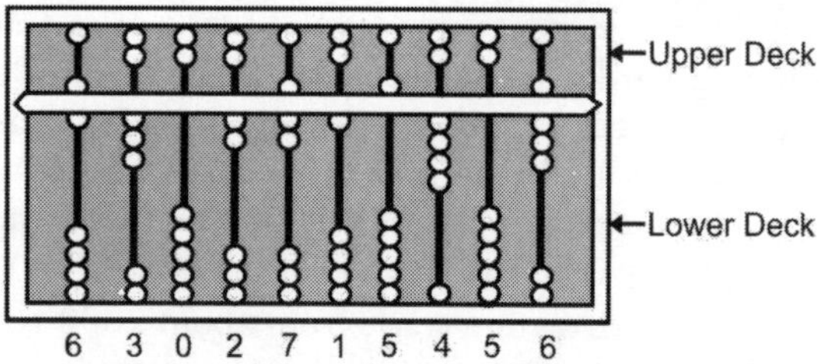

Fig. 1.2: Abacus Storing the Number

Devices to aid computation have evolved from simple recording and counting devices through the abacus, the slide rule, and more recent electronic computers. Even today, an experienced abacus user using a device designed hundreds of years ago can sometimes complete basic calculations more quickly than an unskilled person using an electronic calculator - though for more complex calculations, computers out-perform even the most skilled human.

2.2 FIRST MECHANICAL CALCULATORS

Gears are at the heart of mechanical devices like the Curta calculator. In 1623, Wilhelm Schickard built the first mechanical calculator and thus became the father of the computing era. Since his machine used techniques such as cogs and gears first developed for clocks, it was also called a 'calculating clock'. It was put to practical use by his friend Johannes Kepler, who revolutionized astronomy.

2.3 EARLY MECHANICAL CALCULATING MACHINES

As early as the 1640s mechanical calculators are manufactured for sale. Records exist of earlier machines, but Blaise Pascal invented the first commercial calculator, a hand powered adding machine. Although attempts to multiply mechanically were made by Gottfried Liebnitz in the 1670s the first true multiplying calculator appeared in Germany shortly before the American Revolution.

The example below shows the number 6302 > 15456 entered in the current status of abacus.

Blaise Pascal is usually credited for building the first digital computing machine in 1642. It added numbers entered with dials and was made to help his father, a tax collector. In 1671, Gottfried Wilhelm von Leibniz invented a computer that was built in 1694. It could add, and, after changing some things around, multiply. Leibniz invented a special stepped gear mechanism for introducing the addend digits, and this is still being used.

Fig. 1.3: Pascaline

The prototypes made by Pascal (Pascaline) and Leibniz were not used in many places, and considered weird until a little more than a century later, when Thomas of Colmar (A.K.A. Charles Xavier Thomas) created the first successful mechanical calculator that could add, subtract, multiply, and divide. A lot of improved desktop calculators by many inventors followed, so that by about 1890, the range of improvements included:

- Accumulation of partial results
- Storage and automatic re-entry of past results (a memory function)
- Printing of the results

Each of these required manual installation. These improvements were mainly made for commercial users, and not for the needs of science.

Machines by Gottfried Wilhelm von Leibniz (1671) followed. Around 1820, Charles Xavier Thomas created the first successful, mass-produced mechanical calculator, the Thomas Arithmometer, that could add, subtract, multiply, and divide. It was mainly based on Leibniz's work. Mechanical calculators, like the base-ten addiator, the comptometer, the Monroe, the Curta and the Addo-X remained in use until the 1970s.

2.4 THE SLIDE RULE ENGINEER'S TOOL

The slide rule, a basic mechanical calculator, facilitates multiplication and division. John Napier noted that multiplication and division of numbers can be performed by addition and subtraction, respectively, of logarithms of those numbers. Since these real numbers can be represented as distances or intervals on a line, the slide rule allowed multiplication and division operations to be carried significantly faster than was previously possible. Slide rules were used by generations of engineers and other mathematically inclined professional workers, until the invention of the Facit Calculator and Pocket Calculator till 1968. The engineers in the Apollo program to send a man to the moon made many of their calculations on slide rules, which were accurate to 3 or 4 significant figures.

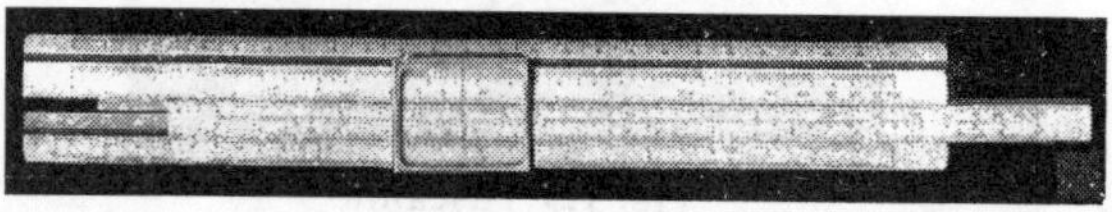

Fig. 1.4: Slide Rule

While producing the first logarithmic tables Napier needed to perform many multiplications and it was at this point that he designed Napier's bones.

2.5 PUNCHED CARD TECHNOLOGY, 1801

In 1801, Joseph-Marie Jacquard developed a loom in which the pattern being woven was controlled by punched cards. The series of cards could be changed without changing the mechanical design of the loom. This was a landmark point in programmability.

Fig. 1.5: Jacquard's Loom

Jacquard's loom was able to weave clothing by reading punched holes stored on small sheets of hardwood. These plates are then inserted into the loom which reads (retrieves) the pattern and creates (process) the weave. Powered by water, this "machine" came 140 years before the development of the modern computer.

2.6 ADA LOVELACE

Shortly after the first mass-produced calculator (1820), Charles Babbage begins his lifelong quest for a programmable machine.

Although Babbage was a poor communicator and record-keeper, his difference engine is sufficiently developed by 1842 that Ada Lovelace uses it to mechanically translate a short written work. She is generally regarded as the first programmer. Twelve years later George Boole, while professor of Mathematics at Cork University, writes An Investigation of the Laws of Thought (1854), and is generally recognized as the father of computer science.

2.7 BABBAGE AND HIS ANALYTCAL ENGINE

While Thomas of Colmar was developing the desktop calculator, a series of very interesting developments in computers was started in Cambridge, England, by Charles Babbage a mathematics professor. In 1812, Babbage realized that many long calculations, especially those needed to make mathematical tables, were really a series of predictable actions that were constantly repeated. From this he suspected that it should be possible to do these automatically.

Fig. 1.6: Difference Engine

He began to design an automatic mechanical calculating machine, which he called a difference engine. By 1822, he had a working model to demonstrate with. With financial help from the British government,

Babbage started fabrication of a difference engine in 1823. It was intended to be steam powered and fully automatic, including the printing of the resulting tables, and commanded by a fixed instruction program.

The difference engine, although having limited adaptability and applicability, was really a great advance. Babbage continued to work on it for the next 10 years, but in 1833 he lost interest because he thought he had a better idea—the construction of what would now be called a general purpose, fully program-controlled, automatic mechanical digital computer. Babbage called this idea an Analytical Engine. The ideas of this design showed a lot of foresight, although this couldn't be appreciated until a full century later.

Fig. 1.7: Analytical Engine

The plans for this engine required an identical decimal computer operating on numbers of 50 decimal digits (or words) and having a storage capacity (memory) of 1,000 such digits. The built-in operations were supposed to include everything that a modern general-purpose computer would need, even the all important Conditional Control Transfer Capability that would allow commands to be executed in any order, not just the order in which they were programmed.

The analytical engine was soon to use punched cards (similar to those used in a Jacquard loom), which would be read into the machine from several different Reading Stations. The machine was supposed to operate automatically, by steam power, and require only one person there.

Babbage's computers were never finished. Various reasons are used for his failure. Most used is the lack of precision machining techniques at the time. Another speculation is that Babbage was working on a solution of a problem that few people in 1840 really needed to solve. After Babbage, there was a temporary loss of interest in automatic digital computers.

2.8 USE OF PUNCHED CARDS BY HOLLERITH

The 1890 census is tabulated on punch cards similar to the ones used 90 years earlier to create weaves. Developed by Herman Hollerith of MIT, the system uses electric power (non-mechanical). The Hollerith Tabulating Company is a forerunner of today's IBM.

Fig. 1.8: Hollerith Machine

Just prior to the introduction of Hollerith's machine the first printing calculator is introduced. In 1892 William Burroughs, a sickly ex-teller, introduces a commercially successful printing calculator. Although hand-powered, Burroughs quickly introduces an electronic model.

A step towards automated computing was the development of punched cards, which were first successfully used with computers in 1890 by Herman Hollerith (left) and James Powers, who worked for the US. Census Bureau. They developed devices that could read the information that had been punched into the cards automatically, without human help. Because of this, reading errors were reduced dramatically, work flow increased, and, most importantly, stacks of punched cards could be used as easily accessible memory of almost unlimited size. Furthermore, different problems could be stored on different stacks of cards and accessed when needed.

These advantages were seen by commercial companies and soon led to the development of improved punch-card using computers created by International Business Machines (IBM), Remington, Burroughs, and other corporations. These computers used electromechanical devices in which electrical power provided mechanical motion — like turning the wheels of an adding machine. Such systems included features to:

- feed in a specified number of cards automatically
- add, multiply, and sort
- feed out cards with punched results

As compared to today's machines, these computers were slow, usually processing 50–220 cards per minute, each card holding about 80 decimal numbers (characters). At the time, however, punched cards were a huge step forward. They provided a means of I/O, and memory storage on a huge scale. For more than 50 years after their first use, punched card machines did most of the world's first business computing, and a considerable amount of the computing work in science.

2.9 ELECTRONIC DIGITAL COMPUTERS

The start of World War II produced a large need for computer capacity, especially for the military. New weapons were made for which trajectory tables and other essential data were needed.

In 1925, unaware of the work of Charles Babbage, Vannevar Bush of MIT builds a machine he calls the differential analyzer. Using a set of gears and shafts, much like Babbage, the machine can handle simple calculus problems, but accuracy is a problem.

2.10 KONRAD ZUSE'S Z-SERIES

Working in isolation in Nazi Germany, Konrad Zuse started construction in 1936 of his first Z-series calculators featuring memory and (initially limited) programmability. Zuse's purely mechanical, but already binary Z1, finished in 1938, never worked reliably due to problems with the precision of parts.

Zuse's subsequent machine, the Z3, was finished in 1941. It was based on telephone relays and did work satisfactorily. The Z3 thus became the first functional program-controlled computer. In many ways it was quite similar to modern machines, pioneering numerous advances, such as floating point numbers. Replacement of the hard-to-implement decimal system (used in Charles

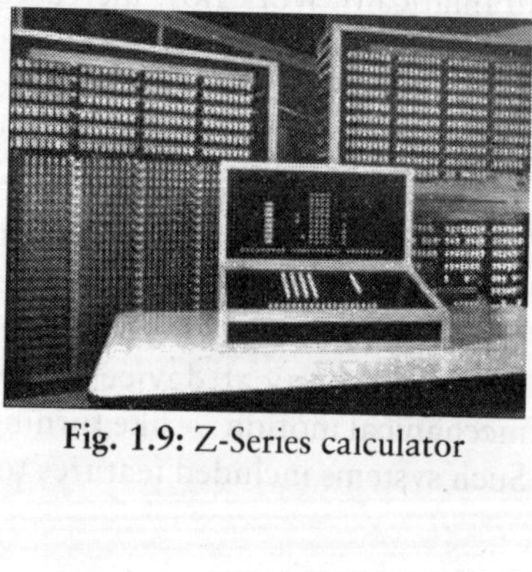

Fig. 1.9: Z-Series calculator

Babbage's earlier design) by the simpler binary system meant that Zuse's machines were easier to build and potentially more reliable, given the technologies available at that time. This is sometimes viewed as the main reason why Zuse succeeded where Babbage failed.

Programs were fed into Z3 on punched films. Conditional jumps were missing, but since the 1990s it has been proved theoretically that Z3 was still a universal computer (ignoring its physical storage size limitations). In two 1937 patents, Konrad Zuse also anticipated that machine instructions could be stored in the same storage used for data—the key insight of what became known as the Von Neumann architecture and was first implemented in the later British EDSAC design (1949). Zuse also claimed to have designed the first higher-level programming language, (Plankalkl), in 1945, although it was never formally published until 1971, and was implemented for the first time in 2000 by the Free University of Berlin — five years after Zuse died.

Fig. 1.10: Enigma

Zuse suffered dramatic setbacks and lost many years during World War II when allegedly British or American bombers destroyed his first machines. Apparently his work remained largely unknown to engineers in the UK and US until much later, although at least IBM was aware of it as it financed his post-war startup company in 1946 in return for an option on Zuse's patents.

2.11 THE ENIGMA AND OTHER CODE GENERATOR COMPUTER

The Enigma, a complex mechanical encoder was used by the Germans and they believe it to be unbreakable. Several people involved, most notably Alan Turing, conceive machines to handle the problem, but none are technically feasible. Turing proposes a "Universal Machine" capable of "computing" any algorithm in 1937. That same year George Steblitz creates his Model Kitchen), a conglomeration of otherwise useless and leftover material, to solve complex calculations. He improves the design while working at Bell Labs and on September 11, 1940, Steblitz uses a teletype machine at Dartmouth College in New Hampshire to transmit a problem to his Complex Number

Calculator in New York and receives the results. It is the first example of a network.

First in Poland, and later in Great Britain and the United States, the Enigma code is broken. Information gained by this shortens the war. To break the code, the British, led by Touring, build the Colossus Mark I. The existence of this machine is a closely guarded secret of the British Government until 1970. The United States Navy, aided to some extent by the British, builds a machine capable of breaking not only the German code but the Japanese code as well.

2.12 COLOSSUS

Colossus was used to break German ciphers during World War II. The British at Bletchley Park achieved a number of successes at breaking encrypted German military communications. The German encryption machine, Enigma, was attacked with the help of electro-mechanical machines called bombes. The bombe, designed by Alan Turing and Gordon Welchman, after Polish bomba, ruled out possible Enigma settings by performing chains of logical deductions implemented electrically. Most possibilities led to a contradiction, and the few remaining could be tested by hand.

Colossus was the first totally electronic computing device. The Colossus used a large number of valves (vacuum tubes). It had paper-tape input and was capable of being configured to perform a variety of boolean logical operations on its data, but it was not Turing-complete.

2.13 ANALOG COMPUTERS, PRE-1940

Nomograms, like the Smith chart serve as analog computers for specific classes of problems. Before World War II, mechanical and electrical analog computers were considered the 'state of the art', and many thought they were the future of computing. Analog computers use continuously varying amounts of physical quantities, such as voltages or currents, or the rotational speed of shafts, to represent the quantities being processed. An ingenious example of such a machine was the Water integrator built in 1936. Unlike modern digital computers, analog computers are not very flexible, and need to be reconfigured (*i.e.*, reprogrammed) manually to switch them from working on one problem to another. Analog computers had an advantage over early digital computers in that they could be used to solve complex problems while the earliest attempts at digital computers were quite limited. But as digital computers have become faster and used larger memory (*e.g.*, RAM or internal store), they have almost entirely displaced analog computers, and computer programming, or coding has arisen as another human profession.

Since computers were rare in this era, the solutions were often hard-coded into paper forms such as graphs and nomograms, which could then allow analog solutions to problems, such as the distribution of pressures and temperatures in a heating system.

Some of the most widely deployed analog computers included devices for aiming weapons, such as the Norden bombsight and artillery aiming computers for battleships. Some of these stayed in use for decades after World War.

Hybrid analog computers, controlled by digital electronics, remained in substantial use into the 1950's and 1960's, and later in some specialised applications.

3.0 Generations of Computers

3.1 ELECTRONIC COMPUTERS, GENERATIONS

Electronic computers developed in five distinct phases known as generations of computer. Each phase is characterized by type of switching circuits.

Generation	*Computing Element (Switch)*	
1st generation	Thermionic valve	1945-56
2nd generation	Transistor (Flip-Flop)	1956-63
3rd generation	Integrated circuits	1964-71
4th generation	Extra Large scale Integrated on a micro chip	1971-till today
5th generation circuits	Parallel processing under development	

There are two different architecture which are known as Von *Neumann architecture* and *Non-Von Neumann architecture*. In the former architecture computer processes data one by one and in the latter it breaks data into several parallel circuits. Super computers are based on Non Von-Neumann architecture.

3.2 FIRST GENERATION COMPUTERS, 1945–1956

First generation computers used Thermionic valves and programming was done in machine languages. All the computers were of very big sizes and expensive programming was very difficult. The first generation computers were different from their ancestors, that they used the concept of stored program.

The first generation of computer systems initially had little or no operating system. The computer was used by programmers one at a time. Each user had the computer to himself or herself while writing debugging, and executing the program. Input output was controlled completely by the user's program. Remember too that early programs were usually written in assembly language and some were machine language.

These first computers also cost many thousands of dollars and required a lot of space, so there may have been only one computer to serve many programmers. As numbers of applications increased, soon became too expensive to have one person tie up the system for such a long period of time.

3.2.1 Vacuum Tube

Fig. 1.11: Vacuum Tube

Also known as *tube*, *Electronic tube*, *electron valve* or *thermionic valve. Nearly* all tubes are of the thermionic type where one electrode, called the cathode, is heated, and electrons are emitted from its surface with a small energy (typically a Volt or less). A second electrode, called the anode (plate) will attract the electrons when it is positive with respect to the cathode, allowing current in one direction but not the other. Tubes are used for amplification of signals.

The first electronic digital computers used hundreds of vacuum tubes as their active components which, given the reliability of these devices, meant the computers needed frequent repairs to keep them operating. The chief causes of unreliability are the heater used to heat the cathode and the connector into which the tube was plugged.

First generation computers were characterized by the fact that operating instructions were made-to-order for the specific task for which the computer was to be used. Each computer had a different binary-coded program called a machine language that told it how to operate. This made the computer difficult to program and limited its versatility and speed. Other distinctive features of first generation computers were the use of vacuum tubes (responsible for their breathtaking size) and magnetic drums for data storage.

3.2.2 The First General-Purpose Electronic Computer ENIAC

By the mid-1940s, the majority of computers were being built out of vacuum tubes rather than switches and relays. Although vacuum tubes were fragile, expensive, and used a lot of power, they were much faster than relays (and much quieter). If we ignore Atanasoff's machine and COLOSSUS, then the first true general-purpose electronic computer was the electronic numerical integrator and computer (ENIAC), which was constructed at the University of Pennsylvania between 1943 and 1946.

Fig. 1.12: ENIAC

ENIAC was the beginning of first generation computers. It was 1000 times faster than a machine using relays and could do in two hours numerical calculations which would have taken 100 engineers an year. This was based on decimal system rather than binary system being used in all other computers, it did not have any memory system. ENIAC was designed to calculate the ballistic course of shells, a project founded by the U.S. Army.

ENIAC was produced by a partnership between the U.S. government and the University of Pennsylvania. Consisting of 18,000 vacuum tubes, 70,000 resistors, 10,000 capacitors 6,000 thousand switches, more than 10 lakhs soldered joints, the computer was such a massive piece of machinery that it consumed 160 kilowatts of electrical power. It was 10 feet tall, occupied 1,000 square feet of floor space, weighed in at approximately 30 tons.

Developed by John Presper Eckert (1919–1995) and John W. Mauchly (1907–1980), ENIAC, unlike the Colossus and Mark I, was a general purpose computer that computed at speeds 1,000 times faster than Mark I.

The machine performed an addition in 200 microseconds, a multiplication in about three milliseconds, and a division in about 30 milliseconds. The ENIAC was a decimal rather than a binary machine. That is numbers were represented in decimal form and arithmetic was performed in the decimal system. Its memory consisted of 20" accumulators, each capable of holding a 10-digit decimal number. Each digital was represented by a ring of 10 vacuum tubes. At any time, only one vacuum tube was in the *ON* state, representing one of the 10 digits. The major drawback of the ENIA was that it had to be programmed manually by setting switches and plugging and unplugging cables.

One of the greatest problems with computers built from vacuum tubes was reliability; 90% of ENIAC's down-time was attributed to locating and replacing burnt-out tubes. Records from 1952 show that approximately 19,000 vacuum tubes had to be replaced in that year alone, which averages out to about 50 tubes a day!

The machine incorporates 20 accumulators (the original plan was for 4). The accumulators and other units are all connected by several data buses, and a set of "program lines" for synchronization. Each accumulator stores a 10-digit number, using 10 bits to represent each digit, and also incorporates circuits to add a number from a bus to the stored number, and to transmit the stored number or its complement to a bus.

A separate unit can perform multiplication (in about 3 milliseconds}, while another does division and square roots; the inputs and outputs for both these units use the buses. There are constant registers, as on the Harvard Mark I: 104 12-digit registers forming an array called the "function table". 100 of these registers are directly addressable by a 2-digit number from a bus (the others are used for interpolations). Finally, a card reader is available to input data values, and there is a card punch for output.

The program is set up on a plugboard—this is considered reasonable since the same or similar program would generally be used for weeks at a time. For example, connecting certain sockets would cause accumulator 1 to transmit its contents onto data bus 1 when a pulse arrived on program line 1; meanwhile several accumulators could be adding the value from that data bus to their stored value, while others could be working independently. The program lines are pulsed under the control of a master unit, which can perform iterations. The ENIAC's clock speed is 100 kHz.

3.2.3 The First Stored Program Computer—EDVAC (June, 1945)

In August 1944, Mauchly and Eckert proposed the building of a new machine called the electronic discrete variable automatic computer (EDVAC). Unfortunately, although the conceptual design for EDVAC was completed by 1946, several key members left the project to pursue their own careers, and the machine did not become fully operational until 1952. When it was finally completed, EDVAC contained approximately 4,000 vacuum tubes and 10,000 crystal diodes. A 1956 report shows that EDVAC's average error-free up-time was approximately 8 hours.

John von Neumann (1903-1957), having joined the ENIAC team, drafts a report describing the future computer eventually built as the "EDVAC" ("Electronic Discrete Variable Automatic Computer" (!)); this is the first description of the design of a stored-program computer, and gives rise to the term "von Neumann computer". The first draft of the report fails to credit other team members such as Eckert and Mauchly; when this version becomes widely circulated, von Neumann gets somewhat too much credit for the design. The final version corrects the oversight, but too late.

In light of its late completion, some would dispute EDVAC's claim-to-fame as the first stored-program computer.

A small experimental machine (which was based on the EDVAC concept) consisting of 32 words of memory and a 5-instruction instruction set was operating at Manchester University, England, by June 1948.

It was, in fact, more of an electronic calculator rather than a computer as understood presently. The computers which were built on the use of binary code and stored program suggested by Dr. John Von Neumann in 1945 are illustrated below:

- **Electronic Delay Storage Automatic Computer (EDSAC):** It was built by M.V. Wilkes at Cambridge University in 1949.
- **Electronic Discrete Variable Automatic Computer (EDVAC):** It was built at Pennsylvania University and was completed in 1950.
- **Lyons Electronic Office (LEO):** The first commercial computer was produced by M.V. Wilkes in 1951.
- **UNIVAC-1:** It was produced in 1951 by Universal Accounting Company (UNIVAC) set up by Eckert and Mauchly.

- **IBM-701 and IBM-650:** These were introduced in 1953 and 1954 respectively by International Business Machines Corporation (IBM). The IBM-650 was the first modern digital computer produced on mass scale.

The first generation computers had the following limitations:

- Slow operating speed and restricted computing capacity.
- High power consumption and short life span.
- Very large space requirement.
- Limited programming capabilities.

3.2.4 Manchestor Mark I (June 1948)

Newman, Freddie C. Williams, and their team at Manchester University, Manchester, England, complete a prototype machine, the "Mark I" (also called the "Manchester Mark I"). This is the first machine that everyone would call a computer, because it's the first with a true stored-program capability.

It uses a new type of memory developed by F. C. Williams (possibly after an original suggestion by Presper Eckert), which uses the residual charges left on the screen of a CRT after the electron beam has been fired at it. (The bits are read by firing another beam through them and reading the voltage at an electrode beyond the screen.) This is a little unreliable but is fast, and also relatively cheap because it can use existing CRT designs; and it is much more compact than any other memory then existing. The Mark I's main memory of 32 32-bit words occupies a single Williams tube. (Other CRTs on the machine are less densely used: one contains only an accumulator.)

The Mark I's programs are initially entered in binary on a keyboard, and the output is read in binary from another CRT.

3.2.5 IBM 604 (1948)

IBM introduces the "IBM 604", a programmable calculator and card punch using vacuum tubes. It can read a card, perform up to 60 arithmetic operations in 80 milliseconds, and punch the results on the same card. The programming is by plugboard.

All machines first mentioned in the chronology from here on are stored-program computers.

3.2.6 UNIVAC I (Universal Automatic Computer)

UNIVAC was built by Remington Rand, became one of the first commercially available computers to take advantage of these advances. Both the U.S. Census Bureau and General Electric owned

UNIVACs. One of UNIVAC's impressive early achievements was predicting the winner of the 1952 presidential election, Dwight D. Eisenhower.

Fig. 1.13: World's first Successful Commecial Computer (UNIVAC)

Characteristics of First Generation Computers

- Operating instructions were made-to-order for the specific task for which the computer was to be used.
- Each computer had a different binary-coded program called a machine language that told it how to operate. This made the computer difficult to program and limited its versatility and speed.
- Use of vacuum tubes.
- Magnetic drums for data storage.
- High Power Consumption.

3.3 SECOND GENERATION COMPUTERS (1956–1963)

Vacuum tube computers continued to be prominent for many years, but vacuum tubes were somewhat problematic. They had an average life span of only about 20 hours, after which they would burn out. When a tube burned out in the ENIAC, for example, searching for the culprit through the computer's 18,000 tubes was a time-consuming process at best. Additionally, vacuum tubes created a great deal of heat when in use, and the heat from a great number of tubes caused some early computers to shut down every few minutes. This is why the development of the transistor in 1947 was a substantial step toward ushering in the second generation of computers, which began in the mid-1950s. In addition to running much more coolly and having longer effective life spans, transistors were faster than vacuum tubes and much smaller, which led to substantial space savings. Bell Labs introduced their first transistor computer in 1955, and more were soon to follow.

In computers became smaller due to a device called *transistor*. The transistors replaced the vacuum tubes, and were a lot better because of the fact that they were small, used little energy, and were much faster than their predecessors.

3.3.1 Transistor

The transistor led to second generation computers that were smaller, faster, more reliable and more energy-efficient than their predecessors. The first large-scale machines to take advantage of this transistor technology were early supercomputers.

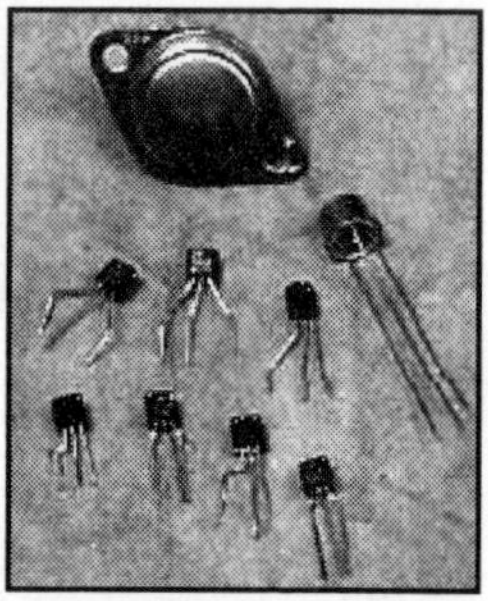

Fig.1.14: Transistor

Transistor is a device composed of semiconductor material that amplifies a signal or opens or closes a circuit. Invented in 1947 at Bell Labs, transistors have become the key ingredient of all digital circuits, including computers. Today's microprocessors contains tens of millions of microscopic transistors.

Prior to the invention of transistors, digital circuits were composed of vacuum tubes, which had many disadvantages. They were much larger, required more energy, dissipated more heat, and were more prone to failures. It's safe to say that without the invention of transistors, computing as we know it today would not be possible.

These second generation computers were also of solid state design, and contained transistors in place of vacuum tubes. They also contained all the components we associate with the modern day computer: printers, tape storage, disk storage, memory, operating systems, and stored programs.

By 1948, the invention of the transistor greatly changed the computer's development. The transistor replaced the large, cumbersome vacuum tube in televisions, radios and computers. As a result, the size of electronic machinery has been shrinking ever since. The transistor was at work in the computer by 1956. Coupled with early advances in magnetic-core memory, transistors led to second generation computers that were smaller, faster, more reliablc and more energy-efficient than their predecessors. The first large-scale machines to take advantage of this transistor technology were early supercomputers, Stretch by IBM and LARC by Sperry-Rand. These computers, both developed for atomic energy laboratories, could handle an enormous amount of data, a capability much in demand by atomic scientists. The machines were costly, however, and tended to be too powerful for the business sector's computing needs, thereby limiting their attractiveness.

3.3.2 Magnetic Storage (1957)

Until 1957, magnetic cores were used as storage devices. In 1957, magnetic tape was used as a much more convenient, and reliable form of saving information. More hard drive space allowed for bigger and more advanced programs.

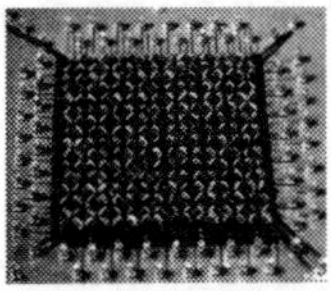

Fig. 1.15: Magnetic Core

3.3.3 Characteristics of Second Generation Computers

Second generation began with the advent of transistorized circuitry, invention of magnetic core to memory storage and high level languages like *Fortran*, *Cobol*, *Algol* etc. Nearly 80% of the computers so far installed were used in business and industry and almost all used batch processing mode.

Second generation is known for:

1. The transistor, which had been invented in 1948 at AT & T Bell Laboratories, gradually replaced vacuum tubes in the design of switching circuits.
2. Cathode-ray-tube memories and delay-time memories were replaced by ferrite cores and magnetic drums as the technologies used in main memories.
3. The use of index registers and floating-point arithmetic hardware became widespread.
4. Machine-independent "high-level" programming languages such as ALGOL, COBOL and FORTRAN were introduced to simplify programming.
5. Special processors (IO processors) were introduced to supervise input-output operations, thus freeing the CPU from many time-consuming housekeeping functions.
6. Computer manufacturers began to provide system software such as compilers subroutine libraries, and batch monitors.
7. Magnetic tapes and disks for secondary storage.
8. Reductions in size and hear generation.

In such computers, transistor circuits called *flip flops* formed the basis of internal binary operation and storage of binary coded data. The use of transistors reduced (i) size (ii) manufacturing and running costs and improved (iii) reliability and processing power.

Transistors were invented in 1948. The first transistorized digital computer, TRADIC, was built by Bell laboratories in 1954. Discrete transistors and diodes were the building blocks: 800 transistors were used in TRADIC. Printed circuits appeared. By this time coincident current magnetic core memory was developed and subsequently appeared in many machines. Assembly languages were used until the development of high-level languages, Fortran (formula translation) in 1956 and Algol (algorithmic language) in 1960.

In 1959, Sperry Rand built the larc system and IBM started the stretch project. These were the first two computers attributable to architectural improvement. The larc had an independent I/O processor which operated in parallel with one or two processing units. Stretch featured instruction look ahead and error correction to be discussed in section 1.2 The first IBM secientfic transistorized computer, IBM 1620, became available in 1960. COBOL (Common Business Oriented Language) was developed in 1959. Interchangeable disk packs were introduced in 1963. Batch processing was popular, providing sequential execution of user programs, one at a time until done.

Some of the second generation computers are:

Model	**Size**	**Application**
IBM 1620	Small	Scientific
IBM 1401	Small to Medium	Business
IBM 7094	Large	Scientific and Business
CDC 1604	Medium to Large	Scientific
CDC 3600	Large	Scientific
RCA 501	Medium	Business
UNIVAC 1108	Large	Scientific and Business

NB: Leo Mark III, ATLAS, ICL 1901 were also in use.

3.3.4 The Digital Equipment Corporation's PDP-1

In 1957, Ken Olsen and Harlan Anderson founded a company called Digital Equipment Corporation (DEC) with the goal of manufacturing and selling high-speed digital circuits. By 1959, the company was well established, and it introduced its first computer, the PDP-1 (the letters stood for "Programmed Data Processor"). The PDP-1 incorporated some of the engineering advances that would later characterize minicomputers, especially in its internal design and attractive packaging. About 50 were produced, its price was $ 120,000. DEC soon designed and began selling other machines as well. The PDP-1 was followed by a series of other more powerful, but less expensive products in the PDP line.

3.3.5 IBM 1401

One important example was the IBM 1401, which was universally accepted throughout industry, and is considered by many to be the Model T of the computer industry. By 1965, most large business routinely processed financial information using second generation computers.

It was the stored program and programming language that gave computers the flexibility to finally be cost effective and productive for business use. The stored program concept meant that instructions to run a computer for a specific function (known as a program) were held inside the computer's memory, and could quickly be replaced by a different set of instructions for a different function.

A computer could print customer invoices and minutes later design products or calculate paychecks. More sophisticated high-level languages such as COBOL (Common Business-Oriented Language) and FORTRAN (Formula Translator) came into common use during this time, and have expanded to the current day.

The computers built in the 1950s and 1960s are considered the second generation computers. These computers make use of the transistors invented by Bell Telephone Laboratories and they had many of the same components as the modern-day computer. For instance, 2nd generation computers typically had a printer, some sort of tape or disk storage, operating systems, stored programs, as well as some sort of memory. These computers were also generally more reliable and were solid in design.

3.3.6 1957: FORTRAN

FORTRAN, an acronym for Formula Translator was the first successful programming language. This language used words and sentences instead of the binary machine codes that had been commonplace in computers prior to the 1950s. FORTRAN could be read by ordinary people with no previous programming experience or knowledge, and it made it easier for computers to be programmed. FORTRAN was extremely useful in that given a single statement, many instructions would be produced. FORTAN created programs that were just as good as those produced by human programmers. Not only was it one of the most successful programming languages ever, but it also dominated many other languages for years.

3.3.7 1959: COBOL

COBOL the short for the Common Business Oriented Language, was invented in 1959. This is a business programming language that allowed for computer programs to be easily read.

3.3.8 1961: IBM 1400 Series

The IBM 1400 series were a major breakthrough for IBM. The first computer in this series is the IBM 1401. The 1401 was a computer system that used transistors instead of the vacuum tubes found in previous IBM computers. The system contained many peripherals, which included, among others, a new high-speed printer. This printer could print 600 lines per minute. The total cost of an IBM 1401 was $ 150,000. 12,000 of these computers were produced.

3.3.9 1962: SpaceWar!

Space War is the first interactive computer game. It was developed by MIT students Slug Russell, Shag Graetz, and Alan Kotok for the PDP-1 computer. In this game, players must battle against enemy spaceships and face obstacles like the gravity of the sun. The players used primitive joysticks to maneuver their ships. SpaceWar has helped to inspired future video games.

3.4 THIRD GENERATION COMPUTERS (1964–1971)

Though transistors were clearly an improvement over the vacuum tube, they still generated a great deal of heat, which damaged the computer's sensitive internal parts. The quartz rock eliminated this problem. Jack Kilby, an engineer with Texas Instruments, developed the integrated circuit (IC) in 1958. The IC combined three electronic components onto a small silicon disc, which was made from quartz. Scientists later managed to fit even more components on a single chip, called a semiconductor. As a result, computers became ever smaller as more components were squeezed onto the chip. Another third-generation development included the use of an operating system that allowed machines to run many different programs at once with a central program that monitored and coordinated the computer's memory.

It seems that the integrated circuit was destined to be invented. Two separate inventors, unaware of each other's activities, invented almost identical integrated circuits or ICs at nearly the same time.

Jack Kilby, an engineer with a background in ceramic-based silk screen circuit boards and transistor-based hearing aids, started working for Texas Instruments in 1958. A year earlier, research engineer Robert Noyce had co-founded the Fairchild Semiconductor Corporation. From 1958 to 1959, both electrical engineers were working on an answer to the same dilemma: how to make more of less.

In designing a complex electronic machine like a computer it was always necessary to increase the number of components involved in order to make technical advances. The monolithic (formed from a single crystal) integrated circuit placed the previously separated transistors, resistors, capacitors and all the connecting wiring onto a single crystal (or 'chip') made of semiconductor material. Kilby used germanium and Noyce used silicon for the semiconductor material.

In 1959, both parties applied for patents. Jack Kilby and Texas Instruments received U.S. patent #3,138,743 for miniaturized electronic circuits. Robert Noyce and the Fairchild Semiconductor Corporation received U.S. patent #2,981,877 for a silicon based integrated circuit. The two companies wisely decided to cross license their technologies after several years of legal battles, creating a global market now worth about $ 1 trillion a year.

3.4.1 Integrated Circuits (1965)

The integrated circuit was invented in 1963. Circuits were built into a like substance called silicon (which is a good (super) conductor of electricity). This was a big break for the computer industry because now big boards lined with transistors which took up a lot of space were replaced with small boards that reduced the size of computers and made them more reliable, and less expensive.

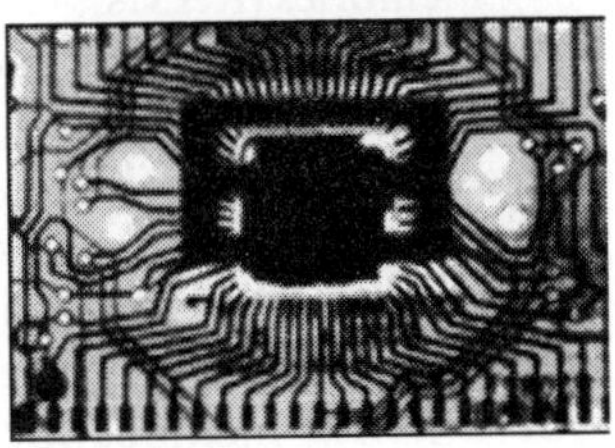

Fig. 1.16: Integrated Circuit

The IC combined three electronic components onto a small silicon disc, which was made from quartz. Scientists later managed to fit even more components on a single chip, called a semiconductor. As a result, computers became ever smaller as more components were squeezed onto the chip. Another third generation development included the use of an operating system that allowed machines to run many different programs at once with a central program that monitored and coordinated the computer's memory.

These computers (released in 1964) used integrated circuits (ICs) built on wafer-thin slices of extremely purified silicon crystal called chip.

> "*What we didn't realize then was that the integrated circuit would reduce the cost of electronic functions by a factor of a million to one, nothing had ever done that for anything before*".
>
> — *Jack Kilby*

3.5 Characteristics of Third Generation Computers

Integrated circuit consisting of transistors, resistors and capacitor synthesized on a single chip, is the characteristic of third generation of computers. This eliminated use of lengthy circuit wires. Silicon transistors were used instead of Germenium. Improved memories up to 4 mega byte range were used instead of 100 kilo bytes of second generation. Many on-line computerized systems became feasible, like airline reservation interactive query and real time closed loop process central system.

Third generation is known for:

1. Integrated circuits (ICs) began to replace the discrete transistor replace, ferrite second generation machines, resulting in a substantial reduction in physical size and cost.
2. Semiconductor (IC) memories began to augment, and ultimately replace, ferrite cores in main-memory designs. The two man types are read-only memories (ROMs) and read-and-write memories, which are usually referred to as random access memories (RAMs).
3. A technique called microprogramming came into widespread use to simplify the design of CPUs and increase their flexibility.
4. A variety of techniques for concurrent or parallel processing were introduced such as pipelining and multiprocessing. These had the objective of increasing the effective speed at which programs could be executed.
5. Efficient methods for automatic sharing of the facilities or resources of a computer system, e.g., its processors and memory space, were developed and incorporated into operating systems.

3.5.1 Miniaturization of Small and Medium Scale Integration

After the invention of the integrated circuit, it took very little time to realize the tremendous benefits of miniaturizing and integrating larger numbers of transistors into the same integrated circuit. More transistors (switches) were required in order to implement more complicated functions. Miniaturization was the key to integrating together large numbers of transistors while increasing hardware speed and keeping power consumption and space requirements manageable.

The transistor was originally a single, discrete device, that you could place individually into a circuit much like any other. Today, some special purpose transistors are still used that way. What allowed for the creation of modern processors was the invention of the integrated circuit, which is a group of transistors manufactured from a single piece of material and connected together internally, without extra wiring. Integrated circuits are also called ICs or chips.

A special material is used to make these integrated circuits. While most materials either insulate from electrical flow (air, glass, wood) or conduct electricity readily (metals, water), there are some that only conduct electricity a small amount, or only under certain conditions. These are called semiconductors. The most commonly used semiconductor is of course silicon.

By careful chemical composition and arrangement, it is possible to create a very small transistor directly on a layer of silicon, using various technologies to manipulate the material into the correct form. These transistors are small, fast and reliable, and use relatively little power. The first integrated circuit was invented in 1959 by Texas Instruments. It contained just six transistors on a single semiconductor surface.

This genration was marked by the use of *small-scale integrated* (SSI) and *medium-scale integrated* (MSI) *circutis* as the basic building blocks. Multilayered printed circuits were used. Core memory was still used in CDC-6600 and other machines but by 1968 many fast computers, like CDC-7600, began to replace cores with solid-state memories. High-level languages were greatly enchanced with intellingent compilers during this period.

3.5.2 Large scale integration (LSI)

LSI came to refer to the creation of integrated circuits that had previously been made from multiple discrete components. These devices typically contained hundreds of transistors. Early computers were made from many of these smaller ICs connected together on circuit boards.

3.5.3 Very Large Scale Integration (VLSI)

As time progressed after the invention of LSI integrated circuits, the technology improved and chips became smaller, faster and cheaper. Building on the success of earlier integration efforts, engineers learned to pack more and more logic into a single circuit. This effort became known as very large scale integration (VLSI). VLSI circuits can contain millions of transistors.

Originally, the functions performed by a processor were implemented using several different logic chips. Intel was the first company to incorporate all of these logic components into a single chip. This was the first microprocessor, the 4004, introduced by Intel in 1971. All of today's processors are (highly advanced!) descendants of this original 4-bit CPU.

3.5.4 Multiprogramming

Multiprogramming was well developed to allow the simultaneous execution of many program segments interleaved with I/O operations. Many high-performance computers, like IBM 360/91, Illiac IV, TI-ASC, Cyber-175, STAR-100 and Cmmp, and several vector processors were developed in the early seventies. Time-sharing operating systems became available in the late 1960s. Virtual memory was developed by using hierarchically structured memory systems.

3.5.4 Famous Computers of Third Generation

Some of the mainframe computers which belong to this generation are:

- IBM-360 series
- ICL-1900 series
- IBM-370/168
- ICL-2900
- Honeywell-6000 series

The IBM-360/370 series is designed to meet the requirements of both scientific computing and business data processing.

IBM 360 (1964–1965): The IBM 360 was introduced by IBM (duh!) in April of 1964, and was finally delivered in 1963. It was not actually a single computer, but was rather a family of six computers and their peripherals. These computers were all mutually compatible and all worked together. The first models of the IBM 360 used transistors, but later these computers made a transition from transistors to integrated circuits.

PDP-8 (1965): The PDP-8 was a successor to the PDP-1. Because it was the first computer to successfully make use of integrated circuits, it was much smaller and cheaper compared to other computers available at the same time. When it first entered the minicomputer market, it sold for about $ 20,000. Five years later, this price was reduced to only $ 3,000. The PDP-8 was relatively simple in design-physically, logically, and electronically. It only had 4,096 words of memory, and its word length was only 12 bits. This machine became the first commercially successful minicomputer because of its reasonable cost, speed, and small size. Minicomputers are medium-sized computers that were much cheaper than the larger and bulkier mainframe computers.

UNIX (1969): At the AT & T Bell Laboratories, programmers Kenneth Thompson and Dennis Ritchie developed a new type of operating system known as UNIX. UNIX is a multi-user operating system able to perform multiple tasks. In addition, UNIX is written in the C language, which allows it to be less machine-specific than other available operating systems. C was specifically designed for UNIX. Because UNIX used C, it was able to be used in any computer system. This was a big achievement at the time!

GUI (1970): (Graphical User Interface, pronounced gooey) was designed by the Xerox Corporation. GUI allowed users to be able to "point and click." In other words, the computer screen was designed to resemble a desktop. It had clickable folders, calculators, etc. Which were represented by images known as icons. Users could click on these icons to move and manipulate the folders and other tools. GUI made using computers much easier and is currently found in modern-day computers.

3.6 FOURTH GENERATION (1971— PRESENT)

The development of micro-processor made it possible to place complete control processing unit of a computer on a single chip. Very large scale integrated circuits having about 50000 transistors on a chip, made it possible to have micro computers of fourth generation. Magnetic core memories were replaced by semiconductor memory and microcomputers became not only interrogative for question answer purpose but became user's friendly since they are menu driven which guides the user. Memory size increased up to 16 megabytes.

After the integrated circuits, the only place to go was down-in size, that is Large Scale Integration (LSI) could fit hundreds of components onto one chip. By the 1980's, very large scale integration (VLSI) squeezed hundreds of thousands of components onto a chip. Ultra-large scale integration (ULSI) increased that number into the millions. The ability to fit so much onto an area about half the size of a U.S. dime helped diminish the size and price of computers. It also increased their power, efficiency and reliability. The Intel 4004 chip, developed in 1971, took the integrated circuit one step further by locating all the components of a computer (central processing unit, memory, and input and output controls) on a miniscule chip. Whereas previously the integrated circuit had to be manufactured to fit a special purpose, now one microprocessor could be manufactured and then programmed to meet any number of demands. Soon everyday household items such as microwave ovens, television sets and automobiles with electronic fuel injection incorporated microprocessors.

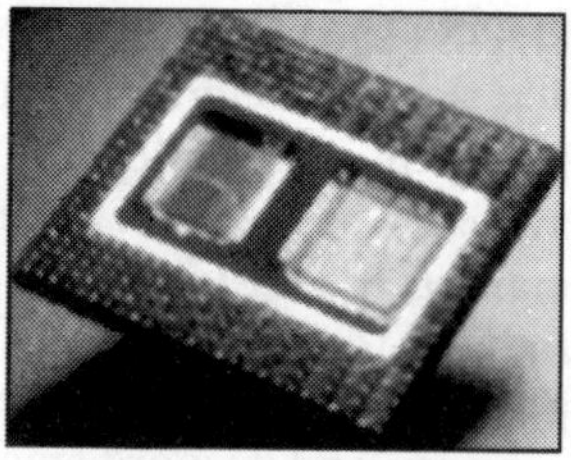

Fig. 1.17: Microchip

3.6.1 Intel Corporation

In 1968, Noyce and another engineer, Gordon Moore, left Fairchild to start their own electronics firm, which they named the Intel Corporation. The company started with 12 employees and their first year revenues were $ 2,672.00. Now, over a quarter century later, Intel's innovations have changed the world.

Intel focussed initially on making semiconductor computer memory-practical and affordable. Within a year, Intel had rolled out its first product—the 3101 64-bit memory chip. Intel continued to successfully develop memory chips, but in 1971 the event happened which changed the world and launched the microcomputer revolution.

3.6.2 Programmable Calculator

A Japanese calculator company named Busicom had approached Intel back in 1969 about designing a set of chips for a programmable calculator and had advanced Intel $ 60,000. Their original design had called for multiple custom chips, but Ted Hoff, a young Intel engineer, thought it was too complex. His solution was to develop a single-chip, general purpose logic device which would retrieve its instructions from semiconductor memory. He envisioned this solution to enable an off-the-shelf processor to handle many different functions, and eliminate a lot of custom circuit design.

Hoff's vision was transformed into silicon by a team of engineers and designers, and within several months the Intel 4004 microprocessor was created. 1/8"wide and 1/6"long, and consisting of 2300 transistors, this revolutionary computer on a tiny chip had as much computing power as its ancient great-grandfather, the room-sized ENIAC. Intel decided to buy the rights to this product back from the Japanese company, which had run into financial problems - and the rest - as they say - is history.

The Intel 4004 was introduced by the end of 1971, sold for $ 200, and was followed less than a year later by the 8008, an 8-bit microprocessor which sold for $ 360. For the first time, affordable computer power was available to everyone.

3.6.3 Development of Microprocessor by Intel

Intel 4004 1971: ICs which have the entire computer circuitry on a single silicon chip are called MICROPROCESSORS. The computers using these chips are called microcomputers or in short micros. These are, in fact, the scaled down versions of mini computers.

In 1971, Intel introduced its 4-bit bus, 108-KHz 4004 chip—the first microprocessor. Microprocessors allow computers think more quickly, and think more powerful thoughts. It made computers much faster, and allowed them to do more powerful things. The Microprocessor was using integrated circuit technology all of which is still used today.

The Intel 4004 microprocessor chip, developed in 1971, took the integrated circuit one step further by locating all the components of a computer (central processing unit, memory, and input and output controls) on a minuscule chip.

A silicon chip that contains a CPU. In the world of personal computers, the terms microprocessor and CPU are used interchangeably. At the heart of all personal computers and most workstations sits a microprocessor. Microprocessors also control the logic of almost all digital devices, from clock radios to fuel-injection systems for automobiles.

Fig. 1.18: Intel 4004 (1971) World's first Micro Processor Chip

Three basic characteristics differentiate microprocessors:

Instruction set: The set of instructions that the microprocessor can execute.

Bandwidth: The number of bits processed in a single instruction.

Clock speed: Given in megahertz (MHz), the clock speed determines how many instructions per second the processor can execute.

In both cases, the higher the value, the more powerful the CPU. For example, a 32-bit microprocessor that runs at 50 MHz is more powerful than a 16-bit microprocessor that runs at 25 MHz.

In addition to bandwidth and clock speed, microprocessors are classified as being either RISC (reduced instruction set computer) or CISC (complex instruction set computer).

The 4th generation computers are marked by the usage of integrated circuits and microprocessors. As the years passed, computers became smaller and smaller, and their prices became lower and lower. Millions of components could be placed onto a single silicon chip. Computers became more efficient and more reliable, and they could perform more and more operations. They began to catch the eye of the general public, and soon more sophisticated software and equipment were designed. Networks became commonplace, and the whole world was connected by the Internet and by the World Wide Web.

The Intel 4004 is a 4-bit microprocessor. Microprocessors are integrated circuits that contain thousands and millions of transistors. The Intel 4004 is the essence of a general-purpose computer. It performed many input and output operations and was also able to read and respond to instructions stored in its memory. The microprocessor was able to group all the components of a computer, including the CPU, or central processing unit, the memory, and the input and output controls on a single chip. The Intel 4004 was first used on March 2, 1972 in the Pioneer 10 space probe.

1972: Intel 8008: In November of 1972, Intel came out with a new 8-bit processor, known as the Intel 8008.

1974: Intel 8080 The Intel 8080 is a re-engineered version of the Intel 8008. It had a larger and more versatile instruction set.

1975: Altair 8800 The Altair 8800 is often called the first personal computer. It was developed by Micro Instrumentation Telemetry

Systems (MITS), a small company centered in Albuquerque, New Mexico. The Altair 8800 was sold as an electronic hobbyist kit for only $397. The Altair computers were of blue box-shaped machines with dimensions of 17 inches by 18 inches by 7 inches. They were very simple, and so in complex that they did not even have a keyboard or display. These computers did not even have enough memory to perform any useful tasks. The Altair 8800 used an Intel 8080 microprocessor and had switches on the front panel for input and rows of neon light bulbs for the output. These lights were known as light-emitting diodes, or LEDs. It is hard to believe, but it is the in-complexity of the Altair 8800 that helped to establish the personal computer industry. Because of its limitations, the Altair allowed for other companies to refine the computer and do develop more software.

One such example of software developed for the Altair 8800 is the development of a version of the programming language, BASIC, for this computer. BASIC was a popular programming language used by many minicomputers and other machines. It was simple and easy to develop. The BASIC programming language designed for the Altair was developed by Bill Gates and Paul Allen, the cofounders of the software giant, Microsoft.

Microsoft was not the only company to develop programs and software for the Altair. In fact, more and more companies began entering the software business by developing software and peripherals for the Altair 8800. The personal computer soon became a business machine.

1976: CP/M: Developed in 1976 by Gary Kildall of the company Digital Research, CP/M was a popular operating system that made it possible for one version of a program to run on a variety of computers.

1976: Apple II: Apple computer was started by Stephen G. Wozniak and Steven P. Jobs. In 1976, the company introduced the Apple II to the personal computer market. The Apple II was appealing to consumers, because of its abundance of appealing software and with its well-written manuals. The computer also was able to be plugged into a standard household outlet.

1983: Microsoft Word: The powerful word processor used in nearly every household today, Microsoft Word, makes its debut as "Multi-Tool Word" in 1983. Microsoft Word can be used for all sorts of word-processing tasks and for creating all sorts of text documents. It is now widely used and widely distributed. Two years after introducing Word to the general public, the company, Microsoft, then introduced its new-famous Microsoft Windows.

1983: Compaq Computer Corporation: In 1983, Compaq Computer Corporation built a portable computer, a clone of the IBM PC. Looking like a sewing machine, Compaq's computer weighed 38 pounds (13 kg). Amazingly, this was considered lightweight at the time! The Compaq personal computer was compatible with the IBM personal computer, and could use software designed for IBM computers. It was an immediate success, bringing in $110 million in its first year of sales.

1984: Apple Macintosh: The Apple Macintosh was designed by Apple Computer and had many of the same features as the Lisa computer, also designed by Apple. The main goal of the Macintosh was to create a computer that would accommodate as many as Lisa's features as possible at a much lower price. It slimmed down on Lisa's high-level languages, and it also used the minimum number of chips and circuit boards needed to operate efficiently. It had graphical user interface and a user-friendly design. The computer contained software in which users could point and click with a mouse. For example, there was MacPaint and MacWrite. The Apple Macintosh used a 32-bit Motorola 68000 central processing unit and had a 9 inch black and white screen. It was first sold for $ 2,500.

Fig. 1.19: Apple's Macintosh

3.6.4 Characteristics of Fourth Generation

The present generation computers emphasize the use of large-scale intergrated (LSI) circuits for both logic and memory sections. High-density packaging has appeared. High-level languages are being extended to handle both scalar and vector data, like the extended fortran in many vector processors. Most operating systems are time-sharing using virtual memories vectorizing compilers have appeared in the second generation of vector machines like the Cray-1 (1976) and the Cyber-205 (1982). High-speed mainframes and supers appear in multiprocessor systems , like the Univac 1100/80 (1976), Fujitsu M 382 (1981). The IBM 370/168 MP, the IBM 3081 (1980), the Burroughs B-7800 (1978), and the Cray X-MP (1983). A high degree of pipelining and multiprocessing is greatly emphasized in commercial supercomputers. A massively parallel processor (MPP) was custom-designed in 1982. This MPP, consisting of 16,384 bit-slice microprocesors is under the control of one array controller of satellite image processing.

Steve Wozniak, a calculator technician at Hewlett Packard, and Steven Jobs, a free-thinking visionary, formed Apple Computer, Inc. after developing the Apple II, one of the first personal computer systems. The Apple II contained the first disk operating system. Apple had several wildly successful years until IBM's PC eventually dominated the Apple II in the corporate marketplace. Apple's groundbreaking Macintosh established new standards for user friendliness and creativity in personal computers, but it failed to retake the corporate market. Today Macintoshes are most widely used in education, publishing, graphics, and multimedia work.

3.7 Fifth Generation

The fifth generation is yet to come in actual shape although people are already working.

The fifth generation is the story of the new super-breed of computers now being planned by the Japanese. These computers will contain bits of information and will be able to think and make decisions and in certain instances better than human beings. Artificial Intelligence (AI) is thus emerging from behind the curtains to take its place in real world. Radically different from the preceding four generations in their construction, capabilities and application, this generation is all set to being momentaneous qualitative changes in human life-styles.

The Japanese call their fifth generation machines, knowledge information processing systems (KIPS). KIPS will work as amplifier of human thought and intelligence. The entire world is likely to become dependent on these machines for greater efficiency.

Many advances in the science of computer design and technology are coming together to enable the creation of fifth generation computers. Two such engineering advances are parallel processing, which replaces on Neumann's single central processing unit design with a system harnessing the power of many CPUs to work as one. Another advance is superconductor technology, which allows the flow of electricity with little or no resistance, greatly improving the speed of information flow. Computers today have some attributes of fifth generation computers. For example, expert systems assist doctors in making diagnoses by applying the problem-solving steps a doctor might use in assessing a patient's needs. It will take several more years of development before expert systems are in widespread use.

Many advances in the science of computer design and technology are coming together to enable the creation of fifth generation computers. Two such engineering advances are parallel processing, which replaces von Neumann's single central processing unit design with a system harnessing the power of many CPUs to work as one. Another advance is superconductor technology, which allows the flow of electricity with little or no resistance, greatly improving the speed of information flow. Computers today have some attributes of fifth generation computers. For example, expert systems assist doctors in making diagnoses by applying the problem-solving steps a doctor might use in assessing a patient's needs. It will take several more years of development before expert systems are in widespread use.

3.7.1 Moore's Law

In 1965, semiconductor pioneer Gordon Moore predicted that the number of transistors contained on a computer chip would double every year. This is now known as Moore's Law, and it has proven to

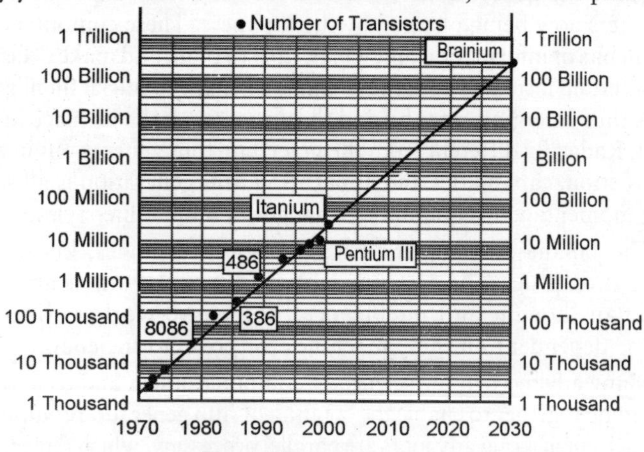

Fig. 1.20: Moore's Law

be somewhat accurate. The number of transistors and the computational speed of microprocessors currently doubles approximately every 18 months. Components continue to shrink in size and are becoming faster, cheaper, and more versatile.

3.7.2 Future Developments

With their increasing power and versatility, computers simplify day-to-day life. Unfortunately, as computer use becomes more widespread, so do the opportunities for misuse. Computer hackers-people who illegally gain access to computer systems often violate privacy and can tamper with or destroy records. Programs called viruses or worms can replicate and spread from computer to computer, erasing information or causing computer malfunctions. Other individuals have used computers to electronically embezzle funds and alter credit histories . New ethical issues also have arisen, such as how to regulate material on the Internet and the World Wide Web. Individuals, companies and governments are working to solve these problems by developing better computer security and enacting regulatory legislation.

Computers will become more advanced and they will also become easier to use. Reliable speech recognition will make the operation of a computer easier. Virtual reality, the technology of interacting with a computer using all of the human senses, will also contribute to better human and computer interfaces. Standards for virtual-reality program languages, called Virtual Reality Modeling language (VRML), currently are being developed for the World Wide Web.

Breakthroughs occurred in the area of quantum computing in the late 1990s. Quantum computers under development use components of a chloroform molecule (a combination of chlorine and hydrogen atoms) and a variation of a medical procedure called magnetic resonance imaging (MRI) to compute at a molecular level. Scientists used a branch of physics called quantum mechanics, which describes the activity of subatomic particles (particles that make up atoms), as the basis for quantum computing. Quantum computers may one day be thousands to millions of times faster than current computers, because they take advantage of the laws that govern the behaviorur of subatomic particles. These laws allow quantum computers to examine all possible answers to a query at one time. Future uses of quantum computers could include code breaking and large database queries.

Communications between computer users and networks will benefit from new technologies such as broadband communication systems that can carry significantly more data and carry it faster, to and from the vast interconnected databases that continue to grow in number and type.

4.0 Architecture of CPU, Input/Output Devices

4.1 WHAT ARE COMPUTERS

Computer is an information processor. It converts information (raw data) into intelligent form for the direct use of human beings.When we talk of computer we actually mean digital electronic computer and that to in most of the applications personal computer.

The two principal characteristics of a computer are:

- It responds to a specific set of instructions in a well-defined manner.
- It can execute a prerecorded list of instructions (a program).

Modern computers are *electronic* and *digital*. The actual machinery — wires, transistors and circuits — is called hardware; the instructions and data are called software.

Computer technology has fascinated the human beings, the most. Computers are all around us and going to stay around us centuries together. Some people have a science fiction view of computers; they think that they have an intelligence of their own and, if we're not careful, will eventually take over the universe.

In fact, the *opposite* is probably more correct. Within the limitations imposed by their designers, computers can do whatever they are programmed to do, but not more. They cannot whatever they are programmed to do, but not more. They cannot (so far) "think" for themselves, but their method of dealing with a problem is similar in many respects to a human's thought processes.

Data and Information

Computer gets information or data through some form of input unit. This information is stored in its memory. But unlike the human brain which usually remembers all the information entering it (although we might not be aware of this) the computer only retains the information it has been programmed to store. Any other information is lost. If the computer has been programmed to compare two bits of data, it will do so and make an appropriate decision. The result is transmitted to the output unit of the computer.

Data is matter in its raw form put into the computer for processing and storage as contrasted with "code" the programs that make the computer work.

Note: Data is a plural noun; the singular form is datum

Data is raw Information, especially information in a form that can be used by a computer. It can include text, numbers, sounds and pictures. A single piece of information is called a Datum.

A computer follows a series of instructions that have been programmed into its memory by the user and operates on data that is presented to it by the user.

The data input may be in form of :

- Typed (linguistic) data or instructions in computer's language. Linguistic Raw data can be:
  (a) Numeric
  (b) Alphabetic
  (c) Alphanumeric.

Then this is coded into data acceptable to computer in its language.

- It may be audible, visual or audiovisual analogic data.

This processing of information is at the heart of any computer system. But remember that information will only be processed in accordance with the program put into the computer, which is quite unable to think for itself, Information presented in an unexpected way will not be dealt with, and if the machine is asked to process information in a way that it has not been programmed for, it will refuse to continue. Information can exist in many forms, so both the input and output units must be able to 'translate' this information into or from the form that the computer can handle.

Basically any computer is supposed to carry out four functions:

- Accepting the data as input.
- Storing the data and instructions in its memory and recalling the same as and when required (Retrieving).
- Processing the data as per instructions to convert it into useful information.
- Communicating the information as output.

Processing of data is often similar to our way of thinking logically. One part of this process is to compare two items of information. If is often necessary to compare the item of information being fed in with one that is stored in the computer memory. The item of information already in the memory may itself have come in as new information and been sorted. Or it may have been given to the computer as part of the working instructions the program.

It gets information or data through some form of input unit. This information is stored in its memory. But unlike the human brain which usually remembers all the information entering it (although we might not be aware of this) the computer only retains the information it has been programmed to store. Any other information is lost. If the computer has been programmed to compare two bits of data, it will do so and make an appropriate decision. The result is transmitted to the output unit of the computer.

4.2 BASIC PARTS OF COMPUTER

Every machine has got three basic units in it. A computer appears to be one single machine, but it is really a group of machines which work together. These machines are called units.

Computer has three basic units:

- Input unit
- Output Unit
- Central Processing Unit

Let us take example of flour mill. In flour mill we put wheat as input and get flour as output. Every machine has got a process to do some useful work. In flour mill the process is of grinding. The unit containing the grinding stones is called processing unit as this is the part of machine that processes the wheat to get converted into flour.

Computers can really do only five things:

- Receive input.
- Process information.
- Produce output.
- Store information (when power is on: Short term Memory)
- Store information (when power is off:Long term Memory)

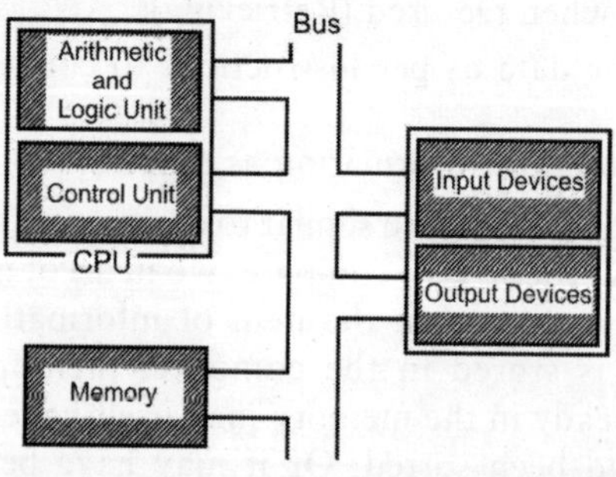

Fig. 1.21: Computer's Functional Block Diagram

4.3 PERIPHERAL DEVICES

Peripheral devices include all the I/O devices used with a computer system. When these devices are under control of the CPU, we say they are online. When they perform their functions independently, not under direct control of the CPU, we say they are offline. The peripheral devices described in the following paragraphs are the ones we commonly use: magnetic ink character readers, scanners, bar-code readers, key-to-online data entry terminals, magnetic tape units, magnetic disk drive units, floppy disk drive units, cathode-ray tube (CRT) terminals, printers, and plotters.

If a peripheral device is disconnected, the computer will still be able to work; only functions performed by this peripheral device will not be available. For example, if you disconnect a keyboard, you would not be able to type in information but you will be able work with the graphical user interface by clicking the mouse. If you remove the hard disk, you will have to start up your computer using floppy disk and you will have to save your data on floppy disks only, but the computer will still be working.

Peripheral devices can reside inside the computer case or outside. For example, a mouse, a keyboard, a monitor are always outside the case and they are connected to the computer via ports which are situated on the back of the case. Disk drives are positioned inside the case and they are connected to the motherboard via disk controller cards which are inserted in the expansion slots on the motherboard.

Peripheral devices are classified according to the purpose they serve:

- Input devices are used for data input
- Output devices are used for data output
- Storage devices are used to store data for long periods of time

Every year new more sophisticated devices appear in the market and prices on existing devices drop down dramatically.

Peripheral devices are connected to the computer through ports, or sockets, that are mounted outside. Some peripherals (e.g. disk drives) are located inside the computer case, so they are connected via expansion slots.

4.4 BASIC COMPONENTS OF A DIGITAL COMPUTER

The block diagram in following figure illustrates the five major operational divisions of an electronic digital computer. Although presently available machines very greatly in the construction details of various components, the overall system concepts remain roughly the same.

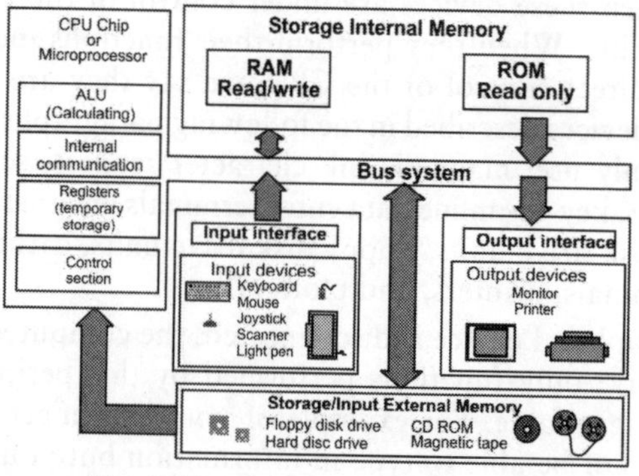

Fig. 1.22: Computer Organization

4.4.1 Input

Input is supplied to the microcomputer with the use of a keyboard, a mouse, or another input device. These input devices may be called peripheral devices.

The input units, as the name implies, pass input information to the computer. The input information may be programs or program related, or it may be data. You will read about input units in detail in unit 2.

4.4.2 Output

Output is the result of a computer process. Output may be viewed on a monitor screen, heard through speakers, printed on printers, and so forth. Output devices may be considered hardware and are also considered to be peripheral devices.

Output units send the information from inside the computer to the outside so we can use it. The output from the computer may appear on a video display be printed out on a printer, be displayed on an LED (light emitting diode) display, be sent to another digital system over a transmission line, or be stored in auxiliary storage such as magnetic tape or magnetic disks. You will read about output units in detail in unit 3.

4.4.3 Proccssing Unit

Processing is done inside the computer in an area called the central processing unit (CPU). Processing is the conversion of input to output.

4.5 CENTRAL PROCESSING UNIT (CPU)

The central processing unit, or processor or microprocessor, as some call it, is the nerve center of the system. It performs the central control functions. All the computational, logical, and operational decisions are made here. It contains the logic circuitry for performing the various computational activities. It controls the operation of all the functional units. It fetches machine instructions from memory, decodes these instructions and insures that the operations called for by the instructions are executed correctly. In order to do all this, it communicates or interfaces with the input and output units and the memory.

CPU is responsible for all events inside the computer. It controls all internal and external devices, performs arithmetic and logic operations. The operations a microprocessor performs are called the instruction set of this processor. The instruction set is "hard wired" in the CPU and determines the machine language for the CPU. The more complicated the instruction set is, the slower the CPU works. Processors differ from one another by the instruction set. If the same program can run on two different computer brands they are said to be compatible. Programs written for IBM compatible computers will not run on Apple computers because these two architectures are not compatible.

Central Processing Unit has basically three component parts :

- ALU
- Control Unit
- Storage (memory)

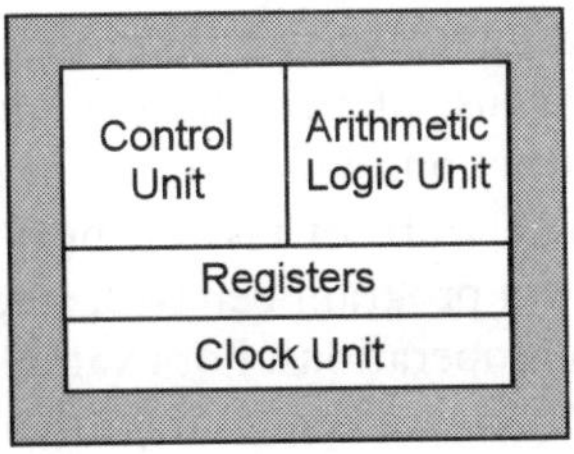

Fig. 1.23: A Simplified Diagram of the CPU

CPU Operation

The fundamental operation of most CPUs, regardless of the physical form they take, is to execute a sequence of stored instructions called a program. Herein we are discussing a very common class of CPUs that conform to the Von Neumann architecture. The program is represented by a series of numbers that are kept in some kind of computer memory. There are three steps that nearly all Von Neumann CPUs use in their operation:

- fetch,
- decode, and
- execute.

Fetching

The first step, or fetch, involves retrieving a sequence of numbers from program memory, whose location is determined by a "program counter". The program counter stores a number that identifes the current location in this sequence. In other words, the program counter marks the CPU's place in the current program.

Decoding

The numbers that the CPU fetch from memory ultimately instruct it as to how to proceed. In the decode step, the number is broken up into parts that have significance to the CPU. Often, some of the digits of the number (for example, the so-called "high" or "low" bits in a binary CPU's memory) indicate which operation (called an instruction) to perform. The remaining digits in the number usually provide information required for that instruction like, for example, operands for an addition operation. These operands generally come in two forms: a memory address, or a constant number. Some types of instructions manipulate the program counter. These are generally called "jumps" and facilitate behavior like loops, conditional program execution (through the use of a conditional jump), and functions in programs. Many instructions will also change the state of digits in a 'flags' register (a single word of fast memory). These flags can be used to influence how a program behaves, since they often indicate the outcome of various operations. For example, if two numbers are added together and produce a result larger than the CPU is designed to handle, then an arithmetic overflow flag may be set.

Executing

After the fetch and decode steps, the execute step is performed. During this step, various portions of the CPU are "connected" (by some switching device, like a relay or a transmission gate) so they can perform the desired operation. If, for instance, an addition operation was requested, an ALU will be connected to a set of inputs and a set of outputs. The inputs provide the numbers to be added, and the outputs will contain the final sum. If, as discussed earlier, the addition operation produces a result too large for the CPU to handle, an overflow flag may also be set.

This process then repeats due to the ever incrementing program counter. In more complex CPUs than the one described here, multiple instructions can be fetched, decoded, and executed simultaneously. This section describes what is generally referred to as a 'single cycle data path,' which in fact is quite common among the simple CPUs used in many electronic devices (often called microcontrollers).

The type of operations that the CPU can do include:

(a) Retrieving data from memory.
(b) Outputting data to memory or ports.
(c) Storing the data temporarily and retrieving it as and when required.
(d) Converting data of one form into another.
(e) Manipulating alphabetic or alphanumeric data like word processing, letter writing, shorting in alphabetic or alphanumeric orders, editing, making catalogues etc.
(f) Math operations *i.e.*, making arithmetic calculations like addition subtraction, multiplication, division and exponentiation etc.
(g) Making logical decisions like comparing two values to find out which one is greater with the help of logical operations such as AND, OR, NOT
(h) Communicating data to far off distances.
(i) Program control operations such as jump to a new spot, skip the next instruction, and "if..then.." statements.
(j) Response to special external impulses ("interrupts").

CPU Architecture

Today most desktop computers and laser printers have 32-bit processors; 64-bit processors are being phased in. Smaller devices like mobile phones, PDAs, or portable video game devices may have 16 or 8-bit processors. Embedded systems such as microwave ovens, calculators, computer keyboards, and infrared remote controls typically have 8-bit or 4-bit processors.

Basically the CPU does everything in binary language *i.e.* with the help of 0 and 1.

A CPU doesn't have to run a whole computer. There are specialized CPU made that are explicitly geared to do specific tasks, such as run the electronics in a car, or a VCR, or microwave oven. If fact, inside the computer itself are a number of these "imbedded" CPUs that run the keyboard, the video controller, the serial ports, the network card, and the disk controller.

The two main features of the CPU are:

- Its speed measured in millions of instructions per second;
- The word size.

4.5.1 Clock Unit and Computers Speed

All activities in the computer system are composed of thousands of individual steps. These steps should follow in some order in fixed intervals of time. These intervals are generated by the clock unit. Every operation within the CPU takes place at the clock pulse. No operation, regardless of how simple, can be performed in less time than transpires between ticks of this clock. But some operations require more than one clock pulse. The faster the clock runs, the faster the computer performs.

CPUs in the same class or generation are distinguished from each other by clock speed, the rate at which information is moved around the chip. Generally, the faster the clock speed, the faster a computer operates although other CPU characteristics can make a faster clock-speed chip slower than expected.

The clock speed of a CPU is defined as the frequency that a processor executes instructions or that data is processed. This clock speed is measured in millions of cycles per second or megahertz (MHz).

The clock rate is measured in megahertz (MHz) or million ticks per second. At the moment of writing of these notes the highest clock rate is 2.8 GHz but by the time you read them probably quicker microprocessors will be on the market. Larger systems are even faster. In "older" systems the clock unit is external to the microprocessor and resides on a separate chip. In most modern microprocessors the clock is usually incorporated within the CPU.

The clock itself is actually a quartz crystal that vibrates at a certain frequency when electricity is passed through it. Each vibration sends out a pulse or beat, like a metronome, to each component that's synchronized with it.

The processor's speed, though frequently being used to denote a computer's performance, does not 100 per cent reflect how a computer truly perform because it measures only the speed of the processor, not the entire computer that has many other components in addition to the processor.

As a result, for example, when a computer is "working" on a spreadsheet, it's Pentium 42.8 GHz processor might be very quick in processing data, but all the time it will be waiting for the hard drive, which is a slower component, to transfer data to it, meaning most of the time the processor is sitting in idle waiting. So a 2.4 GHz Pentium 4 processor might be 50 per cent faster than it's 4.6 GHz counterpart based on processor speed; on the whole, with all of the computer's components accounted for, the two computers might be only 8 or 10 per cent different in performance.

The processors of choice for today's computers are Intel's Pentium 4 and Celeron. Pentium 4 covers the high end of the performance spectrum, and Celeron covers the lower end. The processors come in a variety of speeds measured in GHz (*i.e.* billion hertz, or billion cycles-per-second). For Pentium 4, the speeds go from 4.4 to 2.8 GHz (in year 2004); for Celeron, the speeds go from 0.85 to 4.8 GHz.

In today's computer marketplace there are two main companies that manufacture microprocessor CPUs: Intel Corporation and Motorola Corporation. Intel makes CPUs for IBM and IBM-compatible computers. The older ones had "names" like 8088 (or XT), 8086, 80286 (or AT), and 80386 (or just 386) and 80486 (or just 486). Newer ones have names like PENTIUM II, PENTIUM III and PENTIUM IV.

Motorola makes CPUs for Apple Macintosh, Commodore AMIGA and Atari Mega and Atari ST computers. These are called the 68000, 68020, 68030, 68040 and 68060. In both instances, the higher numbers represent the more advanced and expensive CPUs. Motorola's newer microprocessors are called PowerPCs. Their G3 (Generation 3) chip is used in Apple's new iMac computer, an inexpensive but powerful home desktop computer.

4.6 ARITHMETIC AND LOGIC UNIT (ALU)

The arithmetic and logic unit (ALU) is the part where actual computations take place. It consists of circuits which perform arithmetic operations (*e.g.* addition, subtraction, multiplication, division) over data received from memory and capable to compare numbers.

Computer has to do a large number of calculations and computations. It has to add, subtract, multiply and divide. It has to do very complex calculations. It all does with a very fast speed. Computer has to do a number of computations. It has to take a number of decisions. It has to apply logic. By logic we mean

- to compare two quantities and see which one is bigger.
- to count any element and test whether desired number is reached.
- to decide further route.

All such calculations and comparisons are done in this unit.

While performing these operations the ALU takes data from the temporary storage area inside the CPU named registers. Registers are a group of cells used for memory addressing, data manipulation and processing. Some of the registers are general purpose and some are reserved for certain functions. It is a high-speed memory which holds only data for immediate processing and results of this processing. If these results are not needed for the next instruction, they are sent back to the main memory and registers are occupied by the new data used in the next instruction.

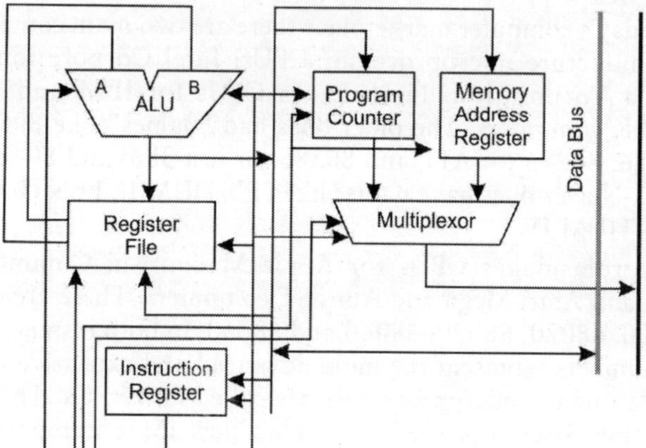

Fig. 1.24: Computer Architecture: Working of the Computer

ALU takes instruction from control unit. It takes data from memory unit and returns information to memory unit. Since it has to do calculations very fast, all components of ALU work very fast. By how fast? Can you imagine that this unit of microcomputer can do 1000000 calculations (say multiplication of 16 digit numbers) per second. This unit has a number of registers and accumulators for short term storage of characters while doing calculations or computations.

ALU comprises of a number of accumulators and registers. It derives all the data from the main memory as directed by the control unit based on the program given to it. This data gets loaded into accumulators in the ALU. For example, if two numbers A and B are to be added, the control unit selects number A from its address in memory and loads it into arithmetic unit. Then the value of B is selected and added to A in the arithmetic unit. The results are placed in the main storage (again under the direction of control unit) or maybe retained in the accumulators themselves for further calculations. Some logical operations of the ALU enable the computer to have the decision-making quality.

4.7 CONTROL UNIT

The control unit supervises all of the CPU's operations. It fetches the software instructions from the memory, and it also coordinates the time and order in which the instructions are carried out. For example, the control unit has a special feature known as the interrupt. An interrupt is exactly what the term means! It is a special signal that interrupts the task the central processing unit is currently performing. It will tell the CPU to put aside this operation and to instead begin work on another operation.

Fig. 1.25: Animation of Control Unit

The control unit directs and controls the activities of the internal and external devices. It interprets the instructions fetched into the computer, determines what data, if any, are needed, where it is stored, where to store the results of the operation, and sends the control signals to the devices involved in the execution of the instructions.

4.8 MEMORY

Memory is the electronic holding place for instructions and data that your computer's microprocessor can reach quickly. When your computer is in normal operation, its memory usually contains the main parts of the operating system and some or all of the application programs and related data that are being used.

Memories are one of the most important parts of a computer system. The term memory refers to any device that can be used for storage. There are two basic ways to store memory:

- Internal (Temporary) Memory
- External (Permanent) Memory

First way is to store the memory within the actual circuitry of the computer, which allows for it to be quickly accessed by the central processing unit. These memory devices are generally used for temporary storage of data and programs that are currently used by the CPU. This type of memory is often used as a shorter synonym for random access memory (RAM). This kind of memory is located on one or more micro chips that are physically close to the microprocessor in your computer.

The second type of memory storage is to store the memory in external storage devices. They include hard drives (also known as hard disks), floppy disks, and CD-ROMs. Programs and information that you are not currently using can be permanently stored on these devices. The term "memory" usually pertains to the devices stored within the circuitry of the computer. They are temporary, meaning that when the computer is turned off, they will be deleted. These devices are connected directly to the CPU and are pretty fast! They will be discussed in the following section. On the other hand, the external storage devices are not considered part of a computer's memory, and these will be talked about in a later section of this website.

4.8.1 Units of Memory

A computer's memory stores all of the data currently being processed as well as the program that controls the processing. These data are stored in cells. In computers used for commercial applications, each memory cell stores one byte of data.

The size of a computer's memory is measured in terms of kilobytes or megabytes. Because "kilo" stands for 1,000 and "mega" stands for 1,000,000, computer memory is measured by the thousands or millions of bytes that can be stored in memory at one time. One gigabyte is equal to 1000 megabyte and one terabyte is equal to 1000 gigabyte.

A computer works by electrical impulses. Hence, a binary number system which uses only two digits, namely, 0 and 1 is a convenient way to represent information inside a computer. We would use the symbol '1' to represent the presence of an electrical pulse and the symbol '0' to represent the absence of it. Information in a computer consists of data (numerical and non-numerical) and instructions which are made up of a large number of characters, namely, decimal numbers 0 to 9, alphabets A to Z, arithmetical operators like (+), (-) etc., relational operators like (=), (<), etc. and many other special characters like (,), (;), etc. Computers use eight binary digits (bits) to represent information internally. This allows upto 2^8=256 different characters to be represented uniquely. A collection of eight bits is called a byte.

Actully in computer usage, the prefix "kilo" actually stands for 1,024 bytes. Because computers are binary devices, most of the numbers found in the internal operations of computers are powers of 4. A kilobytes, then, is 2^{10} = 1,024 bytes. It is usually indicated with the initial K or, sometimes, KB. Thus a computer with a memory size of 512 K is said to have a 512,000 byte main storage. Really, it's 512 × 1024, or 524,248 bytes. A memory with a capacity of one megabyte (abbreviated 1M or 1MB) has

$$2^{20} = 1,048,\ 576$$

Computer memory goes by a number of different names, including immediate Access Storage (IAS), main storage, primary storage, main memory, primary, and RAM.

The following table gives the comparison:

Prefix	*Nominal Value*	*Exponential*	*Exact value*
Kilo	1,000	10^3	1024210
Mega	1,000,000	10^6	1048576220
Giga	1,000,000,000	10^9	1073741824230
Tera	1,000,000,000,000	10^{12}	1099511627776240

The more RAM you have, the faster you computer performs. RAM is measured in megabytes (MB) or gigabytes (GB).

To put this into perspective, consider that a printed page of single-spaced text contains about 3,000 characters. One MB holds about 400 pages of single-spaced text.

4.8.2 Hierarchy of Memory System

Memory in a computer system is required for storage and subsequent retrieval of the instructions and data. A computer system uses variety of devices for storing the instructions and data which are required for its operations. Normally we classify the information to be stored on computer in two basic categories: Data and the Instructions.

The Memory Hierarchy

The hierarchical arrangement of storage in current computer architectures is called the memory hierarchy. Each level of the hierarchy is of higher speed and lower latency, and is of smaller size, than lower levels.

A memory system can be considered to consist of four group of memories. These are:

1. Internal Processor Memories: First, there is the CPU itself. There are special registers in it that can temporarily store data and results of operations. These are fast, but are also very small because of their extremely high price. These consist of the small set of high speed registers which are internal to a processor and are used as temporary locations where actual processing is done. Processor registers provides the fastest possible access, only hundreds of bytes.

2. Cache Memory: It is logically positioned between the internal memory (registers) and main memory. It stores or catches some of the content of the main memory which is currently in use of the processor. Caches are used to temporarily store the information most frequently used by the CPU and by the user. They are also small, but can be extremely fast—almost as fast as the CPU itself.

Level 1 cache is provided to get accessed in just a few cycles, usually tens of kilobytes. Level 2 cache has higher latency than L1 often 512KB or more.

3. Primary Memory or Main Memory or Partially Internal Memory (Also called Immediate Access Memory IAS): It is a large memory which is fast but not as fast as internal processor memory. This memory is accessed directly by the processor. RAM is the main memory of a computer. It is used to temporarily store all information currently in use by the CPU. DRAM (may take hundreds of cycles, but can be multiple gigabytes.

4. Virtual Memory: However, sometimes the RAM is not big enough to store all of the programs in use at one time. This is when the virtual

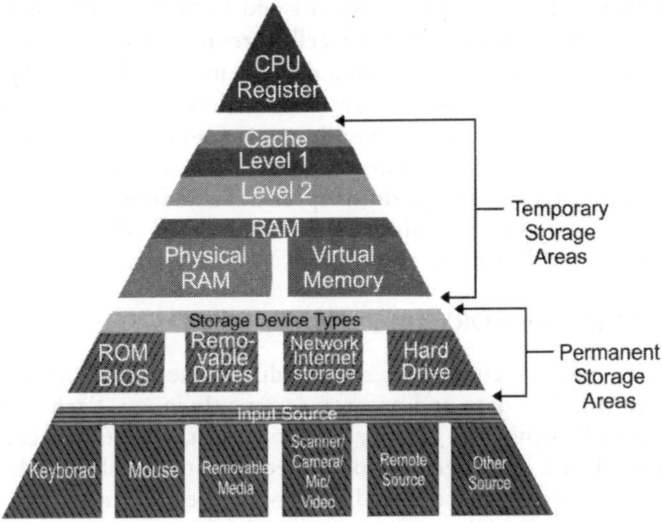

Fig. 1.26: Computer Memory Pyramid

memory feature of a hard drive comes into play. Virtual memory is used to free up space in the RAM so that there will be more room for programs currently in use. It copies areas of the RAM that have not been recently used into the hard drive, thus freeing up space.

5. Secondary Memory or Auxiliary Memory or Backing Store: The word storge or mass media is also used for this Auxiliary memory in fact is much larger in size than main memory but is, slower than main memory. It normally stores system programs (programs which are used by system to perform various operational functions), other instructions, programs and data files. Secondary memory can also be used as an overflow memory in case the main memory capacity has been exceeded (How? The answer is not supplied in the block. You need to refer to further readings to get this answer). Secondary memories cannot be accessed directly by a processor. First the information of these memories is transferred to the main memory and then the information can be accessed as the information of main memory.

4.9 MEMORIES CELLS

It is important to get a basic idea of how memories, in general, work. First, memories are generally laid out in a spreadsheet form—they look like tables with many columns and rows. The cells of these tables are the memory cells. Each cell stores one bit of information. Furthermore, all of the data contained in a memory has an address, which is like your house address. The address pinpoint the exact location of the data, just like the address of your house pinpoints the exact location of where your house is. The instructions telling a computer what tasks to perform specify the address of the data it needs. The data can be easily gathered by using this address to locate the memory cell that it is in.

4.10 MAIN MEMORY OR PRIMARY STORAGE

Primary memory consists of semiconductor memory chips and is used to store the data and programs currently in use. Each storage element of memory is directly (randomly) accessible and can be examined and modified without affecting other cells and hence primary memory is also called Random Access Memory (RAM). Main memory stores a variety of critical information required for processing by the CPU.

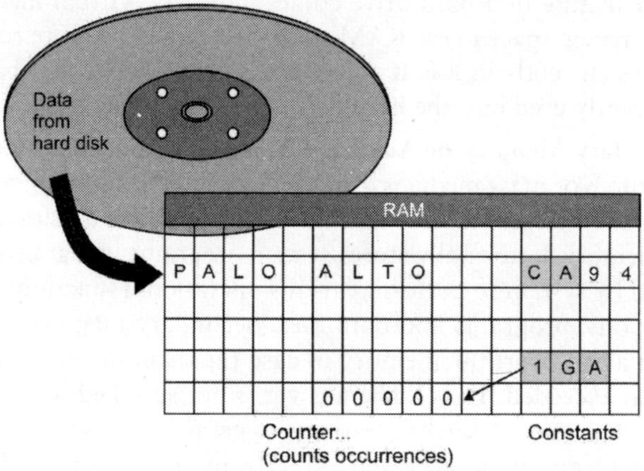

Fig. 1.27: Working of ROM

The memory unit stores all the information in memory cells, also called memory locations, in binary digits. Each memory location has a unique address. The contents of the desired memory locations are provided to the central processing unit by referring to the address of the memory location. The amount of information that can be held in the main memory is known as memory capacity.

4.11 READ ONLY MEMORY (ROM)

ROM is "built-in" computer memory containing data that normally can only be read, not written to. ROM contains the programming that allows your computer to be "booted up" or regenerated each time you turn it on. Unlike a computer's random access memory (RAM), the data in ROM is not lost when the computer power is turned off. The ROM is sustained by a small long-life battery in your computer.

ROM or Read Only Memory is pre-set memory. This means the information on the chip is set by the factory before the user gets it. ROM chips are composed of transistors that are set in a permanent on or off positions (1's or 0's). Since ROM memory is permanent, the access time for the memory is very fast. ROM memory is often used for a computer's BIOS. ROM is very expensive to design and manufacturer.

Non-volatile memory is a general term for all forms of solid state (no moving parts) memory that have a continuous source of power and do not need to have their memory contents periodically refreshed. This includes all forms of Read Only Memory (ROM) such as PROM, EPROM, EEPROM, and flash memory. It also includes Random Access Memory (RAM) that is powered with a battery.

4.11.1 Programmable Read Only Memory (PROM)

A variation of the ROM chip is programmable read only memory. PROM can be programmed to record information using a facility known as a PROM-programmer. However, once the chip has been programmed the recorded information cannot be changed, *i.e.* the PROM becomes a ROM and the stored information can only be 'read'. PROM is also non-volatile storage.

4.11.2 Erasable Programmable Read Only Memory (EPROM)

EPROM or Erasable Programmable ROM was developed to allow programmers to reprogram permanent read only chips. The EPROM works by depositing small electrical charges on the chip that are semi-permanent. The EPROM is erased and reprogrammed by exposing the inside of the chip to Ultra Violet light (UV), through a small window on top of the chip. After the erasure is complete, the EPROM chip may be reprogrammed. The small window on top of the chip distinguishes the EPROM chip from others. EPROM chips were used in the old IBM PCs and XTs for storing the BIOS information.

4.11.3 EEPROM

EEPROM or Electrically Erasable memory was developed to allow the erasure of ROM chips on the fly. Sending a series of special electrical signals through the chip erases EEPROM chips. The EEPROM chip is often referred to as the CMOS BIOS chip in computers. The CMOS chip is constantly supplied with power by the CMOS battery. The CMOS chip is permanently embedded with information about the BIOS, but it also temporarily has the settings for BIOS stored on it.

4.11.4 Electrically Alterable ROM Memory (EAROM)

A recently developed type of memory can be programmed and erased by electrical signals. This type is called Electrically Alterable Read Only Memory (EAROM). It does not require exposure to ultraviolet light to erase its contents as EPROM memory does, and provides an easy means to load and store temporary or permanent information in a form of ROM memory. EAROM memory, similar to EPROM memory, uses small charges of electricity to represent the information contained in it. Information loaded in this memory can be retained for many years without any power supplied; one of its new applications is as a back-up to RAM memory whose contents is lost in a power failure.

A special form of EEPROM is *flash memory*, which uses normal PC voltages for erasure and reprogramming.

4.12 Random Access Memory (RAM)

The term random access means that any word in the memory may be accessed, without having to go through all the other words to get to it. This memory, being monolithic (that is, being contained in an integrated-circuit chip), is much more suitable to the microprocessor. The power requirement for these memory chips is very similar to that of the microprocessor itself.

The only drawback of RAM memory is that it is a volatile form of memory. This means that when the power is removed from the chip, all the memory content is lost. When the power is returned, the content of memory will be unknown.

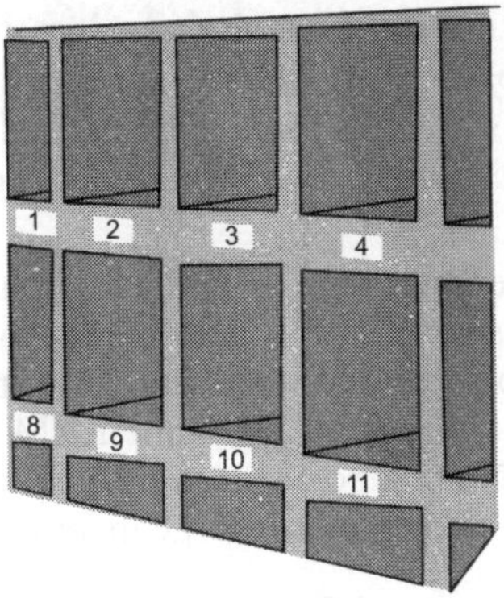

Fig. 1.28: Addresses in RAM

RAM can be of two types:

- Dynamic RAM (DRAM)
- Static RAM

4.12.1 DRAM

Dynamic RAM is a type of RAM that only holds its data if it is continuously accessed by special logic called a refresh circuit. Many hundreds of times each second, this circuitry reads and then re-writes the contents of each memory cell, whether the memory cell is being used at that time by the computer or not. If this is not done, then the DRAM will lose its contents, even if it continues to have power supplied to it. This refreshing action is why the memory is called dynamic.

All PCs use DRAM for their main system memory, instead of SRAM, even though DRAMs are slower than SRAMs and require the overhead of the refresh circuitry. It may seem weird to want to make the computer's memory out of something that can only hold a value for a fraction of a second. In fact, DRAMs are both more complicated and slower than SRAMs.

The reason that DRAMs are used is simple: they are much cheaper and take up much less space, typically 1/4 the silicon area of SRAMs or less.

4.12.2 SRAM

Static RAM chips are also volatile storage devices, but as long as they are supplied with power, they need no special regenerator circuits to retain the stored data. Since it takes more transistors and other devices to store a bit in a static RAM, these chips are more complicated and take up more space for a given storage capacity than do dynamic RAMs. Static RAMs are thus used in specialized applications, while dynamic RAMs are used in the primary storage sections of most computers.

Static RAMs are faster than dynamic RAM a typical Static RAM today has an access time of 80 nano-second which means that it would take 80 nano-second to read from or write into any location. The corresponding figure for a Dynamic RAM is 120 to 200 nsec. However, Static RAMs tend to be more expensive than their Dynamic counterparts and are not used for large memories.

4.13 Secondary or External Storage (Memory)

External memory can be of several types but following types are quite popular:

- Magnetic tapes
- Magnetic disks
- Optical or Laser disk
- Audio cassette

Data is stored on disks and tape in the same way music or television programs are recorded on tape with a difference that in these cases data is stored/recorded digitally. In both cases, a thin plastic film is coated with a thin layer of an oxide of iron—chemically similar to plain old rust. The write/record head of the drive is a sensitive electromagnet, whose magnetic field rapidly changes with the data that is being fed into it. As the magnetic field changes, the head leaves a track of magnetized particles in the film. When the read head passes over these tracks, its field is affected by the fields of the particles, and the data or music can be accurately read or played. The little particles of iron will retain their magnetic tracks for periods of us to several years, but can also be reused, for new data, over and over again.

Oddly enough, ordinary audio cassette tapes can be used as external memory for computers. They are not 'random-access' of course and this makes them slow to use. But cassette tapes and recorders are cheap and widespread, and many home computers depend on them.

Advantages and Limitations of Secondary Storage

- Secondary storage may be used to store programs and data in a machine-readable form for processing at a later date, or for holding data on a temporary basis during the processing of a program. It provides a means of retaining information on a permanent basis.

- Besides providing space for retaining working files, secondary storage may be used to retain copies of files.
- Secondary storage is external to the CPU. A secondary storage device is therefore also a peripheral device, just as is any input or output device.
- Some secondary storage devices permit random access to data, others only serial access.
- Disk devices provide random access. All items of data are equally accessible, or, put another way, it is possible to go direct to any item.

4.14 MAGNETIC DISKS

Magnetic disks can also be used as input or output device and for auxiliary computer's memory. Magnetic disks are getting very popular with modern computer. A magnetic disk is similar to gramophone record but is coated with ferromagnetic materials and rotates at very high speeds of 1800 or 3600 r.p.m. (revolutions per minute). With the help of both the devices, we can have direct access to any information, irrespective of its location on disk.

There are two classifications of drives relating to the operation of the read/write mechanism, these are:

- Fixed-head and
- Moving-head.

On a fixed-head drive, for every usable surface (not usually the very top one or very bottom one of the stack) there is a set of read/write heads equal in number to the number of tracks. The time taken to find an item of data is of the order of 10-15 milliseconds. On a moving-head drive there is only one read/write head per disk surface. The heads are fixed to arms that move in and out in unison between the disks. Access time is slightly slower, 25–100 milliseconds, since the arm has to move to the right position to seek out the specified track. Transfer rates to and from disc and main memory are in the order of 100000 to 2000000 characters per second.

A magnetic disk is a metal or plastic platter that resembles a groove less phonograph record. Let's use this phonograph record analogy to get an idea of how a disk storage device works. Suppose you have a sound tape player and a record player. Suppose, too, that

10 songs are recorded on your favorite tape, and your favorite record has seven pieces of music. Now what must you do if you only want to listen to (or access) the sixth song on the tape and the fourth song on the record? To get to the sixth song on the tape and the fourth song on the record ?

To get to the sixth tape song, you must put the tape on the player and wait until the tape used to record the first five songs has moved through the player. Although your player may "fast forward" the tape quickly past the first five pieces, there's still a delay of several seconds. To get to the fourth song on the record, you can do one of two things. You can place the record on the player, position the pickup arm at the beginning of the first song, and wait until the arm has played the first three pieces. (Following the stored sequence in this way is logically similar to accessing the music on a tape.) Or, you can directly move the pickup arm across the record to the groove where the fourth song begins.

Like the songs on a sound tape, the data records organized on a magnetic tape must be retrieved according to the storage sequence. And like the music, the data recorded on a magnetic disk can also be accessed in a sequence if they have been organized in a way that supports such retrieval. Read/Write heads are tiny electromagnets that can read, write, or erase the polarized spots that represent data on magnetic media. When such heads are fastened to an arm in a disk storage device, they can be moved quickly and directly to any disk location to store or retrieve data. The facts stored on a disk can be read many times. They will remain indefinitely or until the disk surface is erased and reused.

4.14.1 Disks and Disk Drives

Many different kinds of disk are there and but in 1994 only two types are being used with the older computers: Hard disk drives, which are non-removeable, fast, and hold a considerable amount of information; and floppy disk drives, which are slower than hard disks and do not hold as much information, but you can use removable diskettes in the floppy disk drives so that you can transfer and share data between two or more computers. However now a -days a wide variety of disks are available of which details follows:

- Floppy disks 5.25 and 3.5 inch size run by A: and B: drives
- Hard disk run by C: drive

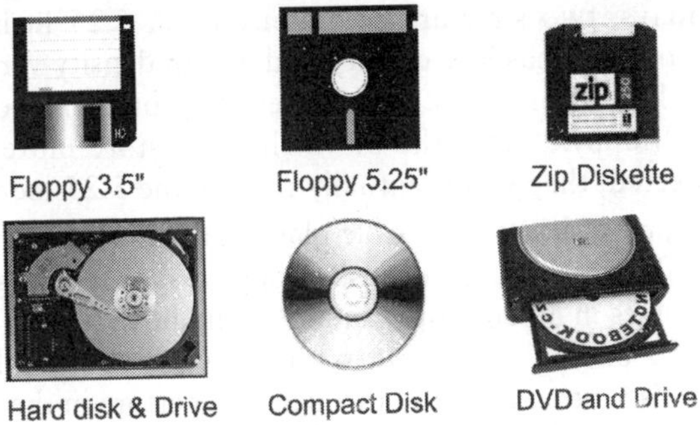

Fig. 1.29: Disk Types

- Compact disks Read Only Memory (CD-ROM)
- Digital Versatile/Video Disk (DVD)

4.15 FLOPPY DISKS

With the advent of the personal computer, a way was needed to distribute software. The Solution was found in the diskette or floopy disk, a small, removable medium so called because the early ones were physically flexible. The floopy disk was actually invented by IBM for recording maintenance information about is mainframes for the service staff, but was quickly seized on by personal computer manufacturers as a convenient way to distribute software for sale.

Unlike hard disks, where the heads float a few microns above the surface, floppy disk heads actually touch the diskettes. As a result, both the media and the heads wear out comparatively quickly. To reduce wear and tear, personal computers retract the heads and stop the rotation when a drive is not reading or writing. About half a second while the motor gets up to speed.

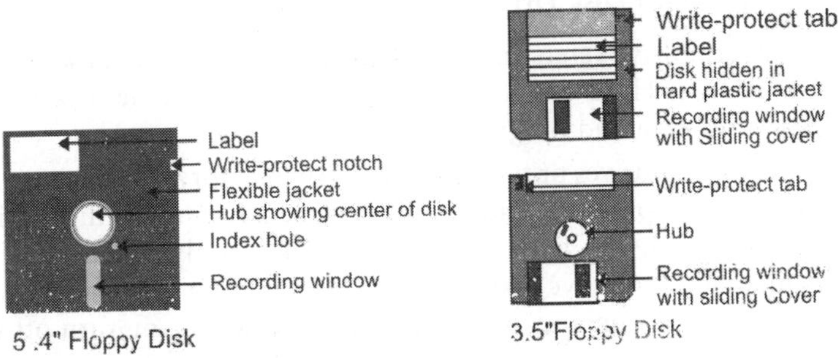

Fig. 1.30: Floppy Disk

Nowadays, two sizes are commonly used, 5.25 inch and 3.5 inch. Each of these has low-density and a high density version. The 3.5 inch diskettes come in a rigid jacket for protection, so they are not really "floppy." Since the 3.5 inch disks store more data are better protected, they will eventually replace the 5.25 inch ones.

The floppy disk uses a flexible plastic base in place of the more conventional rigid, metal-based disk. Floppy disks are changable, and each disk comes in an envelope, as shown in above figure. The disks are mounted on the disk drive with the envelope in place, and information is written and read through an aperture in the envelope. The envelope gives handling protection; in addition, it has a special liner that provides a wiping action to remove wear products and dirt, which would be abrasive to the media and head if left on the surface.

4.16 HARD (MAGNETIC) DISK

The hard disk is the computer's filing cabinet wherein the operating system and all data and program files are stored (See Box). When one switches on the computer, the operating system is automatically loaded into memory from the hard disk making the computer system operational. Not long ago 10 MB equal to around 10,00,000 characters, was a huge amount of storage area and cost the Moon, today disks with higher capacities are relatively cheaper. A 40 GB, (approximately 40,000 MB) hard disk is quite sufficient. However, newer programs are huge and can occupy much hard disk space. If one tries out many games or makes use of development tools, then a higher capacity hard disk may be a good idea. For example, a full installation of Visual C ++5.0, a popular software development program occupies about 500 MB of disk space.

4.16.1 Hard Disk Drives

Ther term drive refers to the box in which you insert diskettes and CD-ROMs. Hard disk drives contain non-removable disk platters that hold data and programs.

The disk drives provide for long-term memory. When you power off your computer the data on the disks remains on the disks. Like a cassette tape, the data leaves a disk drive only when you issue commands to remove that data. Disk data is often called non-volatile memory. The contents of RAM disappears when you turn off your computer.

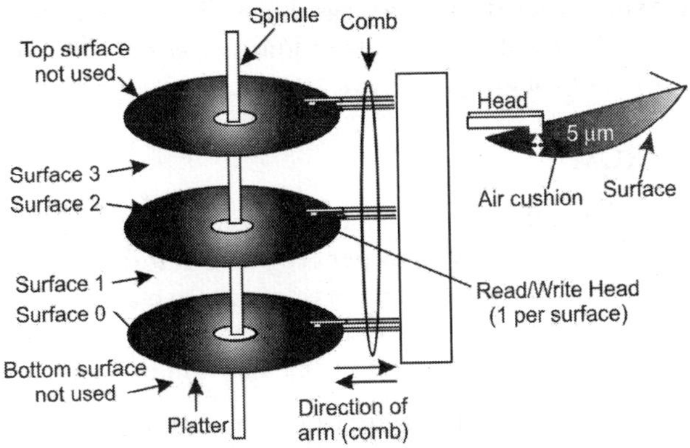

Fig. 1.31: Hard Disk Platters

This disks and RAM work in tandem with each other. When you need to run a program, such as a word processor, you'll instruct the computer to load that program from a disk drive into RAM memory. Once loaded into RAM, your CPU can then execute the instructions in that program.

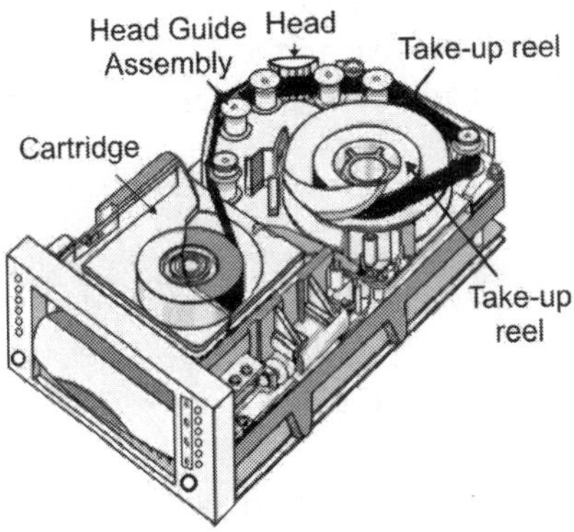

Fig. 1.32: Hard Disk

Average seek time: The average amount of time it takes a disk drive to locate a specific track and sector.

4.16.2 Access Time

Time interval between the instant that a piece of information is requested from a memory or peripheral device and the instant the information is supplied by the device. With hard disks or compact discs, maximum access time is measured as the time it takes to move from one end of the disc to the other, find a piece of information, and transfer that information to RAM.

4.17 CD-ROM

In 1986, Philips and Sony introduced compact disk technology to the consumer market as a way to deliver audio information in digital format. The compact disk (CD) is a thin piece of clear polycarbonate plastic with a metal layer covered with a protective layer of lacquer, with the aluminimum layer measuring 120 mm in diameter.

CD (Compact Disc): A thin platter that has computer data or music recorded on it in optical form. See CD-ROM.

CD-R: A CD which can be written on by the user. A CD-RW can be written, erased and re-written.

CD-ROM (Compact Disc Read only Memory): The term is often used for the CD platters as well as for the drive which reads them.

A CD-ROM: Drive is a necessary component of almost any multimedia system. (The only way around it is to use laser disks, and they are more expensive and less popular. Standard CD-ROM drive are currently quadspeed to 64 speed drives are available.

Fig. 1.33: CD-ROM and CD Driver

A single CD-ROM can hold the entire text of an encylclopedia and video clips, enhanced with sound, to illustrate many of the items in the encyclopedia. The newest and most popular titles are interactive in nature. These products enable the user to react to the presentation and determine how it will proceed. This type of product stimulates the senses, and some studies indicate that interactive electronic books can maintain the user's attention better than traditional books.

There are nearly three miles of pits (the spaces between the pits are called lands) that wind in a spiral from the center of the CD to the edge. These pits measure from 1 to 3 microns long and are about 1/10 micron deep and 1/2 micron wide. A laser in the CD player reflects off the aluminimum as the disk spins. The laser focus is affected by the pits, and this difference in focus is interpreted into a series of 0s and 1s, making up the binary information of the digital audio samples.

The speed refers to the rate at which data can be transferred from the CD to the computer or monitor. A single speed CD-ROM can transfer 150 killobytes of data per second Most CD drives can only play the disks, but prices are coming down on drives that can write as well as play.

The disk spins at a constant linear velocity, which means that the rotational speed of the disk may vary from 530 rpm when it is reading near the center of the disk to about 200 rpm when it is reading near the outside of the disk. This allows the data to be read at a sustained data rate of 150k per second This is sufficient for 16-bit 44.1 stereo audio, but is not sufficient for reading large computer data files. There are now CD-ROM players that read information up to 8 times. This speed (1200 K per second), and that will probably continue to increase in speed

4.17.1 CD-ROM and Information

The basic unit of information in a computer is a byte. A byte is eight bits-one or zeroes that are used to store information.

- **Text:** one page of text can require 500 to 1,000 bytes.
- **Pictures:** A single full-screen can require 300,000 to 1,000,000 bytes.
- **Audio:** One minute of music can require 2,500,000 bytes
- **Video:**One minute of video can require 10,000,000 bytes.

CD-ROMs can hold over 650,000,000 bytes of data as compared to a 3,5 inch diskette, which can only hold 1,400,000 bytes. Having so much storage space on a CD-ROM means that you can deliver impressive amounts of content for a wide variety of applications.

Presenting multimedia in this form has many advantages, which explains the explosive growth of this medium. Some of these advantages include:

- **CD-ROM** is inexpensive to duplicate; depending on volume, the perunit cost of a disk can be less than a dollar.
- **CD-ROM** is easy to distribute, with the format having found a home in many types of retail environments : book stores, record stores, software stores, mail-order houses, and so on.

- **CD-ROM** has an enormous user base. At the time of this writing, approximately 17 million CD-ROM drives are in use here in the U.S. The vast majority of new desktop computers now being sold feature a CD-ROM drive, which for now and the future ensures a sizable target audience for CD-ROM content.
- That content can be of several types. The most common incarnation is that of a consumer product title that is meant to be bought and used on the merits of its content. That includes games, entertainment titles, reference works, clip, art collections, how to disks, and whatever else might be successful subject material for a CD-ROM production.

4.18 DIGITAL VERSATILE/VIDEO DISK (DVD)

To provide for the next generation of software, a new format has been created that is abbreviated DVD. This sometimes stands for digital video disk and sometimes digital versatile disk, depending on whom you ask.

DVD uses the same physical format as CD-ROM but the similarities end there. The physical and logical formats are considerably different. It is likely that within the next few years, DVD may revolutionize not only how we store data on the PC, but also home audio and video as well. A future version of DVD called DVD-RAM may in fact even replaced your VCR! This is still quite new technology and it is changing rapidly.

4.19 ZIP DRIVE

A Zip drive is a removable (external) storage device that securely stores all your important computer data magnetically. Iomega Zip drives and Zip disks are durable, Zip portable, easy-to-use, and extremely efficient - one 250MB disk can store the same amount of data as 173 floppy disks!

Zip drives are available in 100 MB, 250 MB and 750 MB—the fastest and are generally used to back up and protect important data. Each Zip drive is easy-to-use, secure, and extremely durable. You can choose between built-in (internal) or portable (external) versions of the reliable Zip drive, depending on your needs. And each Zip drive comes with (Iomega) Automatic Backup software and Iomega Sync to automatically backup and synchronize your file revisions.

4.20 MINI AND MICRO MINI USB DRIVES

They are also called '*micro-mini*' or *pen drives* and are very convenient as they are portable and look like a keytag. Mini and Micro Mini USB Drives plug into any computer's USB port and are Active Disk Powered to allow applications to conveniently launch and run directly from the mini drive. They are available in sizes of 128 MB, 256 MB, 512 MB and 1 GB. It takes less than one minute to transfer data equal to 256 MB which may reqiure more than one to two hours with the help of hundreds of floppies.

Micro Mini Drives are the size of a thumbnail making them one of the smallest USB drives (year 2003). While Mini Drives are small, the Micro Mini Drives are 1/2 their size and still hold the same amount of data with fast USB 2.0 speeds. You can keep your data close to you by using our dog tag style neck chain or key clip, personalize it with 3 changeable colored caps, and use the iilluminated extension cord to reach the back of your computer easier.

Mini Drives are a little larger than Micro Mini Drives, but are still extremely small! The newly redesigned Mini Drives have a mini window enabling you to personalize them by adding a family photo or a label of your choice and are as small as a large paper clip.

4.21 OTHER COMPONENTS OF COMPUTER

Power Switch: The power switch is located somewhere on the CPU, usually in the front or on the side. You press this to start your computer. But do not press the power switch to turn off your computer unless nothing else works.

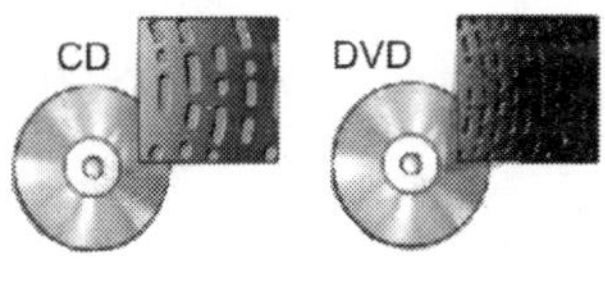

Fig. 1.34: DVD as Compared to CD

Motherboard: All of the computer's components are hooked up to a motherboard, which is the main base for the computer's circuitry and components.

Video controller: is a board in the computer that controls the monitor. It translates the data in the video memory into symbols on the monitor. It can thus draw letters and graphs on the screen.

Fig. 1.35: Mini Drive

Parallel/Serial: ports allow the computer to send data to and receive data from printers, modems, etc.

Network Interface Card A NIC connects the computer to local area networks (LANs) like CLUNet. Networks are a high speed method of transferring data from one computer to another.

6. Expansion Card: A circuit board that fits into a computer expansion slot to add a certain function (like a modem, sound card, or SCSI interface).

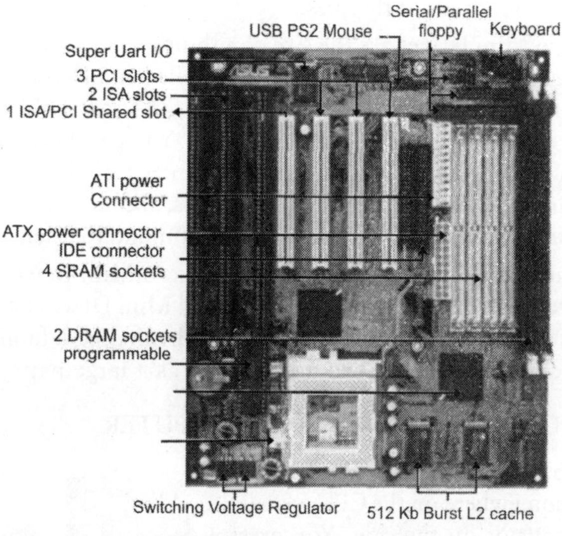

Fig. 1.36: Motherboard detail

4.22 INPUT DEVICES

Input/Output devices are the devices that connect you to your computer. Input devices let you input data and other information into your computer and they also let you give your computer special instructions so that it will know what to do. For example, you can type in data by using a keyboard, or you can input data in picture form by using a scanner. On the other hand, output devices display the results of your computer's computations. Examples of these would be your computer's monitor, which displays all of the programs you're running, as well as the printer, which will print out a hard copy of the information!

Input devices are necessary to convert our every day information into a form which can be understood by computers. Most computing activities are based on the daily collection, preparation and input of large amounts of data. Once collected, the data is either input to the computer system in one step using a removable external storage device like floppy, CD-ROM or pendrive etc.

Entering Input Data

Input data may be entered by:

- Direct-entry method,
- Or it is converted into a machine-readable form and, as a second operation, input to the computer. A great deal of manpower can be tied up with preparing data for input (data preparation) and in inputting the data. In many organizations these activities represent a significant proportion of the overall cost of computing.

4.23 TYPES OF INPUT UNITS

Input devices can be broadly categorized into:

- Online data input devices
- Offline data input devices

Data input can similarly be categorized as:

- Source data input
- Offline data input
- Online data input

Offline Data Input/Entry

Data entered directly into a computer system from online terminals can be accumulated in batches for later processing. But that data for batch processing may be transferred from source documents to a machine-readable medium through the use of offline devices. Offline devices are those devices which don't consume any time of central processor and thus saves the processor time which is otherwise very valuable.

For decades, card punch machines were the most popular offline data entry device, and punched cards were used as the machine-readable input and storage medium. A batch of cards was punched, sorted into the sequence used by the master file being updated, and then carried to the computer, where it was processed. A few punched card systems are still in use, but the card essentially passed into history in December 1984, when IBM closed the last of its punched card manufacturing plants.

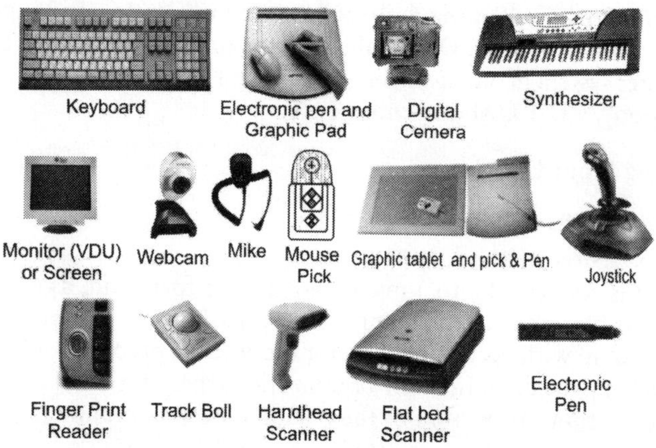

Fig. 1.37: Input Units

Online Data Input Devices

Millions of online data entry devices are now in use, and additional millions are being installed every year. By far the most popular device for interactive processing and for the online entry of data for batch processing is the visual display terminal (VDT or VDU). These are:

- Dumb,
- Smart, and
- Intelligent VDTs.

Smart VDTs have a microprocessor and some internal storage, and they can edit, store, and consolidate input data. But they are not user-programmable.

Intelligent VDTs are user-programmable, and there's very little difference between them and many personal computers.

Functions of Input Unit: Digitizing

Input Unit is actually a device which provides man to machine communication with computer. Input of any form is first converted into binary electronic signals which can be understood by central processing unit. This process is called digitizing.

Input data may be linguistic, mechanical graphical, audio, visual or audio-visual. Linguistic input means a written program in computer's programming language which is keyed into computer either by typing on its keyboard, or by punched card or by some recorded means like floppy disk, magnetic tape or cassettes. Audio input is given to computer through a Voice Data Entry (VDE) system which has got a microphone. The Visual information is recorded in the similar manner as far television video and then inter-faced to computer. The graphical data is put in computer with the help of electronic data tablets (digitizers). Mechanical input means like joystick for tracker ball and micro-mouse are used for playing fun on the visual display unit.

4.24 LIST OF INPUT DEVICES

- Keyboard
- Pointing Devices
  - Mouse
  - Joystick
  - Trackball
  - Light pen
- Data Glove
- Picks
- Stylus
- Reading Devises : Data lapturing at source
  - OMR
  - OCR
  - MICR
  - Bar Code Readers
- Graphics tablet/digitizer
- Scanner
- Modem

4.25 KEYBOARD

To input information in the form of words and numbers, the most widely used device today is the keyboard. Keyboards are considered the most important input device to the average PC user. They serve as a translation of "key hits" into a seemingly understandable input.

Fig. 1.38: Keyboard

A computer's keyboard looks like a typewriter. Each letter has its own key. To type a letter or number, you press its key. The letters and numbers that you type show up on the computer's monitor. Letter and number keys are arranged like those on a typewriter but there are also some extra keys and particular key combinations that help us give instructions to the computer.

A computer keyboard all of the expected alphabetic, numeric, and special characters found on a typewriter. Unlike the typewriter, which records the data on paper, each character keyed is encoded into the computer system's data coding scheme and is stored in memory.

Underneath the keys there is a pressed a metal contact touches a cross-point on the grid and each cross-point has a particular pattern of electronic pulses so that the computer known which key has been pressed. The keyboard translates key pushes and releases into a binary code which it sends to the keyboard controller.

Most keyboards have an ASCII (American Standard Code for Information Interchange) layout. This layout includes a numeric keypad, special function keys, special control keys, as well as typing keys.

Numeric Keypad: The numeric keypad is the group of 17 keys on the very right of your keyboard. It includes all of the numbers from 0-9, a plus sign, a minus sign, and a few other various keys. If you look closely, you'll realize that the arrangement of the keys on a numeric keypad on your keyboard is quite similar to the arrangement of the keys on a calculator! Because many computers are used for business purposes and to input numbers into a computer, the numeric keypad was designed. This made it easier and faster to input numbers.

Function Keys: The function keys can be found in a row at the very top of your keyboard. They are labeled "F1," "F2," "F3," all the way up to "F13." These function keys are mainly used to simplify the use of applications. They perform special commands, depending on which application is open. For example, pressing F7 in a word document opens the spell check. In almost all applications, pressing F1 opens a help menu. Also, the function keys can be combined with other keys, like the control keys to perform different tasks!

Control Keys: The control keys are special keys like del (delete), ctrl (control), alt (alternate), shift, and enter. The delete (or backspace) key is used to erase characters in a word document. The esc (escape) key is generally used to exit out of a program or application, while when you strike the enter key, a function or command is carried out. The tab key is primarily used to indent in a word document. The shift key is often used in conjunction with other keys. For example, pressing shift and a letter will make that letter capital, while pressing shift and a number will display the character shown right above that number. Finally, there are the ctrl and alt keys. Like the shift key, these are also used in conjunction with other keys, and they usually perform functions similar to those of the function keys. There are also the cursor movement keys. These are the arrows shaped like an inverted T located between the typing keys and the numeric keypad. These keys are used to move the cursor around on your screen. For example, if you were in a word document, the blinking cursor will move up and down to different lines if you press the up and down cursor movement keys.

There is no standard computer keyboard, although many manufacturers imitate the keyboards of PCs. There are actually three different PC keyboards: the original PC keyboard, with 84 keys; the AT keyboard, also with 84 keys; and the enhanced keyboard, with 101 keys. These three differ somewhat in the placement of function keys, the Control key, the Return key, and the Shift keys.

There are several different types of keyboards for the Apple Macintosh. All of them are called ADB keyboards because they connect to the Apple Desktop Bus (ADB). The two main varieties of Macintosh keyboards are the standard keyboard and the extended keyboard, which has 15 additional special function keys.

QWERTY Keyboard

The standard layout of letters, numbers, and punctuation is known as a QWERTY keyboard because the first six keys on the top row of letters spell QWERTY. The QWERTY keyboard was designed in the 1800s for mechanical typewriters and was actually designed to slow typists down to avoid jamming the keys. Another keyboard design, which has letters positioned for speed typing, is the Dvorak keyboard.

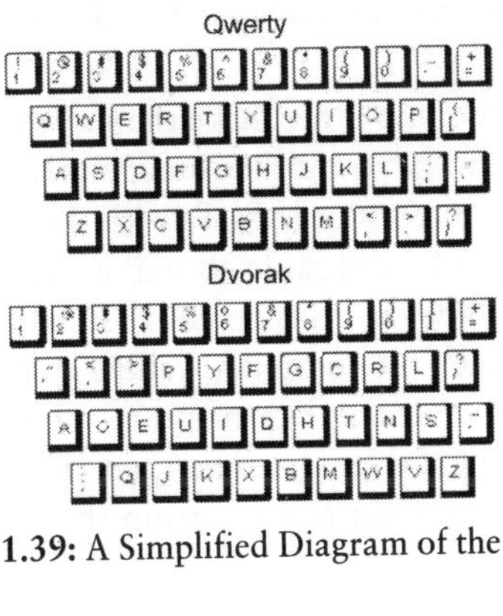

Fig. 1.39: A Simplified Diagram of the CPU

Dvorak Keyboard

In the past few years, considerable interest has been accorded the DVORAK keyboard which arranges keys differently. The DVORAK keyboard is designed for rapid data entry, or typing, and we should expect to see many more such keyboards in the near future. It gets is name from its inventor, August Dvorak.

Unlike the traditional QWERTY keyboard, the Dvorak keyboard is designed so that the middle row of keys includes the most common letters. In addition, common letter combinations are positioned in such a way that they can be typed quickly. It has been estimated that in an average eight-hour day, a typist's hands travel 16 miles on a QWERTY keyboard, but only 1 mile on a Dvorak keyboard.

Wireless Keypad: The wireless keypad is a keypad that does not need to be connected to the computer via wires! It can operate anywhere within a 3 meter radius of the Base Station, and it even includes a folding palm rest! The wireless keyboard is designed for as much comfort to the user as possible. It allows for more movement and can be placed at a comfortable location. This keyboard also includes special multimedia keys and it connects to the computer via a PS/2 or USB connection.

4.25.1 The Speed of Entering Data through Keyboard

For a small business, the keyboard supplied with the computer is normally sufficient to enter the data into the computer. A reasonably efficient operator can enter 60 to 100 records per hour, of approximately 100 characters each, into the computer. As a rule of thumb, an equal amount of time may be set apart for the generation of various reports .

Where volumes of input are more than 500 transactions per day it may be desirable to have another independent terminal or have a multiterminal computer like IBM-PC/AT. In case of such higher volumes, it may also be desirable to make use of an ofline device, shich is a specialised device to enter and record data at a faster rate without being a load on the computer. Where graphice is an important application, an input device like a mouse or an elecltronic tablet may be needed. Larger computer centres generally operate on multiple shifts to increase the output per day from the computer.

4.26 POINTING AND POSITIONING TOOLS FOR INPUT

These are devices that the operater can move, which in turn signal the cursor on the screen to move in the same direction and speed. Their principal advantages are ease of use and speed.

4.26.1 Pointing Devices

Pointing devices, such as a *mouse* or *trackball*, are used for other traditional keyboard functions, like sending commands and positioning the cursor. Both of these devices are designed to move a pointer around the screen and point to specific characters or objects. The most common type of mouse has a ball on its underside that allows it to roll around on the desktop. The trackball remains stationary on the desk while the user moves the protruding ball to control the pointer on the screen. Both devices have one or more buttons that can be used to send signals to the computer, conveying messages.

As you slide the mouse acrohss your desktop, a pointer echoes your movements on the screen. You can click the mouse_press the button while the mouse is stationary_or drag it_move it while holding the button down.

Examples of other pointing devices:

- Mouse
- Joystick
- Trackball
- Picks Stylus

The touch pad or graphics tablet is particularly popular with artists and designers. The touch screen can respond when the user points to or touches different screen regions.

4.27 MOUSE

A device that controls the movement of the cursor or pointer on a display screen. A mouse is a small object you can roll along a hard, flat surface. Its name is derived from its shape, which looks a bit like a mouse, its connecting wire that one can imagine to be the mouse's tail, and the fact that one must make it scurry along a surface. As you move the mouse, the pointer on the display screen moves in the same direction. Mouse contains at least one button and sometimes as many as three, which have different functions depending on what program is running. Some newer mice also include a scroll wheel for scrolling through long documents.

In particular, the mouse is important for graphical user interfaces because you can simply point to options and objects and click a mouse button. Such applications are often called point-and-click programs. The mouse is also useful for graphics programs that allow you to draw pictures by using the mouse like a pen, pencil, or paintbrush.

How a mouse works

1. There is a ball on the underside of the mouse that spins as you move the mouse around.
2. Inside the mouse there are two rollers. In most mice, these rollers look like black bars. If you open up the disk on the bottom of the mouse and take the ball out, you will be able to see these rollers. One of the rollers detects the horizontal movement of the mouse, while the other detects the mouse's vertical movement. Working together, the two rollers can determine which direction the mouse is moving!
3. When the ball rolls, it causes the rollers to roll. When the rollers roll, they cause a special disk to spin. The disk is circular with rectangular holes spaced out along its outer edge.
4. On one side of the disk there is an infrared LED, while on the other side of the disk, opposite the LED, there is an infrared sensor. When the disk spins, the light from the LED can go through the rectangular holes periodically.
5. The sensor is able to sense these pulses of light.
6. A processor directly built into the mouse converts these pulses into binary digits that the computer can understand, and the information is transferred to the central processing unit by through the mouse's cord.

Wireless Mouse

The wireless mouse, like the wireless keyboard is specifically designed for comfort and mobility. It can be positioned anywhere within a 3 meter radius of the Base Station. This maximizes the comfort of the user. In addition, the mouse contains a scrolling wheel and three programmable buttons.

4.28 TRACKBALL

One day, someone came up with the idea to flip the mouse over, and they called it a trackball. A trackball allows for more detailed movement of the mouse pointer. The mouse itself is motionless, but

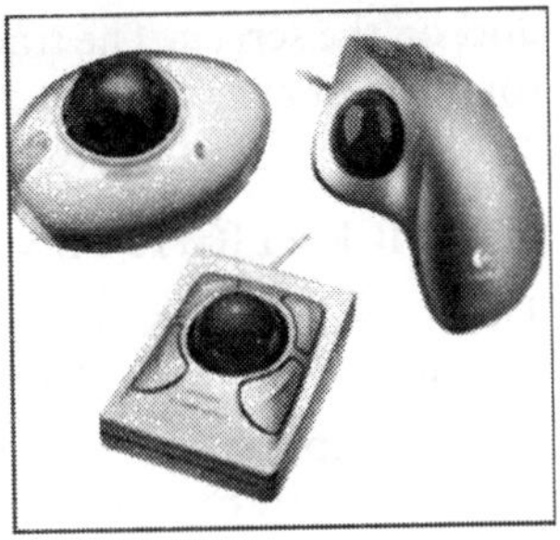

Fig. 1.40: Trackball

the motion comes from a giant ball in the center that rotates by touch. The technology behind the ball is a pure adoption of the opto-mechanical design. Trackballs have been modified into many shapes and sizes and are rarely seen in the old "blocky" design.

A wide variety of desktop and panel mount trackballs or trackball assemblies are available. These devices replace the mouse in computer systems where rapid, accurate pointing functions are desired (CAD/CAM, graphics, workstations, kiosks, etc.), and in industrial/military computing and control environments where a mouse cannot be used due to space limitations, harsh environments, or other factors.

4.29 GAME PAD

The Game Pad consists of controllers that are specially designed for various computer games. Like the mouse and keyboard, it, too, works anywhere within a 3 meter radius of the Base Station and is designed for comfort. Its sleek design and customizable controls allow for the greatest comfort for its users and its target precision and character profile sharing make it easy to use. In addition, the Game Pad allows for multiple players!

4.30 JOYSTICK

Fig. 1.41: Joystick

Children can play with computers in a simple way by use of a joystick or tracker ball. While playing video games like SPACE-INVADERS, the user need to move the space craft quickly across the screen. Although this can be done by pressing a key on the keyboard but it is neither convenient nor comfortable for small children to use the keyboard. This type of action can be better done with the help of joystick and tracker ball which are not only easier to use, but provide a better control.

As you move the stick around in its socket the movements are translated into binary instructions with the help of electrical contacts in the base of joystick. When the movement of firing comes, a button provided at the base is pressed that sets off an explosive pattern of dots on the screen. The tracker ball also does the same thing but it is round in shape. Both of these devices allow you to move objects around the screen easily.

4.31 PICKS: LIGHT PEN

These are devices used to point at objects or positions on the screen. The most common is the light pen, which is very easy to use.

An input device that utilizes a light-sensitive detector to select objects on a display screen. A light pen is similar to a mouse, except that with a light pen you can move the pointer and select objects on the display screen by directly pointing to the objects with the pen.

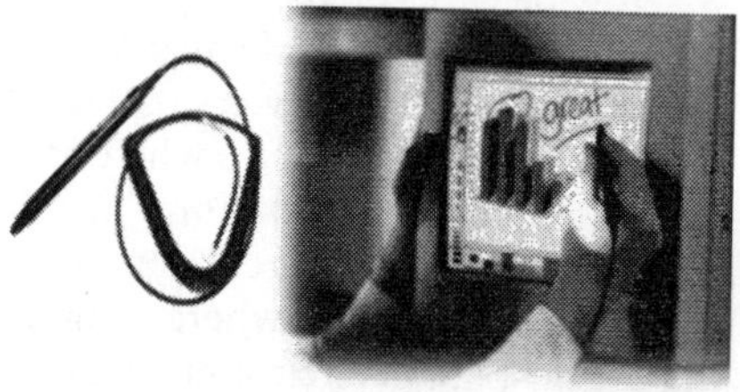

Fig. 1.42: Light Pen and Graphic Pad

The lightpen is one of the most common manual input device on graphic displays (and even on some alphanumeric ones), for use in identifying ("picking") information displayed on the screen for subsequent computer processing. Handwritten text analysis is critical in a pen-based computer. This keyboardless machine accepts input from a stylus applied directly to a flat-panel screen. Many people think the real future of pen-based technology is in personal digital assistants (PDAs), which serve as pocket-sized organizers, notebooks, appointment books, and communication devices for people on the go.

Light pens are the input devices of choice for Point of Sale applications used in restaurants, hotels, and retail establishments.

4.32 GRAPHIC INPUT DEVICES

Analog devices: This category includes a variety of devices that can be used to measure, point to, or code cordinate positions by converting analog signals (*e.g.*, speed, direction, distance, rotation) to digital signals. They are often used to trace drawings so that the designs can be entered into th graphics systems. There are two basic approaches to analog input device.

Flat-surface devices: There are digitizer tablets that trace drawings. They can use a number a different techniques, such aselectromagnetic, touch sensitive, sonic, incremental, and light detector techniques. Each differs in speed, accuracy, and cost. The electromagnetic digitizer is the most common and precise.

Fig. 1.43: Touch Screen

Touch Terminals (Touch Screens): Touch terminals have touch-sensitive screens, which dispense with the keyboard and allow data to be input by touching the screen with the fingertip. The surface of the screen consists of a number of programmed "touch points" each of which triggers a different input when selected by the user. A user is normally provided with a menu of prompts and does not need to be an expert or even familiar with how a computer operates. Small touch sensitive boxes are provided which act something like the function keys. This form of input is particularly suitable to applications designed for use by the general public, for example, for searching library catalogue files. Although these are user-friendly, touch screens cannot be used to enter large amount of data.

Touch screens were developed to give the user the opportunity for direct visual pointing. Unlike most pointing devices, the user doesn't have to navigate a pointer across the screen to click the desired target. Instead the user can directly touch the screen to indicate a desired action. Touch Screen technology has two options for its application: capacitive and optical.

4.33 TOUCH PAD

Touch Pads were developed as a pointing technology for laptops. The touch pad is contained in a small area and can easily be adopted into the surface of a computer laptop. The touch pad uses a capacitance method to determine the exact location of the user's touch.

Fig. 1.44: Touch Pad

4.34 DIGITIZER TABLETS (GRAPHIC TABLETS)

Also known as digitizer, data-pads, data-tablets or hand print terminals, these devices are pad-like in shape, ranging in size from 25 cm square to as large as 150 cm by 120 cm. They may be used to hand-print characters directly as input to a computer (typically transmitted via a VDU or paper tape) or, in the case of some microprocessor-based tablets, they may be special-purpose designed, for example, as a drafting tool for surveying or meteorological work.

This looks similar to a drawing board on which any illustration or drawing can be placed, and then a scanning head (usually called a "puck") is moved to a desired position and a button pressed. This transmits and coordinates of the puck position to the system, allowing essentially "analog" information to be input to the system digitally. Some allow the puck movement to create a string of coordinates ("drawing a line") if that option is chosen. Information once entered is stored by the computer;quite often this implies a three dimensional description of a number of objects which may be displayed from different angles. The components may be built into structures which in turn may be displayed. This facility is being increasingly used in architects offices to store details of building components such as doors and window frames for use in computer produced plans.

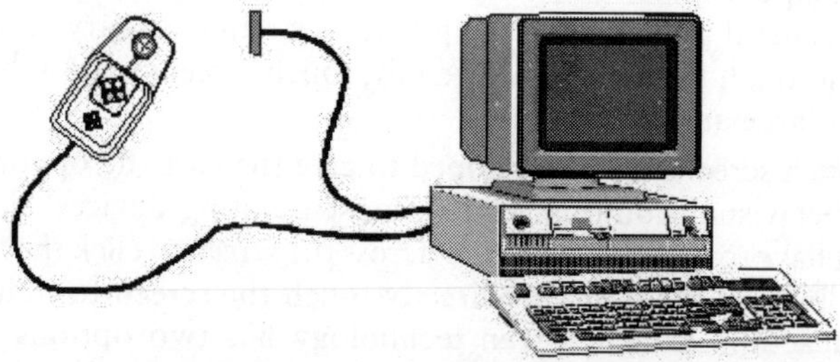

Fig. 1.45: Digitiser attached to a PC

The sensing technique by which data is captured varies. It may be based on the pressure of the pen registered beneath the surface of the pad. A pad may also have a touch-sensitive keyboard for the input of non-written characters.

4.35 SCANNERS

Scanners and digitizers have been developed to electronically store printed data. Some systems can identify printed characters. Others break down printed images into bits of data, store them, and then translate the bits into an image on a monitor or printer. The simplest scanners can only read specially shaped characters, such as those on checks or grocery products.

All scanners are essentially opto-electronic devices which convert any type of printed or written information (including photographs into digital pulses which can be manipulated by computer into any mode).

Scanners consist of two components:

- Data-capturing subsystem
- Image-processing subsystem

Data-capturing subsystem illuminates the document so the image can be recorded. Then an image-processing subsystem uses sophisticated software to convert the optical images into digital form and store the data in a file. Unfortunately, scanners often make errors that require manual correction. High intensity LASER beam is created in such devices for scanning the pictorial details. So the data files invariably require some editing at a CAD terminal.

Scanning a piece of artwork (say, a photo) requires controlling the scanner from your computer with some kind of software, which also handles transferring the digital data from the scanner to your hard drive.

4.35.1 Types of Scanners

Fig. 1.46: Flatbed Scanner

Flatbad Scanners: Also called desktop scanners, these resemble 'photostat'. The most common type of scanner is a flatbed scanner, designed for scanning relatively large, flat prints and pages. Scanning a document here is as easy as photocopying. The document or the picture to be scanned rests on a glass plate and the scanning head moves underneath it. This type of scanner is versatile and can scan flat originals of various sizes—even small 3D objects. These are also most probably the best scanners for you. By adding a transparency adapter with a separate light source, you can also use this type of a scanner for scanning slides, x-rays, and other transparent 'original'. In fact, for scanning photographs, a 300 dpi scanner is generally sufficient.

Sheetfed Scanners: The Sheetfed Scanner is the most commonly used in the large copy machines to copy a large number of documents at one time. Although, a few independent feed scanners have taken the place of flatbed scanners for low-quality scanning needs. The idea is the same as the rest of the scanners, except instead of a motorized carriage for the CCD, there is a motorized carriage for the paper.

Personal sheetfed scanners look more like a fax machine than a copier. They occupy less space on your desk and are also more expensive.

3. Hand-held Scanners: Hand-held scanners work almost exactly like flatbed scanners. The main difference is that the user becomes the motorized carriage. Therefore, the user must move the scanner across the image at a certain rate so the CCD can scan the item. The advantage of a hand-held scanner is the price. However, if the user is pushing for quality, the flatbed scanner is definitely the correct choice.

Fig. 1.47: Hand -held Scanner

4. Barcode Scanners: With the 1980 introduction of the first hand-held bar code laser scanning device, symbol changed the way the bar code was read.

Barcodes are made up of bars of different width and spacing. They are used in supermarkets (*e.g.* universal product codes (UPC)), mail sorting, libraries, etc. Optical Character Recognition Reader (OCR) reads barcodes.

Fig. 1.48: Barcode Reader

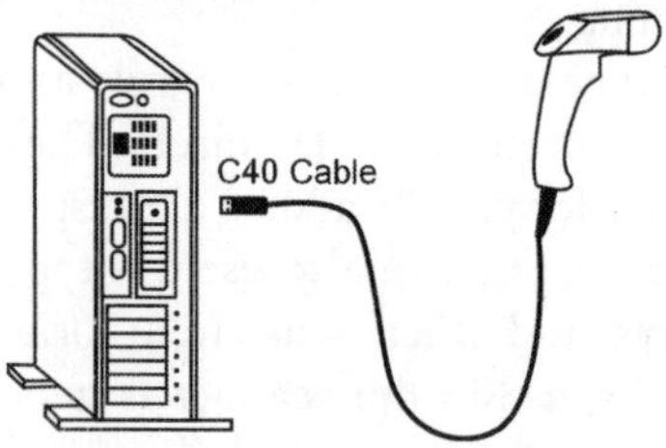

Fig. 1.49: Barcode Reader: Connection

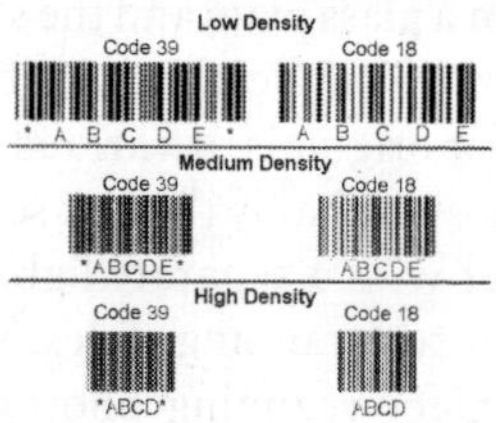

Fig. 1.50: Barcode Chart

4.36 READING TOOLS

The following reading devices provide the computer with limited ability to "read" directly from paper, converting printed information into bit patterns that can be processed by the computer:

- Optical-mark readers use reflected light to determine the location of pencil marks on standardized test answer sheets and similar forms.

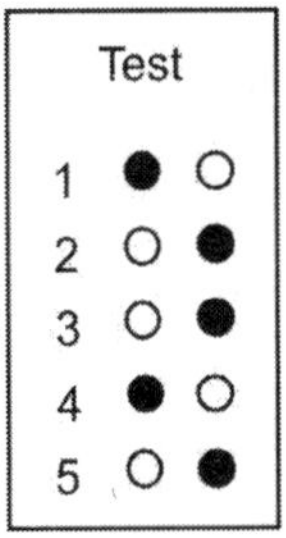

Fig. 1.51: OMR

Fig. 1.52: Optical Marks

OMR is short for optical mark recognition, the technology of electronically extracting intended data from marked fields, such as checkboxes and fill-infields, on printed forms. OMR technology scans a printed form and reads predefined positions and records where marks are made on the form. This technology is useful for applications in which large numbers of hand-filled forms need to be processed quickly and with great accuracy, such as surveys, reply cards, questionnaires and ballots. A common OMR application is the use of "bubble sheets" for multiple-choice tests used by schools. The student indicates the answer on the test by filling in the corresponding bubble, and the form is fed through an optical mark reader also abbreviated as OMR, a device that scans the document and reads the data from the marked fields. The error rate for OMR technology is less than 1 per cent.

OCR

Often abbreviated OCR, optical character recognition refers to the branch of computer science that involves reading text from paper and translating the images into a form that the computer can manipulate (for example, into ASCII codes). An OCR system enables you to take a book or a magazine article, feed it directly into an electronic computer file, and then edit the file using a word processor.

All OCR systems include an optical scanner for reading text, and sophisticated software for analyzing images. Most OCR systems use a combination of hardware (specialized circuit boards) and software to recognize characters, although some inexpensive systems do it entirely through software. Advanced OCR systems can read text in large variety of fonts, but they still have difficulty with handwritten text.

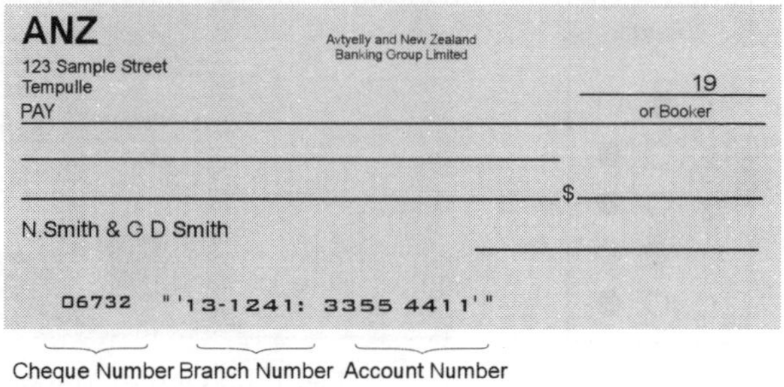

Fig. 1.53: Magnetic-ink

- Bar-code readers use light to read universal product codes (UPCs), inventory codes, and other codes created out of patterns of variable width bars.
- Magnetic-ink character readers read those odd-shaped numbers printed with magnetic ink on checks.
- Wand readers use light to read alphabetic and numeric characters written in a specially designed typeface found on many sales tags and credit card slips. In many stores wand readers are attached to point-of-sale (POS) terminals. These terminals send information scanned by the wand to a mainframe computer.

When wand readers are used to recognize words and numbers at a POS terminal, the computer is performing optical character recognition (OCR).

4.37 SOUND: MICROPHONES

The human voice, or any sound for that matter, travels through media, such as water or air, as waves. These waves hit another object and cause vibrations; in the case of people, these waves hit our ears and are turned into signals that are sent to the brain. The brain then processes these signals into the sounds with which we are familiar. Computers, on the other hand, use a microphone that acts like an electronic ear.

4.38 MIKE AND SOUND CARD

As computers get more sophisticated, they will be able to recognize the user's voice. The microphone will eventually replace the keyboard and mouse as the most used input device.

A piece of computer hardware which digitizes and allows a sound to be played back on a computer. Sounds are usually sent into a sound card via the computer's microphone. They are played back using the computer's speakers.

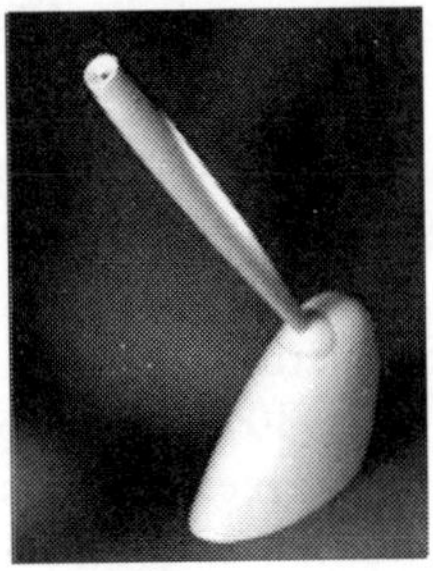

Fig. 1.54: Mike

Mike: As the sound is transformed into a marching column of numbers, the software has something it can bite into. Somehow, the software must recognize which words are represented by all of these numbers.

4.39 DIGITAL CAMERA

With a digital camera, you can improve photos after they're shot, make your own prints at home, and share them easily over the Internet. Computers are being used to create multimedia presentation. Low cost cameras like the one pictured below are being used to add movies to presentations or for video conferencing across the Internet.

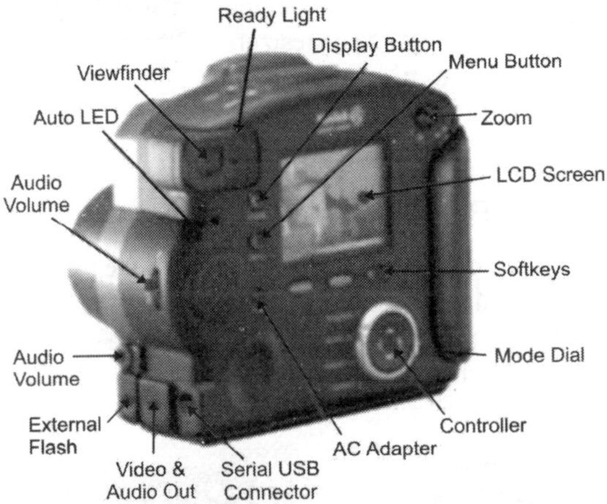

Fig. 1.55: Camera Details

Digital cameras let you take lots of shots in a way that's easily transmitted from a far, while using reusable memory instead of consumable film. The technology requires just enough fiddling to be of interest more to relatively serious amateurs than casual snapshooters. But as with so many things digital, capabilities are increasing as prices drop, making digital cameras more attractive to more people.

You can transfer the shots you like to your computer and alter them in various ways—cropping, adjusting color and contrast, adding textures and other special effects, and so on—then make them into prints, cards, and T-shirts, or send them with your e-mail. You do all this with the software that comes with the camera.

Digital cameras share many features with ordinary cameras—an electronic image sensor, autofocus, and zoom functions.

Digital cameras are typically categorized by how many pixels, or picture elements, the image sensor contains. Cheap, 'kid' cameras have 307,200 pixels or less and are priced below 5000/-. One-megapixel cameras run around 12500 to 25000, two-megapixel models, 25000 to 37500.

4.40 SYNTHESIZER: THE MUSICAL INPUT

It is an electronic instrument that can create sounds from different musical instruments such as horns and woodwinds to other sounds that cannot be duplicated with musical instruments. Synthesizers usually consists of a piano type of keyboard along with controls to choose the instruments.

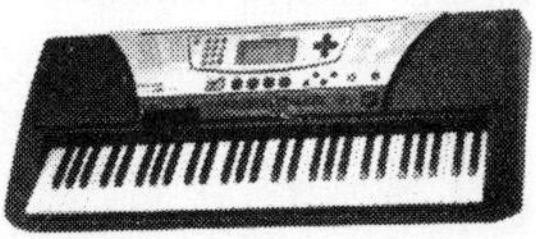

Fig. 1.56: Synthesizer

4.41 VIDEO DIGITIZER

Video digitizers are collection of circuits that can capture input from a video camera, video cassette recorder, television, or other video source and convert it to a digital signal that can be stored in memory and displayed on computer screens.

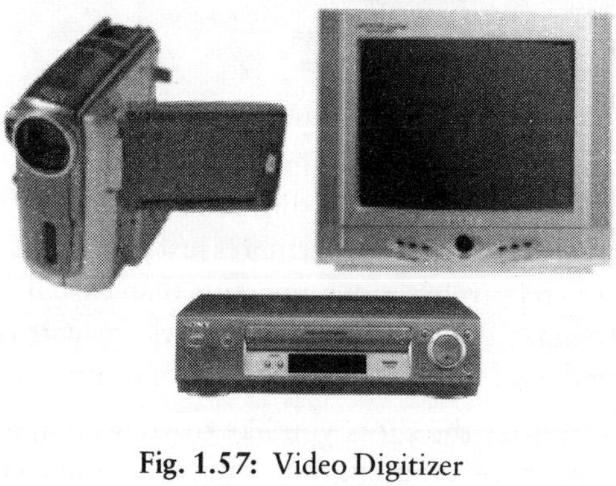

Fig. 1.57: Video Digitizer

4.42 OUTPUT DEVISES

Output devices serve the reverse function of input devices. They let the computer tell you what it has processed. Without output devices the user would not be able to communicate with the computer. So just like input devices, output devices are a necessity to function.Output devices allow computers to send information to the outside world. The most common output devices are monitor screens and printers.

Output devices may adopt different shapes and styles. The output may be in form of typed, printed or sketched matter or may be visible or audic type. But we can broadly classify them as

- Hard Copy Device
- Soft Copy Device

By hard copy we mean output in direct usable form that is in printed or plotted form. Whereas soft copy is in magnetic/audible form that cannot be used directly.

Terminals can be generally classified as soft copy or hard copy. Hardcopy devices produce a permanent record on media such as paper sheets or paper tape. Softcopy devices do not produce a permanent record, but some softcopy devices incorporate a printer for hard copy. Hard copy devices are very slow in operation as compared to soft copy devices.

Following hard copy devices are very popular:

- Printers
- Plotters
- Photographic output
- Punched card or Punched paper

Following soft copy devices are very popular:

- Visual Display soft copy devices are very popular
- Liquid crystal Display (LCD)
- Audio Response Unit

A variety of different types of computer output equipement are used to provide hard copy of the graphics information created on a display. This information may be output when it is created or when it is retrieved from storage. The output may be generated directly from the user's terminal or workstation while it is online, or it may be generated offline from a remote terminal.

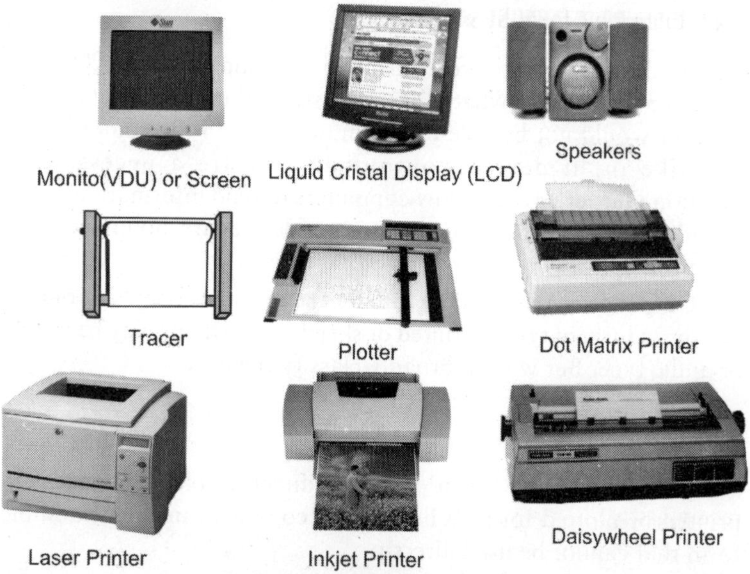

Fig. 1.58: Output Units

There are three basic classes of graphics output equipent, each of which includes several different types. The selection of a particular type of output equipment depends on user requirements, such as speed, quality, size, and cost.

4.42.1 Function of Output Units

Output units works just in reverse manner than input units. We feed data into computer by typing on the keyboard of computer. Similarly information coming out of computer can be typed on the sheet of paper by an output unit known as dot-matrix printer in a similar manner as it happens in teleprinters. The output coming from the CPU is in the form of electronic binary signals which needs conversion in the same form which can be easily understood by human beings *i.e.* graphical, alphanumeric in human language, audio visual. This function is performed by Output Units. When the data has been processed by CPU and converted into information, the results are in the form of electronic signals. Output unit converts them into a form which could be easily understood by us. The output can be given:

(a) On the screen by visual display unit
(b) In printed form by printers

(c) In graphic form by plotters
(d) In audible sound form by speech synthesizers
(e) Through other output units

Following are some popular output devices:

1. Visual display unit.
2. Plotters
3. Printers
4. Speech synthesizers
5. Coders
6. Robotic structures etc.

4.43 MONITOR

A Video Monitor, or Video Display Unit (VDU) or Video Display Terminal (VDT), makes it possible for a computer user to see input characters as they're typed, but it also serves as an output device for receiving messages from the computer.

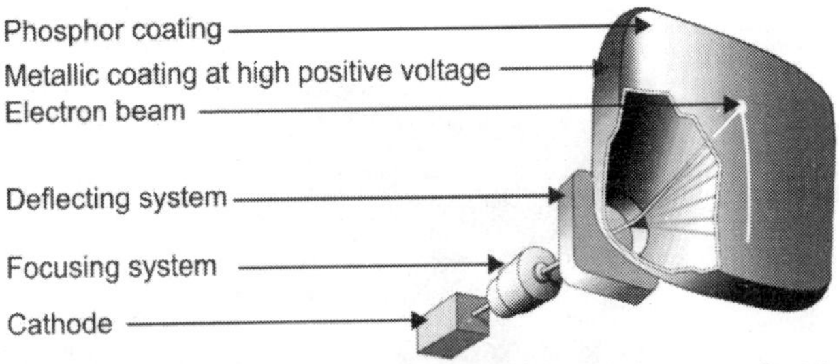

The vector display draws lines by controlling a single beam of electrons and "writing" the image on the screen. The random-scan display connects endpoints whose coordinates are stored in the computer and presented as a display file

Fig. 1.59: Video Monitor (VDU)

Images on a monitor are composed of tiny dots, called pixels (for picture elements). A square inch of an image on a typical monitor is a grid of about 72 pixels on each side. Such a monitor is said to have a resolution of 72 dots per inch (dpi). The higher the resolution, the closer together the dots.

Monitors fall into one of two classes:

- Television-style CRT (cathode ray tube) monitors and
- Flat-panel LCD (liquid crystal display) monitors.

Both types are available in either color or monochrome models in a variety of sizes. Monochrome monitors display only one color, usually black on white, or green or amber on black.

A cable connects the monitor to a video adapter (video card) that is installed in an expansion slot on your computer's motherboard. This system converts signals into text and pictures and displays them on a TV-like screen (the monitor).

The computer sends a signal to the video adapter, telling it what character, image or graphic to display. The video adapter converts that signal to a set of instructions that tell the display device (monitor) how to draw the image on the screen.

A Computer Display System Consists of a:

- Monitor and a Graphics card also called a video display adapter or called a video card
- Accelerated graphics card

Graphics card (video card) is an expansion card providing data path from main board to monitor.

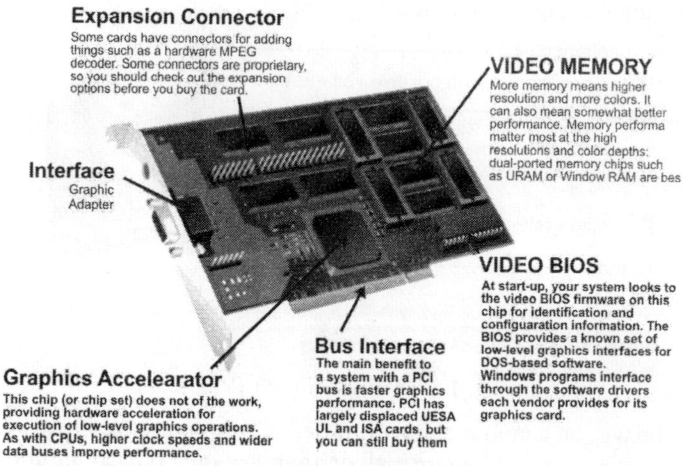

Fig. 1.60: Graphic Adapter

Accelerated graphics card - increases speed at which images are displayed when connected to a PCI bus can move data between the micro processor and graphics card as fast as the microprocessor can process it.

4.44 CRT MONITORS

CRT or Cathode Ray Tube monitors are composed of an "electron gun" and a specially coated screen. When the monitor receives the instruction from the video adapter, the electron gun "shoots electrons" onto the screen according to the instructions. So your saying, "that"s great so why do we see the stuff we do". Well, when the electrons hit the screen, they react with a chemical surface made of phosphors. The reaction of the electrons with the phosphors produces a small glow. The combined glow of the entire screen produces the visible image. The beams of electrons generally travel across the screen in a vertical manner.

4.45 LIQUID CRYSTAL DISPLAYS (LCD)

LCD screens have two sheets of glass. Sandwiched between these two sheets is a type of liquid containing crystalline structures. These crystals can move around by sending in electrical pulses. They either let light through, or they block light, depending on their location. This can create images and pictures.

There are two main types of LCD displays:

- Passive matrix screens and
- Active matrix screens.

Both these screens have a table-like display. Like with cathode-ray monitors, the LCD screens are divided into tiny sections called pixels. The pixels are arranged in a system of rows and columns. There are transistors at the beginning of every row and every column. To activate a pixel, the transistors send electric pulses down the rows and columns to that pixel. The main difference between the passive matrix screens and the active matrix screens is that the active matrix screens use more transistors for each pixel than passive matrix screens. This is what allows the former to have the brightest overall picture and the widest viewing angle. This is also what makes the active matrix screen much more expensive than the passive matrix screen!

Since its advent in 1971 as a display medium, liquid crystal displays have moved into a variety of fields, including miniature televisions, digital still and video cameras and monitors and today many believe that the LCD is the most likely technology to replace the CRT monitor. The technology involved has been developed considerably since its inception, to the point where today's products no longer resemble the clumsy, monochrome devices of old. It has a head start over other flat screen technologies and an apparently unassailable position in notebook and handheld PCs where it is available in two forms:

- Low-cost, dual-scan twisted nematic (DSTN)
- High image quality thin film transistor (TFT) also known as AMLCD.

4.46 VIDEO DISPLAY CARD

If there's one area where a potentially fast computer can be slowed to a crawl, it's the video display. Two computers equipped with identical PENTIUM CPUs, identical SD-RAM, identical RAM caches, identical hard drives, can look like the tortoise and the hare when only one of them is equipped with a high speed video display card.

Video display cards contain their own RAM chips. Today's bare minimum standard is 1 MB of RAM on the video card. These RAM chips hold the image that the monitor is actively displaying.

4.47 PRINTERS

Output displayed on a monitor is immediate, but it's temporary. A printer allows a computer user to produce a hard copy on paper of any information that can be displayed on the computer's screen.

Printers can be classified as:

- Serial or Character Printers
- Line Printers
- Page Printers

All of them print the characters in the American Standard Code for Information Interchange (ASCII) character set which is standardized by the American National Standards Institute (ANSI). This is a set of 128 printable and special control characters.

4.47.1 IMPACT PRINTERS

Impact printers can be further grouped as

- Serial or character impact printers
  - (a) Dot matrix Printer
  - (b) Daisy wheel Printer
  - (c) Golf ball Printer
  - (d) Cylinder Printer
- Line Printers
  - (a) Chain
  - (b) Band
  - (c) Drum

Impact printers include line printers and dot-matrix printers. The following impact printers form images by physically striking paper, ribbon, and print hammer together, the way a typewriter does:

4.48 DOT-MATRIX PRINTER (DMP)

The dot-matrix printer presses pins into a ribbon to produce the desired pattern. Due to the low quality of print, dot matrix printers are mainly used for fast low-quality printing.

Dot-matrix printer speeds usually range between 30 and 600 cps. The printing head contains a vertical array of pins. As the head moves across the paper, selected pins "fire" against an inked ribbon to form a pattern of dots on the paper. This gives a dot-matrix printer graphics capability—a feature lacking in daisy-wheel devices. The capital letters are formed by using 5-dot rows and 7-dot columns. Such a pattern is called a 5 × 7 matrix. Though there are 9 pins in the print head. The bottom 2 pins are used to form the descenders of lower-case letters such as p, g, q, and *y*. Many inexpensive dot-matrix printers don't provide descenders, but print quality is improved when they are used. A5 × 7 matrix produces readable text, but many other matrix sizes are used. The print quality is also improved when larger matrices are used.

Fig. 1.61: Dot-Matix and Dot-matrix Printer

4.49 LINE PRINTERS

High-speed line printers use impact methods to produce line-at-a-time printed output. Some are dot-matrix devices, but others typically use rapidly moving chains or bands of print characters or some form of a print drum to print lines of information on paper. From 300 to over 3,000 lines can be printed each minute depending on the printer used.

In the drum printer, raised characters extend the length of the drum. There are as many bands of type as there are printing positions.

Each band contains all the possible characters. The drum rotates rapidly, and one revolution is required to print each line. A fast-acting hammer opposite each band strikes the paper against the proper character as its passes. Thus, in one rotation, hammers of several positions may "fire" when the a row appears, several others may strike to imprint D's, etc. At the end of the rotation, the line has been printed.

Fig. 1.62: Line Printer

4.50 NON-IMPACT PRINTERS

These use a variety of new printing technologies that individual picture elements to be printed from computer signals. They can provide fast, high quality outputs. These include thermal and xerographic printers, which are often used to provide convenient but not necessarily high-quality hard copy during the design process.

Non-impact printers are categorized under five types.

- Thermal
- Inkjet
- Sprayed Ink
- Bubble Jet
- Electrostatic
- Electrographic
- Electrophptpgraphic: Laser

Non-impact printers are mainly of the following types:

4.51 INKJET

Sprayed ink printers use a simple method of spraying ink onto the papers to represent the desired character. An inkjet printer system produces characters by spraying small ink droplets onto paper. A high-frequency electrical generator causes a piezoelectric crystal to vibrate and throw off tiny ink droplets supplied from a reservoir. The droplets are then charged by an electrode and guided by horizontal and vertical deflection plates in the precise manner to form the desired characters on paper. Such printers are capable of speeds over 45,000 lines per minute with software control of styles and size of characters.

This is one of the fastest printing methods and also one of the more costly machines. However, for large volumes of printed output, the cost is comparable to that produced by line printers. Color printing is also available in some models.

A key characteristic of inkjet printers is their ability to print in color. This is one of the reasons why they are so popular in the home and office. Printers use the three primary subtractive colors of cyan, magenta, and yellow. Then, they use a process known as dithering. Dithering is when the primary colors are positioned close together so that different combinations of the cyan, magenta and yellow can create a whole spectrum of additional colors! Some printers are equipped with three colors, while others with four. Those with three contain only cyan, magenta, and yellow. These printers cannot switch between black and white text and color documents. Other printers have four colors-cyan, magenta, yellow, and black. These can easily switch between black and white documents and color documents!

Fig. 1.63: Inkjet Printer

4.52 LASER PRINTER

The laser xerographic print mechanism shown in Figure: 1.64 uses a laser produced light to charge a surface in the latent image of the characters to be printed. A special powder is attracted to the charged surface areas corresponding to the latent characters and fused into place to produce visible characters. These printers are capable of very high speeds, up to 15,000 lines per minute.

Fig. 1.64: Laser Printer

Laser-printer's output is dry, crisp, and water-resistant. By contrast, inkjet output emerges wet, takes several seconds to dry, and isn't very water-resistant.

4.53 PLOTTERS

This is the most common class of output equipment used for graphics applications, particulary for drafting sytems. Plotters are usually used by architects, surveyors, engineers, pattern makers.

A plotter is an output device used to produce high quality drawings. It holds pens on paper and can draw curved lines. Text is drawn as images.

Plotters can draw in black and white or in colour. Color plotters would have a set of color pens. A plotter selects the pen from the predefined position so if the pens are in the wrong order the wrong colour will be selected. Plotter can be connected to the port normally used by a printer.

Fig. 1.64: Flat bed Plotter

Plotters have built-in microprocessors that convert the characters and instructions from the computer into a series of coordinates which the plotter then draws. They require lots of memory to store highly accurate images.

One of the quality factors is resolution—the smallest movement a pen can make on a paper (a typical movement can vary from 0.001 to 0.005").

There are two basic types of plotters:

1. Pen Plotters: The pen plotter an outgrowth of NC technology uses stepping motors to control the motion of the pen relative to the drawing board, it traces the image as an outline drawing stored in the computer. Because of this, pen plotter cannot handle a lot of intricate detail or solid-color areas efficiently. Although they are also relatively slow, they can be inexpensive and can be equipped to use multiple pens for color plotting.

2. Electrostactic plotters: These are faster and can handle more intircate designs. An electorstatic charge is used to produce an image on paper, and then a tonger is applied to th paper to create the printed image. This is accomplished either by useing a matrix of needles to "write" the electrostatic image or by using a photoconductor which is written on by a light from a CRT or laser, asin copying machines.

Plotters usually come in two designs:

- flat bed and
- drum.

3. Flat bed: Flat bed plotters can fit on a desk but there is a restriction on the paper size. They draw with the arm that moves to the left or right and a pen-holder moves from top to bottom or vice versa.

4. Drum plotters: Drum plotters are much bigger than flat bed plotters and they usually use rolls of paper. They move the paper past a set of pens. Pens move to the left or right in only one horizontal plane while the paper moves forward and backward perpendicular to the paper's motion. Drum plotters must have extremely accurate paper feed mechanisms.

4.54 SPEECH RECOGNITION AND SOUND OUTPUT SYTEMS

Even if your computer doesn't understand what you say, the newest software can at least recognize it. Today's voice recognition software comes closer than ever to the goal of making keyboards optional. We aren't there yet, but for a variety of users, voice recognition software is actually well worth the money.

Audio Devices

Sound effects and music are vital to the computer's success. Computer games are all equipped with amazing sound effects and music, while popular songs can be downloaded off the Internet. There are a few devices that make this possible:

Sound Cards

Sound cards are a key component to any computer system. They let you listen to music while using your computer and they also bring games to a new level! They add all sorts of sound effects to your computing experience.

Sound cards perform four basic functions with sound. They can play music or other sounds (*i.e.* from MP3s, from CDs, games, *etc.*), they can record sounds (by the use of a microphone or other similar device), they can synthesize sounds, and finally, they can process sounds.

You might be wondering how sound cards can take sounds like your own voice and save them as files to your computer's hard disk.

Well, first, you will need some sort of device, like a microphone. When you speak into the microphone, a device inside the microphone will convert your voice to electrical signals that your sound card can understand. Your voice is in analog form, and not digital. This merely means that it is not in the form of the binary digits (bits) of 1 and 0. In order for the sounds to be saved and used later by the computer, the sound card must convert the analog signals into digital signals. This is done by a special device called the analog-to-digital converter (ADC).

The digital output created by the ADC then enters the digital signal processor, or DSP. Because sound files are usually really really big, they need to be compressed so that they won't take up too much room in your hard disk. This, along with many other different computations, is done by the DSP.

Finally, the output created by the DSP is transferred to the computer's central processing unit. It is sent to your hard disk, and is usually saved as a wave file.

Playing sounds saved as wave file in your hard disk is a fairly straightforward task. The computer's CPU retrieves the file from the hard disk and ships it off to the sound card. The file is decompressed by the DSP, and is then converted back to analog form by the digital-to-analog converter (DAC), and the music is ready for us to listen to!

Musical Instrument Digital Interface (MIDI)

Almost all sound cards have a MIDI synthesizer or sampler. MIDIs let MIDI computers communicate with MIDI instruments. At first, MIDI sound cards made use of FM synthesis. FM synthesize can produce sounds by taking two or more sound waves and them combining them to make an approximation of a particular sound. It can be pretty realistic, but now there is a newer technology called wavetable synthesis. Wavetable is much more realistic than FM synthesis because it creates sounds by taking samples of real instruments. For example, if it wanted an example of an A flat on a piano, it would take a sample of a piano playing A flat. The sample sound is played in a loop and can realistically recreate the sound of the original instrument!

Speakers

Speakers are the amazing devices that are used to play all sorts of music and sounds from your computer! They take the electric signals stored CDs and when you want to listen to certain music, your sound card will retrieve it from your computer's hard disk, where it is stored as electric signals. Your speakers will take these signals and convert them to sound waves, which will travel to your eardrums!

Speakers bring life to your computer system. Without them, the sounds effects of high-tech games cannot be heard, you couldn't listen to music or to DVDs. Sound brings your computer system to a whole new level, and you can only hear the sound waves because of your speakers!

4.55 MODEMS

The modem is a device that allows a computer to be connected to your telephone line. Modems may be installed inside the computer's case or connected to the computer's exterior serial port.Once connected, you can send or receive faxes and electronic mail. Many people are purchasing modems today so they can connect their computers to the Internet. The Internet is a vast connection of about 100 million computers worldwide. It provides instantaneous access to thousands of companies, news services, libraries, and individuals anywhere in the world.

Today, the industry standard modem transmits data at 57,000 bits per second or bps (NOT bytes per second). They are often called 56K modems. Less expensive models, that transmit data at 28,800 bps.

4.56 MULTIFUNCTION DEVICES

These are the devices which incorporates two or more input output units and their qualities. As with printers, the main types are inkjet and laser, with similar pros and cons. When deciding which to buy, you'll want to consider how the machine will be used. If you plan to scan and print color photos or graphics, an inkjet multifunction device will give you the versatility you need. But if you only plan to print, scan, or copy black-and-white text or graphics, a laser multifunction device can be a good choice. A multifunction device's print quality can be expected to match that of a regular printer using the same ink cartridges or toner and having the same resolution options.

Case: The case houses the microchips and circuitry that run the computer. Desktop models usually sit under the monitor and tower models beside. They come in many sizes, including desktop, mini, midi, and full tower. There is usually room inside to expand or add components at a later time. By removing the cover off the case you will may find plate covered, empty slots that allow you to add cards. There are various types of slots including IDE, ASI, USB, PCI and Firewire slots.

Notebook computers may have room to expand depending on the type of computer. Most Notebooks also have connections or ports that allows expansion or connection to exterior, peripheral devices such as monitor, portable hard-drives or other devices.

Cards: Cards are components added to computers to increase their capability. When adding a peripheral device make sure that your computer has a slot of the type needed by the device.

Sound cards allow computers to produce sound like music and voice. The older sound cards were 8 bit then 16 bit then 32 bit. Though human ear can't distinguish the fine difference between sounds produced by the more powerful sound card they allow for more complex music and music production.

Color cards allow computers to produce color (with a colour monitor of course). The first color cards were 2 bit which produced 4 colors [CGA]. It was amazing what could be done with those 4 colours. Next came 4 bit allowing for 16 [EGA and VGA] colours Then came 16 bit allowing for 1064 colours and then 24 bit which allows for almost 17 million colours and now 32 bit is standard allowing monitors to display almost a billion separate colours.

Video cards allow computers to display video and animation. Some video cards allow computers to display television as well as capture frames from video. A video card with a digital video camera allows computers users to produce live video. A high speed or network connection is needed for effective video transmission.

Network cards allow computers to connect together to communicate with each other. Network cards have connections for cable, thin wire or wireless networks. For more information see the section on Networks.

Cables connect internal components to the Motherboard, which is a board with series of electronic path ways and connections allowing the CPU to communicate with the other components of the computer.

5.0 Hardware and Software

Hardware comprises all of the physical parts of a computer, as distinguished from the data it contains or operates on, and the software that provides instructions for the hardware to accomplish tasks. The boundary between hardware and software is slightly blurry-firmware is software that is "built-in" to the hardware. Firmware is usually the province of computer programmers and computer engineers in any case and not an issue that computer users need to concern themselves with.

5.1 COMPUTER HARDWARE

The hardware are the parts of computer itself including the Central Processing Unit (CPU) and related microchips and micro circuitry, keyboards, monitors, case and drives (floppy, hard, CD, DVD, optical, tape, etc.). Other extra parts called peripheral components or devices include mouse, printers, modems, scanners, digital cameras and cards (sound, colour, video) etc.

A peripheral is a type of computer hardware that is added to a host computer in order to expand its abilities. More specifically the term is used to describe those devices that are optional in nature, as opposed to hardware that is either demanded, or always required in principle.

The term also tends to be applied to devices that are hooked up externally, typically through some form of computer bus like USB. Typical examples include joysticks, printers and scanners. Devices such as monitors and disk drives are not considered peripherals because they are not truly optional, and video capture cards are typically not referred to as peripherals becuase they are internal devices.

5.2 SOFTWARE

Software refers to the programs that run on computers and those which run the computer. The difference between hardware and software is that you can touch and hold hardware, but not software. For example, if you have an installation CD for Microsoft Office, the CD itself is hardware, but the programs on the CD are software.

Software fills the communication gap between humans with problems to solve and computers that can only understand zeros and ones.

Software is the set of instructions that are given to the computer. Some of the instructions are put into the computer when it is being manufactured, that is, when it is made in the factory in its memory that is ROM. Other instructions are put into the computer's RAM later by the programmer. You will write some of these instructions yourself. You will also use some instructions that have been written by other people.When we run a program actually we put the sets of instructions stored in the hard disk to RAM.

Software can also refer to the data that computers process, but generally it means programs. A program is a set of instructions that direct the computer. The CPU interprets those instructions and either activiteis one of the hardware devices (such as the printer) or processes data given to the program.

Software gives "intelligence" to the computer. Software provides the computer with information is called data and it is arranged into groups, or programs. Software is connected to the computer's input.

Examples of software programs or applications would be the Operating System (DOS, Windows 9x/Millenium/XP, O/S2, UNIX, MacOS 9.x/10.x and various others), Wordprocessor (typing letters), Spreadsheet (financial info), Database (inventory control and address book), Graphics program, Internet Browser, E-mail and many others.

As well any document that you create, graphic you design, sound you compose, file you make, letter you write, e-mail you send or anything that you create on your computer is referred to as software. All software is stored in files.

Software is stored on a disk or tape whether that disk is a floppy, hard disk, CD, tape or one of the dozens of other storage devices available.

There are millions of different pieces of software available for almost every conceivable need. Software is available commercially through stores and mail order and also available on the Internet. Software is also available through an Open Source license which allows anyone to use the Open Source software free of charge as long as the license is maintained. If you can't find the application that you need software development companies can custom design software for you.

The largest software companies offer packages of software or suites that include many of the programs that the average person or business needs. Software packages or suites contain programs that work together and share information, making it easier to combine that information in versatile ways. For example, when writing a letter you can get the mailing address from an address book, include a letterhead from a graphics program and included a financial chart from a spreadsheet and combine this collection of information in the body of the letter.

5.2.1 Types of Software

There are three major categories of software:

1. System Software
    - Operating systems
    - Translators
        - (a) Assemblers
        - (b) Compilers
        - (c) Interpreters
    - Simulators
    - Emulators
2. Application Software
    - User applications programs
    - Applications packages.
3. Utility Software
    - Utilityware
    - Middleware
    - Groupware
    - Shareware
    - Charityware
    - Crippleware
    - Freeware

According to another classification the four basic types of software are:

- Commercial,
- Middleware
- Shareware and
- Open source software.

Software can be categorised as open system or proprietary.

Open system software can be modified for use with any hardware. It is not the exclusive property or design of a particular vendor. For example, UNIX is the first operating system that works on minicomputers, mainframes and PCs. Its inner workings are available to public and anyone can adapt it or develop applications to work with it.

The proprietary products are designed for particular systems and cannot be used with other hardware. Its inner workings are protected information. For example, Apple computers and software for them have been proprietary products for many years. Only recently Apple licensed its MacOS to other hardware companies to build Macintosh clones.

5.3 SYSTEM SOFTWARE

By system we mean computer system. The user of a computer has at his disposal a large amount of software provided by the manufacturer. Much of this software will be programs which contribute to the control and performance of the computer system. Such programs are given the collective name system software. Any one of these programs is a systems program, *e.g.* compilers assemblers, DOS, UNIX, Interpreter and DBMs are systems programs.

5.4 OPERATING SYSTEM SOFTWARE

Also known as system software, the operating system acts as your computer system's manager. The operating system has two main functions:

- Presenting a user interface — the way in which the OS accepts commands from its human user. Interfaces can be text-based (DOS) or graphical (Windows, Macintosh).
- Managing system resources — the OS grants access to the CPU, RAM, disks, and other resources, making sure that no program is allowed to perform an operation that makes the system unstable.

5.4.1 Operating System: The Resource Manager

An operating system is primarily a resource manager, and the primarily resource it manages is computer hardware. It provides many features including:

- defining the "user interface"
- sharing the hardware among users.
- allowing users to share data among themselves
- scheduling resources among users
- facilitating input/output.
- recovering from errors.

The Key Resources an Operating System Manages are

- Processors
- Job Sheduling and Monitoring

- Memory
- Device (Input/ Output)
- Data/ Information

The job manager is responsible for reading in your program, determining the resource needs, ensuring that these resources are available, scheduling the job for execution, monitoring the job throughout execution, and releasing the resources used when the job terminates.

Ther memory manager is responsible for assingning memory and releasing memory as a job executes .

The processor manager is responsible for assigning computation time or processor time.

The I/O manager is responsible for assigning and performing all I/O operations to specific devices.

The information manager handles requests for filing an auxiliary storage a specific group of date items under an assigned file name . When the data items are needed, the information manager finds the file and gives the location to the I/O manager.

5.5 TRANSLATORS

A translator is a program which converts statements written in one language into statements in another language. Two types of translator have already been introduced but there are three basic types of translators:

- Assemblers
- Compilers
- Interpreters.

5.6 COMPILERS AND INTERPRETERS

The compiler translates the whole of the high level language source program into a machine code object program prior to the object program being loaded into main memory and executed. Contrast this with the interpreter which deals with the source program one instruction at a time, completely translating and executing each instruction before it goes into the next.

(a) If a compiler is used, the same program need only be translated once. Thereafter, the object program can be loaded directly into main storage and executed.

(b) If an interpreter is used, the source program will be translated every time the program is executed. Executions carried out in this way may be ten times slower than the execution of the equivalent object programs!

5.6.1 Compiler Translation

Naturally, a source program written in a high-level language must also be translated into a machine-usable code. A translating program that can perform this operation is called a compiler. Compilers, like advanced assembly programs, may generate many lines of machine code for each source program statement. A compiling run is required before problem data can be processed. With the exception that a compiler program is substituted for an assembly program, the procedures are essentially the same as those shown in (Fig. 8.2). The production run follows the compiling run.

The compiler:

(a) Translates the source program statements into machine code.
(b) Includes linkage for closed sub-routines.
(c) Allocates areas of main storage.
(d) Produces the object program on disk or magnetic tape.
(e) Produces a printed copy (listing) of the source and object programs.
(f) Produces a list of errors found during compilation *e.g.* the use of 'words' or statements not included in language vocabulary; or violation of the rules of syntax.

5.6.2 Interpreter Translation

An alternative to using a compiler for high-level language translation is often employed with personal computers. Instead of translating the source program and permanently saving the object code produced during a compiling run for future production use, the programmer merely loads the source program into the computer along with the data to be processed. An interpreter program stored on the computer's operating system disk or permanently wired inside the machine then converts each source program statement into machine language form as it's needed during the processing of the data. No object code is saved for future use.

The next time the instruction is used, it must once again be interpreted and translated into machine language. For example, during the repetitive processing of the steps in a loop, each instruction in the loop will have to be reinterpreted every time the loop is executed. The interpreter eliminates the need for a separate compiling run after each program change to add features or correct errors. But a previously compiled object program should obviously run much faster than one which has to be interpreted each step of the way during a production run.

5.7 THE ASSEMBLER

(a) Translates symbolic operation codes into machine code, and symbolic addressed into actual machine addresses.
(b) Includes the necessary linkage for closed subroutines.
(c) Allocates areas of main storage.
(d) Will indicate invalid source language instructions.
(e) Produces the object program on disk or tape.
(f) Produces a printed listing of the object program together with comments. Notice that one symbolic instruction is translated into one machine instruction (*i.e.* one for one) which is one feature which distinguishes a low level language from a high level language.

5.8 EMULATORS

To emulate means to imitate one system with another such that the imitating system accepts the same data, executes the same program, and achieves the same results as the limited system.

Emulation was the name given to the technique introduced in the IBM System/360 machine series for aiding in the conversation problem (to System/360 from previous IBM computers in the 1400 and 7000 series). An emulator is a package that includes both special hardware and a complementary set software, which runs in the manner of an interpretive routine simulator program but is 5 or even 10 times as fast as a purely software similator. In terms of software emulator means a translaters which translates one HLL to another, say a program written in PASCAL to FORTRAN.

5.9 SIMULATORS

Simulation is a process that employs a computerized model of certain significant features of some physical or logical system. The object of the process of simulation is to provide an experimental model for the accumulation of data on the target system. The process of simulation comprises the steps of experiment definition, modeling, computer implementation, validation, and data gathering.

Simulators are programs developed to apply the process of simulation. The term "simulator" usually implies the incorportion of some model elements (job-shop simulator: computer simulator) even though the tool may be quite flexible and thus useful for many modeling adaptations.

Simulators are distinguised from "simulation languages," which are genereal purpose and contain no model basis.

5.10 PROGRAMMING LANGUAGES

Another type of software is programming languages. Programming languages are used to create all other software whether it is system or application software. Modern programming environments have graphical user interfaces (GUI) featuring symbolic icons and drag-and-drop technique and English-like statements. Delphi and Visual Basic are the most popular examples of such environments. Programming in visual environment creates an impression that it is very easy to write programs. But becoming a good programmer requires talent, ability to learn, logic, creativity, patience and experience.

5.11 APPLICATIONS SOFTWARE

Application software is, quite simply, software that performs a specific function for you. Applications programs may be provided by the computer manufacturer or supplier but in many cases the user produces his own applications programs called user programs. (eg. payroll programs, stock control programs etc.) A single applications program is often called a job. Sometimes a job may be divided into smaller units called tasks. A job may comprise program + data. Most applications programs can only work if used in conjunction with the appropriate systems programs.

Common applications include: Word processors, Databases, programs that let you organize data into useful informationPersonal productivity applications such as calendars, address books, personal graphics, Internet/World Wide Web browsers such as Netscape Navigator and Microsoft Internet Explorer Spreadsheets, programs that perform calculations for you. There are many other types of applications, including games, tax software, children's educational software, flight simulators, and thousands more.

5.12 GENERAL CLASSIFICATION OF APPLICATION SOFTWARE

Application software is of two types:

- Standardised
- Customised

Once again, what distinguishes standardised applications software from customised software is the former's portability. They are canned software packages one can buy off the shelf, and run on all compatible systems. Examples of standardised applications software packages are spreadsheets, wordprocessing and Data base Mangement packages.

Customised software packages on the other hand, are programs specially written for carrying out specific jobs like financial accounting inventory management and invoicing, Railway and airways enquiry and reservations. As the name suggests, they are custome-written, tailor-made programs that are not portable across systems/jobs.

5.12.1 Customised User Applications Programs

User applications programs are programs written by the user in order to perform specific jobs for the user. Such programs are written in a variety of programming languages according to circumstances, but all should be written in a systematic way.

For many applications it is necessary to produce sets of programs which are used in conjunction with one another and which may also be used in conjunction with service programs such as sort utilities.

For some applications it may be necessary to compare available software or hardware before writing the applications software. For example, the user may wish to compare the performance of two different COBOL compilers. Such a comparison may be made using a benchmark.

5.12.2 Standard Applications Software

Standard applications software may be

- general-purpose commercial software, and
- specialised software for particular applications.

A number of pre-written programs are available in market, which can be used for variety of purposes.

5.13 GENERAL PURPOSE COMMERCIAL SOFTWARE

The software you buy in stores—Microsoft Office, WordPerfect, CORELDraw, and the like—is commercial software. You are required to pay for commercial software before you can legally use it. Commercial software is usually distributed by software publishing companies, and usually comes in attractive packaging.

Software is also categorized by how it is used:

- Document production
- Graphics
- Presentation
- Numeric analysis
- Data Management
- Information and Reference
- Connectivity
- Education and Training
- Entertainment
- Accounting and Finance
- Business
- Document Production Software assists you with composing, editing, designing, and printing documents.
    - **Word processing software:** Correspondence, reports, papers, novels, memos.
    - **Desktop Publishing:** Newsletters, brochures, magazines.
    - **Web Authoring:** Web pages for the internet.
- Graphics used to create, edit, and manupulate.
    - Photos
    - Paintings
    - Drawings and 3-D Objects
    - Animations and videos

Bitmapped images if you want to use a computer to create sketches and paintings.

- Vector graphics Images composed of lines and filled shapes.

Digitizing the process of converting videos into a format that can be stored on a computer disk.

Rendering creating a wire frame drawing when creating a solid 3-D image by covering the wire frame and applying computer generated highlights and shadows.

Presentation software provides all the tools to combine text, graphics, graphs, animations, and sound into a series of electronic slides.

Output the presentation:

- On overhead transparencies
- As paper copies
- As 35 mm slides
- On a color monitor
- On a computer projection device.

Numeric Analysis Software. Constructs numeric models of physical and social systems then analysis those models to predict trends and understand patterns.

- Spreadsheet Software
- Graphing Software
- Statistical Software

Spreadsheet software performs calculations based on numbers and formulas that you enter.

Top selling packages:

- Microsoft Excel
- Lotus 1-2-3
- Quatro Pro

Graphing software transforms complex data into meaningful graphs that allow you to visualize and explore data.

Statistical software helps you analize large data sets to discover patterns and relationships.

- Useful for summerizing survey results, test scores, lab results, census data.
- Examples of top selling packages: SSPS, JMP, Data Desk.

Data Management Software helps store, find, update, organize and report information.

- File management software for simple lists.
- Database software for related files or databases.

Information and reference software provides you with a collection of information and a way to access that information. Data management software contains no data.

Reference software:

- Generally shipped on CD-ROM
- Web sites often provide updates for dated information
- Popular software in this category include, encyclopedias, mapping, and trip planning.

Connectivity software connects your computer to a local computer network or to the internet. Includes.

- Basic communications software
- Remote control software
- E-mail
- Web Browsers

Communications software interacts with your computers modem to dial and establish a remote connection.

Remote control software accesses an unmanned computer. (Procomm Rapid Remote, pcANYWHERE, LapLink)

E-mail software manages your computer mailbox. (Outlook Express, Netscape Communicator.)

Web Browser software accesses information on the web. (Netscape, Internet Explorer.)

Web accessaries for Internet Explorer 5.0 (Free Download) Accessories to make your browser a more powerful tool.

Education and training software helps you learn new skills. Basic skills are taught through levels of play that adapt to age and ability.

Advanced learning includes:

- Foreign languages
- Exam preparation
- ❐ Music instruction

Exam preparation software is available for tests such as SAT, GMAT, and LSAT.

Education and training software is sometimes referred to as edutainment software, bcause it blurs the lines between learning and game playing.

Game software includes software toys, simulations and software designed for hobbies and leisure.

Action, adventurerole playing, classic, puzzle, simulations, strategy wargames.

Generally divided into six categories:

1. Adventure role playing realistic 3-D graphics, players interact with the environment, weapons and monsters galore.
2. Action games similar to arcade games.
3. Simulation software broad interests including building a city, playing a sport, flying airplanes.
4. Accounting and finance software keeps records of monetary transactions and investments.
5. Personal finance software, gearing towards individual finances.
6. Online banking, a way to use your computer and modemm to access your banking functions.

Business software helps organizations accomplish specific routine tasks.

Horizontal Market Software used by many kinds of businesses.

- Accounting applications
- Payroll applications
- General ledger systems

Vertical market software designed for specific businesses:

- Medical billing and insurance
- Hotel management applications
- Law firm time tracking systems

Following general purpose commercial softwares are extremely popular:

- Word processing
- Databases
- Spreadsheets
- Graphics
- Communication
- DTP
- Games

World Processing: Simple editing programs are found in the operating systems of many computers and these are sufficient for very simple tasks. To be able to create more elaborate documents with figures, Greek symbols, equations etc. one needs one of the more complete word processors.

The essence of a word processor is that it gives you the ability to use many different fonts, to tailor the appearance of the page and to include drawings or graphs in the text. Add to that the ability to cut and paste pieces of text from one page to another or from one article to another and it becomes an essential piece of software for most people using a computer.

Proprietary programs such as Word, Word Perfect and AmiPro are all available in DOS and Windows versions.

Spreadsheet: This is a program that stores data in tabular form and uses it to perform calculations. Formulae may be included as part of the table so that derived parameters are automatically updated.

To clarify this a little, imagine a group practice with the patient fees, time of visits and the office expenses for each physician entered on an individual basis. A spreadsheet would automatically tally the totals, allocate percentage expenses, determine monthly income and could generate averages on a daily, weekly or monthly basis. It could also translate this into the number of patients seen on each day of the week and the average length of time for the interviews and express this as a graph for each day of the week. Programs such as Lotus 123 and Microsoft's Excel fall into this category.

More than any other type of personal computer software, the spreadsheet has changed the way people do business. Spreadsheet software allows the user to take control of numbers, manipulating them in ways that would be difficult or impossible otherwise.

A spreadsheet program can make short work of tasks that involve repetitive calculations: budgeting, investment, management, business projections, grade books, scientific simulations, checkbooks, and so on.

A spreadsheet can also reveal hidden relationships between numbers, taking much of the guesswork out of financial planning and speculation.

Spreadsheet applications:

- Cost analysis
- Budgets
- Sales projections

Inventory, tax, and profit planning

- Student marks
- Engineering tables
- Decision support systems etc.

Database: This program may be thought of as a card file where you store information about items using a standard format. Thus, if you want to catalogue your slides for lectures you might decide you need fields for the name of the slide, the date when used, the description of the slide, the type of slide (Patient, Graph, Biopsy etc.) and some yes/no aspects such as whether it is a good slide or mediocre. Once you have the database set up, it may be sorted, searched or manipulated in a way that a card file would not permit. Program's such as Borland's D-Base, Lotus Approach or Microsoft's Foxpro are well known.

Graphics: A graphics program is particularly useful for those who are interested in designing their own slides or who want to include sophisticated graphics into their texts, as in a newsletter. The graphics programs also let you import pictures from a prepared library and manipulate the pictures in endless ways, rotating, bending, stretching, flipping them back to front and changing the colors.

Most graphics programs have a "slideshow" feature. This allows you to project a series of slides, just as you would with the more familiar 35 mm slides. To do this the computer is hooked up to a data projector or a color palette that goes on top of an overhead projector.

Special effects such as fading the slides in and out as you change from one to the other are available. As the quality of the projectors improves, this will be an increasingly popular way of giving talks - not to mention the ability to change your slides on the airplane!

Communications: If you want your computer to communicate with others a communication program is a must. These are of various types. When you buy a fax/modem a communication program will be included which lets you place phone calls and send faxes. A full function fax program can turn your computer into a convenient and efficient Fax machine. You can usually recognize such programs by their names: Winfax, Bitfax etc.

A "Terminal" program is usually part of the Windows environment and allows you to turn your computer into a machine that emulates standard machines used to communicate over network lines or on the Internet.

Other programs allow you to hook up your laptop to your home computer by phone lines from a remote location and to operate your home computer as if you were at home.

The year of the LAN happened sometime in the 1980s, and it was Novell's NetWare that made it so. NetWare is no lightweight desktop OS. NetWare was an OS that systems administrators could rely on. Versions of this OS are still in use in businesses everywhere.

Desk Top Publishing

Desktop publishing software places your words on pages with fancy lettering and pictures. Newspapers, award, and even books are often made with this tool.

Desktop publishing software lets you work with picture and words at the same time. Newspapers and magazines that you see today are made using desktop publishing.

Games: Flight Simulator: To work its magic, Microsoft's simulation of an airplane's cockpit employed low-level graphics routines. It became a mainstay of software suites used to test compatibility with the IBM PC standard. It was also one of the best-selling games of all time.

5.12.1 Advantages of Commercial Software

- **Documentation:** Commercial software almost always comes with printed manuals for easy reference. In addition, many commercial packages put the entire manual online in a hypertext-based Help system.
- **Technical support:** Almost all commercial packages provide a phone number (in many cases, a toll-free 800 number) you can call if you have questions. You're more likely to know people using the same software. If you need help with Microsoft Word or WordPerfect, you'll probably know someone else who is familiar with it.
- **Virus-free:** Commercial software is pressed onto brand-new installation CDs or floppy diskettes. Viruses never have a chance to infect your software.

5.12.2 Disadvantages of Commercial Software

Cost: Full-featured commercial software is expensive. For example, you can expect to pay well over $ 600 for a new installation of Microsoft Office Professional (which contains a processor, database, spreadsheet, presentation manager, and e-mail client). You have to pay for it first. There's nothing like plunking down big bucks for a piece of software and then finding it doesn't fit your needs. Many stores won't take software returns once you've broken the seal on the envelope containing the installation disk or CD.

5.13 Specialised Applications Software

These handle particular problems. There are enormous numbers of such problems. A few examples:

- Acconting package: Tally
- Payroll processing
- Inventory management
- Class scheduling
- Human resource management
- Project management
- Personal information management (diary, addresses, etc.)
- Insurance claims processing
- Preparing for seminars/presentations
- Games

5.14 APPLICATION PACKAGES

These are similar in concept to service programs except that application packages are generalized programs for solving business problems as opposed to programs for carrying out computer systems tasks.

Many users have the same type of problem to put onto a computer and thus manufacturers and specialist software writers have written standard programs to solve these problems and sell them to the many users who want them. Examples are: Stock Control, Sales Invoicing, Network Analysis.

Most Important Old Software Products

The 20-year story of personal computing often seems to be dominated by hardware. But it's the software that makes the hardware worth owning: Many early buyers of Apple IIs walked into stores and asked for the VisiCalc machine.

VisiCalc: Written in 1979 by first year Harvard Business School student Dan Bricklin and Bob Frankston of MIT, VisiCalc was a godsend </art/9509/sec7/art5.htm>to Wall Street users who had bought the first microcomputers two years earlier. Running initially on the Apple II and nearly single-handedly creating the demand for the machine, VisiCalc established spreadsheets as a staple application, setting the stage for Lotus 1-2-3 on the IBM PC in 1985.

WordStar: While writing programs on the Altair, Michael Shrayer hit upon the idea of writing the manuals on the same machine. Electric Pencil was born, the first microcomputer word processor. But the first program to exploit the market potential was Seymour Rubinstein's 1979 masterpiece, WordStar. Other programs took up WordStar-compatible keyboard commands—including the last major upgrade of Electric Pencil.

dBase III: Wayne Ratliff's creation, first intended to manage a company football pool, was the first serious database management system for CP/M. DBase II, in its DOS incarnation, was a massive success. Ashton-Tate, which acquired dBase from Ratliff, began to lose the lead when it released the bug-ridden dBase IV in 1988. A Windows version (under the ownership of Borland) didn't appear until 1994, much too late. The DBase language survives in the form of Xbase, supported by vendors such as Microsoft and Computer Associates. Foxpro 2.6 and Microsoft's Access are the most popular databases currently.

AutoCAD: Autodesk's AutoCAD started life as a CP/M (Control Program for Microcomputers) application, later moved to DOS, and eventually made the transition to Windows. It brought CAD from minis and mainframes down to the desktop, one of the first programs to make that now-common migration. AutoCAD quickly became—and remains—an industry standard.

Lotus 1-2-3: VisiCalc may have sold Wall Street on the idea of electronic spreadsheets, but 1-2-3 was the spreadsheet that Main Street wanted, too. When the IBM PC and XT took over the world, Lotus's simple but elegant grid was without question the top spreadsheet to run on them, adding graphics and data-retrieval functions to the paradigm established by VisiCalc. By the early 1990s, Lotus could brag that 1-2-3 was the top-selling application of all time.

Quicken: This checkbook-balancing program may be better suited to the needs of its users than any other program on this list save VisiCalc. Scott Cook's Company grew from humble beginnings in the mid-1980s to become Microsoft's multibillion dollar dance partner (until the Department of Justice cut in). Once you start balancing your checkbook in Quicken, you don't ever go back.

SideKick 1.0: Besides being the first PIM (personal information manager), its pop-up notepad, calendar, and calculator made Borland International's SideKick the model for TSRs—an application type that was relatively rare in 1984. Pop-up mini-apps became commonplace in the DOS era, but Windows' task switching killed the TSR market in the 1990s. Excel for the Macintosh VisiCalc and Lotus 1-2-3 started the spreadsheet revolution, but they were character-based. Microsoft Excel for the Macintosh made the benefits of graphical spreadsheets obvious. Microsoft ported Excel to Windows, but Lotus was slow to convert 1-2-3 to Windows. There's a lesson here: Today, Excel for Windows is the best-selling spreadsheet.

PageMaker: This is the program that launched a million newsletters. PageMaker's paste-up metaphor also made sense to people who had worked in traditional design and production departments. QuarkXPress might now have a larger share in higher-end publishing, but with Adobe's money and name behind Aldus, PageMaker promises to remain a competitive desktop publishing system for a long time to come.

LANtastic: For people who thought Novell NetWare was for corporate MIS gurus, Artisoft's affordable network-card-and-software package was an easy and popular way to link PCs and share resources. With the addition of NetWare server functions in Artisoft's new LANtastic Dedicated Server, LANtastic keeps a foothold in the future.

Adobe Type: Desktop publishing was still a bit of a toy when Adobe made Type 1 PostScript fonts available on the Macintosh. Thanks to these fonts and the enhanced line spacing and printing control that PostScript provides, the Mac became a tool on which to run a publishing business.

Lotus Notes 3.0: Notes is the most innovative and powerful of the numerous contenders in the leading-edge groupware category. Not just E-mail, Notes is brilliant at capturing corporate group-think, thanks to its unique, replicated message system. Notes has become the standard applications development environment in every company that's ever uttered the word reengineering.

5.15 INTEGRATED OFFICE SUITES

Application suites (pronounced "sweets") are collections of application software sold together in one package. When different types of software are sold together, we say that they are "bundled."

Suite or Software Suite is a number of applications packaged together and sold as a unit. Several, or even all, of the general purpose software can be integrated into a single integrated software package.

A software suite is a collection of full featured, stand alone programs. The programs usually share a common command structure and have similar interfaces to make them easy to learn. Programs in the suite pass data back and forth easily. Some opular software suites are Microsoft Office, Lotus SmartSuite, and cprel office.

An integrated package usually requires less storage space than three or four individual programs. Because the package uses one set of commands, you can learn an integrated package more easily than you can learn individual productivity progeams. And of course, passing data form one module to another is very easy.

An integrated package does have one significant disadvantage. The modules may not be as powerful or versatile as the individual programs.

Software suites and integrated packages have been steaily increasing in popularity. Users apperciate the comparatively low cast and flexiblity.

Application suites often include word processor, spreadsheet, database, and presentation software all in the same package.

Following integrated packages (suites) are famous:

MS Office XP

- MS WORD 2002 Wordprocessor
- MS EXCEL 2002 Spreadsheet
- MS ACCESS 2002 Database
- POWERPOINT 2002 Presentation Graphics
- MS OUTLOOK EXPRESS E-mail. Calender. Contact

Corel Wordperfect Office

- WORDPERFECT 10 Wordprocessor
- QUATTROPRO 10 Spreadsheet
- PARADOX 10 Database
- COREL PRESENTATION Presentation Graphics
- COREL CENTRAL E-mail. Calender. Contact

Lotus SmartSuite Millenium

- LOTUS WORDPRO Wordprocessor
- LOTUS 123 Spreadsheet
- LOTUS APPROACH Database
- LOTUS FREELANCE GRAPHICS Presentation Graphics

5.16 ADVANTAGES OF INTEGRATED PACKAGES

(a) The main advantage is the saving of programming effort and expense on the part of the user. Development costs are effectively shared between the users.

(b) The user gets a well tried and tested program which he is able to use with confidence.

(c) Relatively quick implementation results from the use of packages.

5.17 APPLICATION PROGRAMMING INTERFACE

An application programming interface (API) is a description of the way one piece of software asks another program to perform a service. The service could be granting access to data or performing a specified function. APIs exist for most enterprise software, and they are essential in operating systems, which control most of a computer's basic functions.

You often have to rely on others to perform functions that you may not be able or permitted to do by yourself, such as opening a bank safety deposit box. Similarly, virtually all software has to request other software to do some things for it.

To accomplish this, the asking program uses a set of standardized requests, called application programming interfaces (API), that have been defined for the program being called upon. Almost every application depends on the APIs of the underlying operating system to perform such basic functions as accessing the file system. In essence, a program's API defines the proper way for a developer to request services from that program.

Developers can make requests by including calls in the code of their applications. The syntax is described in the documentation of the application being called. By providing a means for requesting program services, an API is said to grant access to or open an application.

Building an application with no APIs, says Josh Walker, an analyst at Forrester Research Inc. in Cambridge, Mass., "is basically like building a house with no doors. The API for all computing purposes is how you open the blinds and the doors and exchange information." APIs also exist between applications.

SAPAG's enterprise applications include APIs, called BAPI, that offer other applications access to business data. When an industry settles on a data standard, a common API allowing access to applications that process that data often follows, Walker says.

5.18 WHAT IS SHAREWARE

Shareware is try-before-you-buy software. Shareware is not free. But you can work on it for some time and decide if you want to buy it.

You can use the software for a period of time specified by the author. If you want to continue using it, then you should pay the author a registration fee (varies from software to software). Or you should remove the software from your system. This way you get to see what a program has for you, and the author also gets paid for his hard work. We don't give complete versions of programs as it is unethical and illegal to do so.

Some shareware programs nag you every time you run them. Once the trial period is over, they may stop working or keep nagging you to register. And if you have saved some files in these programs, you may not be able to access them.

On this CD, we bring you a whole lot of shareware. We also bring you some freeware, which are programs that are completely free. Check out the Hottest Freeware of the Month.

Shareware a way of marketing software in which you can try out the program in question for a limited time period, usually 30 days, before plunking down the cash. Some sites dedicated to shareware include ClNet's Shareware.com, WinSite, and TuCows. More information about the shareware concept is available from the Association of Shareware Professionals (ASP).

Freeware differs in that you are usually not required to pay the author anything to use it, however the software is still copyrighted, not in the public domain. Some authors only allow you personal use of their fonts, and in this case use for commercial purposes requires payment of some sort. Consult the included documentation for each individual font, or contact the author for details.

What else? Sometimes the authors are not really interested in payment for the stuff, but ask that you send them a postcard, hence postcardware, or a thank you E-mail, or just that you smile more often!

Shareware is based upon the "try-before-you-buy" system. If you like a shareware program a friend is using, you are free to make yourself a copy to try out. You are also legally allowed to download shareware from the Internet and install it on your computer. Once it is installed, you have a grace period (usually two weeks to thirty days) to decide if it is a product you want to keep.

If so, you are legally and morally required to send the author a registration fee as payment. In some cases, sending in the registration fee entitles you to extra goodies. For example, the author may mail you printed manuals upon receiving your registration, or he/she may send you other programs he/she has written. There are many variations on the shareware concept.

5.18.1 Crippleware

Crippleware is shareware that has been "crippled"—*i.e.* it is not fully functional. In order to un-cripple the software, you have to send in the registration fee, at which time the author sends you a key that unlocks the remaining features. It's common for crippleware to allow you to try almost all of the program's features, leaving out one or two key items—such as the ability to print.

5.18.2 Charityware

Charityware asks you to make a donation to the author's favorite charity, rather than sending the author payment. Postcardware does not ask for payment, but does ask you to send a picture postcard from your hometown, allowing the author to track whom is using his/her program.

5.18.3 Freeware

Freeware is yours to use and copy without any payment. Shareware is almost always self-marketed by the author; by not contracting with a professional publisher, he/she avoids packaging, marketing, and distribution costs. In other words, there's a lot less overhead associated with shareware.

The most common way to get shareware is to download it online. Shareware is sometimes sold in stores for a few dollars per disk, but the disks are rarely packaged attractively and usually contain no printed documentation.

5.18.4 Advantages of Shareware

As mentioned earlier, shareware has a lot less overhead than commercial software. A full-featured shareware program usually costs a half to a fifth as much as comparable commercial software. Try before you buy. If you download a new piece of shareware and, after installation, you find it doesn't meet your needs, all you have to do is delete it from your hard disk. No money lost.

5.18.5 Disadvantages of Shareware

Lack of documentation: Shareware rarely comes with printed manuals (sometimes you get them after registering). In some cases, you get hypertext Help files or text files which offer extensive notes on the program's use. In other cases, however, all you get is a README file that begs you to send the author money.

Lack of Technical Support: Don't count on a phone number you can call when you have questions. In fact, you'll be lucky if the author provides an e-mail address. You're on your own! It's hard to find others using the same program. Commercial software is the standard in home and office today, because it is so well marketed. If you have questions, you may have trouble finding a friend or co-worker who can help.

Chance of Catching a Virus

If you download your shareware from reputable archives such as shareware.com and Ziff-Davis, you'll probably never get infected. Still, there is a slight chance. Always virus-check new files before you install them on your computer.

5.18.6 Which Should you Choose?

Commercial software is the best choice for novice users. You'll have a printed manual to guide you through the installation, and a tech-support number to call if you experience problems. For that matter, anyone (novice or not) who believes in keeping it simple will prefer commercial software.

Experienced users, on the other hand, should consider shareware. There are thousands of outstanding shareware programs that cost only a fraction as much as their commercial counterparts. Also, if you want to try out a lot of different programs, shareware is a good way to do it since you can test-drive them without first paying a fee.

5.19 GROUPWARE

A class of software that helps groups of colleagues (workgroups) attached to a local-area network organize their activities. Yet another kind of client/server application is called groupware. This does not mean that everybody in your office wears the same T-shirt with the company logo on it. Groupware is a fuzzy new type of application that is designed to let people work together efficiently on collaborative projects.

Groupware is sometimes called workgroup productivity software. Groupware includes but is not limited to:

- Electronic mail
- Document routing and workflow
- Compound document managment (multimedia)
- Scheduling and calendar management
- Imaging
- Forms management

Typically, groupware supports the following operations:

- Scheduling meetings and allocating resources
- E-mail
- Password protection for documents
- Telephone utilities
- Electronic newsletters
- File distribution

Groupware is ideal for applications that center around various types of documents that are created, reviewed, updated, and used by several users.

For example, customer service in most companies revolves around a call form created when the service call is first received. Several people may be involved in adding information to the call form, reviewing it, passing it on to the individuals who can best handle the situation, tracking the call so it doesn't fall through the proverbial cracks, referring the call to higher powers if necessary, and so on until the customer is satisfied and the call form is stamped "closed." Every time the customer is contacted, the call form must be updated. With groupware, all of the people involved in processing the form can access it electronically and pass it back and forth.

The best-known groupware is Lotus Notes. In fact, at this writing, Notes is really the only player in the client/server groupware game, though that will change as other players including Microsoft enter the field. For now, if you want to do groupware, do it with Notes.

Lotus Notes includes a powerful scripting language that lets you create completely customized applications. It's more than just a fancy e-mail system; what it is can be difficult to say.

5.20 MIDDLEWARE AND MESSAGING

Middleware is a layer of software or functionality that sits between one system (usually a client) and another system (usually a server) and provides a way for those systems to exchange information or connect with one another even though they have different interfaces. Messaging is one of the methods that has become integral to the way that middleware is implemented.

Middleware and messaging may be employed within an organization to tie together its LAN and legacy systems, its diverse clients and back-end databases, and its local and remote systems. Middleware is also coming into widespread use on the Internet as a way to run sophisticated client/server applications with integral transaction processing, security, and management, something that is difficult to do with current Web protocols such as HTTP (Hypertext Transfer Protocol).

Here are some basic definitions of middleware:

- Middleware helps applications work together across multiple platforms or operating system environments.
- Developers write applications that interface with the middleware layer rather than recreating the functions that the middleware layer can provide.
- Middleware is used in the enterprise to make client/server applications work across the enterprise network.
- Middleware is used on the Internet to tie users together with servers at locations that are potentially worldwide.

5.21 UTILITY SOFTWARE

Computer software that is mostly concerned with maintenance and organization of the computer itself. The distinction between 'utility' and 'system' software is not clear-cut, but probably relates to how 'essential' the software is. Utility software covers those tasks which make the computer easier and more reliable to use, but it is not essential. For example, virus checking software might be classified as a 'utility,' while the operating system is very much an example of system software.

Utilities are system software that augment the operating system and give user a way to control use of hardware resources.

In computers, a utility is a small program that provides an addition to the capabilities provided by the operating system. In some usages, a utility is a special and non-essential part of the operating system. The print "utility" that comes with the operating system is an example. It's not absolutely required to run programs and, if it didn't come with the operating system, you could perhaps add it. In other usages, a utility is an application program that is very specialized and relatively limited in capability. A good example is a search-and-replace utility. Some operating systems provide a limited capability to do a search-and-replace for given character strings. You can add a much more capable search-and-replace utility that runs as an application program. However, compared to a word processor, a search-and-replace utility has limited capability.

Utility programs are a set of programs that perform such tasks as:

1. Copying file media (*e,g.*,disk-to-disk) or transferring data from one medium to another (e.g., tape-to-disk or disk-to-printer)
2. Sorting records on a file into some specified order (*e.g.*, sorting customer records into ascending, account-number sequence)
3. Merging the records from two files into one or selecting those records from one file that match those of another file.
4. Printing the contents of primary memory (this is called a memory dump).
5. Manipulating files by erasing them or checking disk directories for file names, contents, or sizes.
6. Formatting tapes or disks.

These are common tasks at a computer center and, in theroy, any system programmser could write the programs for them. However, it is usually more economical for businesses to acquire a comprehensive utility program to perform these "housekeeping chores."

5.22 UTILITIES AND SERVICE PROGRAMS

Utilities, also called service programs, are systems programs which provide a useful service to the user of the computer by providing facilities for performing common tasks of a routine nature. Common types of utility programs are:

- Disk fragmenter
- Virus scanner and remover
- Sort
- Editors
- File copying
- Dump
- File maintenance
- Tracing and debugging

5.23 VIRUS' VACCINE

Viruses are sneaky little critters—pieces of program code—that attach themselves to existing programs. When run, they may be harmless; for example, popping up a message that says "Happy Birthday!" On the other hand, their behavior may be quite malicious, erasing your hard drive or destroying your File Allocation Table (FAT).

But don't be scared—if you know how viruses work and how to avoid them, you can enjoy years of trouble-free Net-surfing. There is a complete chapter on virus chapter 83 which deals with other types of harmful programs such as bombs, worms, and Trojan horses are also covered.

5.24 THE NORTON UTILITIES

Before Peter Norton rolled up his sleeves, bit twiddlers were on their own when it came to recovering lost clusters and managing other disk catastrophes. It's almost the end of the millennium, and most of us still reach for Norton Utilities when something goes wrong with a disk.

5.25 PERSONAL INFORMATION MANAGERS

Most individuals have hectic schedules, a fact that has resulted in a large market for personal information managers (PIMs), This programs typically focus on four essential functions contact management calendear and scheule management, task and project managemet, and mail merge and print. Some PIMs also offer financial tracking and time and billing capabilities.

A personal inventory manager (PIM) or organizer is a program which allows you to maintain a to do list and a list of phone numbers and addresses. It also includes a notebook, a calendar and an appointment planner. The computer can prompt you with various types of alarms when the appointments are due.

Other programs will scan and remove computer "viruses" from your machine. Viruses are evil little programs deliberately designed to make your computer crash and burn. They may be unwittingly transmitted with other information that you are downloading from, say, the Internet and you will have no clue that they are in your machine until it starts to behave peculiarly.

Disk compression programs squeeze all the available space out of your disk drives and are very useful for the inevitable day when you have too many programs and too little space. Some of the operating systems have their own compression programs, for example, doubles pace in DOS.

Having enough RAM (Memory) is also a problem and as you use the computer more, the likelihood is that you will run out. There are ways of making more memory available and programs such as Quarterdeck's QEMM can do this for you. Some operating systems, such as Windows, have a memory manager built in to the program.

1. **Contact Management:** Contact managemet software is designed to help you keep track of your contacts. Information maintained by the software includes addresses, telephone and fax numbers, a notepad, sutomatic telephone dialing using a modem, and search and sort capabilities.
2. **Calender and Schedule Management:** Most PIMS offer a varity of calendar formats: yearly,monthly weekly, and daily. The amount of detail provided should increase as the time span covered decreases. A PIM usually has an alarm that can be set to derve as an appointement reminder, and the program warns the user of possible schedule conflicts. Recurringntasks are scheduald automatically, and a flexible array of calendar printing options is usually available.

3. **Task and Project Management:** Busy people tend to have several projects going at once. Sometines keeping deadlines ateaight and prioritizing tasks can be overwhelming. The job of task management software is to organize and present tasks both graphically and in the form of "to do" lists. A PIM should enable the user to view tasks by project, deadline data and priority. The PIM should also have the capability of automatically rescheduling overdue projects.
4. **Mail Merge and Print:** PMIs interface with different word processing and fax programs. The capability to merge documents and produce mailing lables is essential. Expanded telecommunications capabilities are appearing in the newest PIMs on the market. Users need to be aware of which programs are compatible with a particular PIM.

Popular PIMs

Some PIMs are designed for home use, some for business use and others for a workgroup. popular PIMs include Sidekick by Starfish Software, LOTUS Organizer, Schedule+ by Microsoft and ACT by Symantec Corporation. PIMs used for business and workgroups include ECCO professional by Arabesque Software, Full Contact by FIT Software, and InfoCentral by WordPerfect.

5.26 PLUG-INS

Even if you're new to the Net, you've probably been hearing about multimedia on the Web — listening to audio, watching animations and videos, even playing in 3-dimensional space. Life online can be a much richer experience when you aren't restricted to just words and pictures. Sound and movement make information come alive.

To experience multimedia online, you should have a computer with a sound and video card. Then what you need are special pieces of software called plug-ins. A plug-in extends the capabilities of your web browser, like Netscape Navigator or Microsoft Explorer, turning your computer into a radio or television.

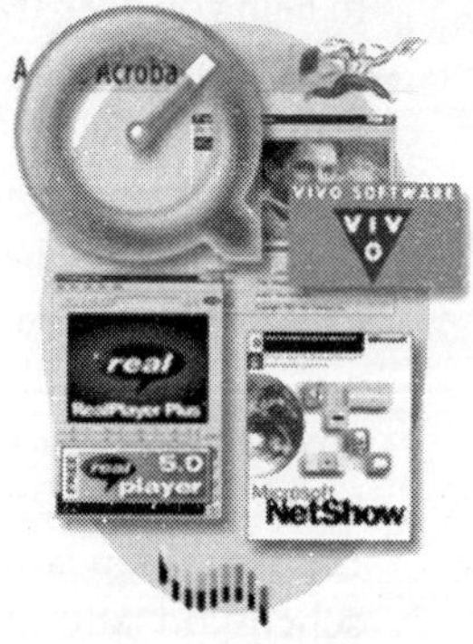

Fig. 1.66: Plugins

First, a couple of terms. The term need to be explained "plug-in" is used in two ways on the Internet. The technical definition of a plug-in is a small add-on piece of software that conforms to Netscape Navigator standard. Other browsers however, including Microsoft Explorer 3.0, support many Netscape plug-ins. But Explorer actually uses a different software standard, called an ActiveX control, instead of plug-ins. Just to make life more confusing, there are plug-ins for Navigator that allow it to run some ActiveX controls.

Evolution

The earliest known library was a collection of clay tablets in Babylonia in the 21st cent. B.C. Ancient Egyptian temple libraries are known through the Greek writers. Diodorus Siculus describes the library of Ramses III, c.1200 B.C. The extensively cataloged library of Assurbanipal (d. B.C.) in Nineveh was the most noted before that at Alexandria. The temple at Jerusalem contained a sacred library. The first public library in Greece was established in 330 B.C., in order to preserve accurate examples of the work of the great dramatists. The most famous libraries of antiquity were those of Alexandria, founded by Ptolemy I, which contained some 700,000 Greek scrolls. The library at Pergamum, founded or expanded by Eumenes II, rivaled those at Alexandria.

The first Roman libraries were brought from Greece, Asia Minor, and Syria as a result of conquests in the 1st and 2nd century B.C. Caius Asinius Pollio established (c.40 B.C.) the first public library in Rome, but the great public libraries of the Roman Empire were the Octavian (destroyed A.D. 80) and the Palatine (destroyed c.A.D. 190) and the more important Ulpian library, founded during the reign of Trajan. In addition to these public collections, there were many fine private libraries by the time the Roman Republic was ended in 27 B.C. of these there remain only fragments of one at Herculaneum.

The early Christian libraries were in monasteries; the Benedictines amassed a fine collection at Monte Cassino. The Romans had brought book collections to the British Isles, but important early monastic libraries were founded in York, Wearmouth, Canterbury, and elsewhere in England and Ireland by Anglo-Saxon monks. Some of the finest manuscript illumination was produced in these libraries. On the continent, St. Columban and other missionaries founded monastic libraries in the 6th cent. Most of the ancient Greek and Latin texts that have survived until modern times were preserved in medieval European monastery libraries.

The Arabs in the 9th to 15th century collected and preserved many libraries, and the Jews and the Byzantines also developed fine libraries during the medieval period. In the 14th and 15th century Charles V of France, Lorenzo de' Medici, and Frederick, duke of Urbino, all formed fine libraries; part of the Urbino library is now in the Vatican Library. In the 15th century the Vatican Library, the oldest public library in Europe, was formed. In 1475, Platina, as its first librarian, made a catalog that included 2,527 volumes. In 1257 the Sorbonne library at Paris was founded, and in 1525 the erection of the Laurentian Library in Florence, designed by Michelangelo, was begun. Many of the great university libraries (*e.g.*, Bologna, Prague, Oxford, and Heidelberg) were opened in the 14th century.

In the United States a circulating library, the Library Company of Philadelphia, was chartered in 1732 on the initiative of Benjamin Franklin. A public library had, however, been opened in Boston as early as 1653 (see Boston Public Library). Other early subscription libraries included the Boston Athenaeum, the New York Society Library, and the Charleston (S.C.) Library Society. In 1833 the first tax-supported library in the country opened at Peterborough, N.H. The American Library Association was formed in 1876, and this organization spurred improvements in library methods and in the training of librarians.

Libraries in the United States and Great Britain benefited greatly from the philanthropy of Andrew Carnegie, who gave more than $65 million for public library buildings in the United States alone and strengthened local interest by making the grants contingent upon public support. Among the innovations of the late 19th century were free public access to books (involving elaborate classification schemes) and branch libraries or deposit stations for books in many parts of cities; in the early 20th century travelling libraries, or bookmobiles, began to take books to readers in rural or outlying areas.

Notable Libraries

Among the chief modern public and university libraries are the Bibliothhque nationale and the Mazarine, Paris; the British Museum, London; the Bodleian Library, Oxford; the Vatican Library, Rome; the Ambrosian Library, Milan; the Laurentian Library, Florence; the Russian State Library, Moscow; the Huntington Library, San Marino, Calif. (see under Huntington, Henry Edwards); the Library of Congress, Washington, D.C.; the New York Public Library; the libraries of Chicago, Columbia, Harvard, Princeton, Yale, and other major American universities; and the Newbery and John Crerar libraries in Chicago.

There are several sorts of libraries in the United States and elsewhere that exist apart from the public and university systems. Three major categories of these are private libraries, usually housing special collections, *e.g.*, the Pierpont Morgan Library in New York City of rare books in the humanities and the Folger Shakespeare Library in Washington, D.C. (see under Folger, Henry Clay); presidential libraries, which contain the papers of past presidents not held in the Library of Congress, *e.g.*, the Jimmy Carter Library, Atlanta, Ga., the Dwight D. Eisenhower Library, Abilene, Kans., the Gerald R. Ford Library, Ann Arbor, Mich., the Rutherford B. Hayes Library, Fremont, Ohio, the Herbert Hoover Library, West Branch, Iowa, the Lyndon Baines Johnson Library at the University of Texas, Austin, the John Fitzgerald Kennedy Library, Boston, the Franklin D. Roosevelt Library, Hyde Park, N.Y., and the Harry S. Truman Library, Independence, Mo.; and industrial libraries formed by many corporations to house research works relevant to their business.

Modern-style library

In its traditional sense, a library is a collection of books and periodicals. It can refer to an individual's private collection, but more often, it is a large collection that is funded and maintained by a city or institution. This collection is often used by people who choose not to, or can not afford to purchase an extensive collection themselves. However, with the collection or invention of media other than books for storing information, many libraries are now also repositories and/or access points for maps, prints or other artwork, microfilm, microfiche, audio tapes, CDs, LPs, video tapes and DVDs, and provide public facilities to access CD-ROM databases and the Internet.

Thus, modern libraries are increasingly being redefined as places to get unrestricted access to information in many formats and from many sources.

Etymology of the Word

The word is derived from Latin liber, which means "book". Derivations from the Greek Bibliotheke (from Biblos, book) are used in at least German, French, Norwegian, Spanish, Swedish, Danish, Polish, Portuguese, Romanian, Russian, Dutch, and of course Modern Greek. Other languages, such as Icelandic, Finnish, Estonian and Persian, use words that derive from their own words for book (Bokasafn, Kirjasto, and Raamatukogu and Ketabkhaneh, respectively). Some European languages use a cognate of library to mean bookshop.

One of two traditional Research Rooms in the main branch of the New York Public Library, now equipped with modern computers

History

The first libraries were only partly libraries, being composed for the most part of the unpublished records that make up archives. Archaeological findings from the diggings of the ancient city-states of Sumer have revealed temple rooms full of clay tablets in cuneiform script. These archives were made up nearly completely of the records of commercial transactions or inventories, with only a few documents touching theological matters or legends. Things were much the same in the Papyrus based government records of ancient Egypt.

Private or personal libraries made up of non-fiction and fiction books, (as opposed to the state or institutional records kept in archives) first appeared in classical Greece. The first ones appeared some time near the 5th century before our era. They were filled with parchment scrolls and later on papyrus scrolls. There were a few institutional or royal libraries like the Library of Alexandria, which were open to an educated public but on the whole, collections were private. In those rare cases where it was possible for a scholar to consult library books there seems to have been no direct access to the stacks. In all recorded cases the books were kept in a relatively small room where the staff went to get them for the readers, who had to consult them in an adjoining hall or covered walkway.

Little is known about early Chinese libraries, save what is written about the imperial library which began with the Qin Dynasty. One of the curators of the imperial library in the Han Dynasty is believed to have been the first to establish a library classification system and the first book notation system. At this time the library catalog was written on scrolls of fine silk and stored in silk bags.

The Geisel Library at UCSD, with its unique architecture is a landmark. In Persia many libraries were established by the

Zoroastrian elite and the Persian kings. Among the first ones was a royal library in Isfahan. One of the most important public libraries established around 666 AD in south-western Iran was the Library of Gundishapur. It was a part of a bigger scientific complex located at the Academy of Gundishapur.

During the Early Middle Ages, after the fall of the Western Roman Empire and before the rise of the large Christian monastery libraries, Islamic libraries knew a period of great expansion in the Middle East, North Africa, Sicily and Spain. Like the Christian libraries they mostly contained books which were of a codex or modern form instead of scrolls. By the 8th century first Iranians and then Arabs had imported the craft of paper making from China, with a mill already at work in Baghdad in 794. By the 9th century completely public libraries started to appear in many Islamic cities. They were called "halls of Science" or dar alilm. They were each endowed by Islamic sects (many of which have disappeared or become less important in our times) with the purpose of representing their tenets as well as promoting the dissemination of secular knowledge. The libraries often employed translators and copyists in large numbers, in order to render into Arabic the bulk of the available Persian, Greek and Roman non-fiction and the classics of literature. After but a few centuries most of these libraries were destroyed by Mongolian invasion and violent strife between the sects in the larger cities of the Islamic world. But by then much of their contents had been copied in the more tranquil settlements in Spain and Sicily, and from there they often made their way to Christian Europe. One of the oldest libraries in the Middle East which is still operational and expanding is the Central Library of Astan Quds Razavi in the Iranian city of Mashhad with a history of over six centuries.

The first public libraries in the west were established under the Roman Empire as each succeeding emperor strove to open one or many which outshone that of his predecessor. Unlike the Greek libraries readers had direct access to the scrolls, which were kept on shelves built into the walls of a large room. Reading or copy was normally done in the room itself. The records give only a few instances of lending features. As a rule Roman public libraries were bilingual: they had a Latin room and a Greek room. Most of the large Roman baths were also cultural centers, built from the start with a library, with the usual two room arrangement for Greek and Latin texts.

The medieval library arose very directly from the fact that books were valuable possessions, were therefore likely to be stolen, and were far too expensive for most people to own. Its architecture derived from the need to chain these books, first to lecterns and later to armaria and shelves, in areas that were illuminated by sunlight. Early libraries were located in monastic cloisters associated with scriptoria and were collections of lecterns with books chained to them. Shelves built above and between back-to-back lecterns were the beginning of bookpresses. The chain was attached at the fore-edge of a book rather than to its spine. Book presses came to be arranged in carrels (perpendicular to the walls and therefore to the windows) in order to maximize lighting, with low bookcases in front of the windows. This stall system (fixed bookcases perpendicular to exterior walls pierced by closely spaced windows) was characteristic of English institutional libraries. In Continental libraries, bookcases were arranged parallel to and against the walls. This wall system was first introduced on a large scale in Spain's El Escorial.

As books became cheaper, the need for chaining them lessened, but as the number of books in libraries increased, so did the need for compact storage and access with adequate lighting, giving birth to the stack system, which involved keeping a library's collection of books in a space separate from the reading room, an arrangement which arose in the 19th century. Book stacks quickly evolved into a fairly standard form in which the cast iron and steel frameworks supporting the bookshelves also supported the floors, which often were built of translucent blocks to permit the passage of light (but were not transparent, for reasons of modesty). With the introduction of electrical lighting, the use of glass floors was largely discontinued, though floors were still often composed of metal grating to allow air to circulate in multi-story stacks.

Ultimately, even more space was needed, and a method of moving shelves on tracks ("compact shelving") was introduced to cut down on otherwise wasted aisle space.

Types of Libraries

Many libraries are comfortable places for reading and studying. Libraries can be divided into categories by several methods: By the Entity (institution, municipality, or corporate body) that supports or perpetuates it

- Private libraries
- Corporate libraries

- Federal/Central/State/Municipal
- Academic libraries
- historical society libraries
- by the type of documents or materials it holds
- digital libraries
- picture (photograph) libraries
- slide libraries
- tool libraries
- by the subject matter of documents it holds
- architecture libraries
- fine arts libraries
- military libraries
- law libraries
- By the users it serves
- military communities
- By traditional professional divisions:
    - **Academic libraries:** These libraries are located on the campuses of colleges and universities and serve primarily the students and faculty of that and other academic institutions.
    - **School libraries:** Most public and private primary and secondary schools have libraries designed to support the curriculum.
    - **Public libraries or public lending libraries:** These libraries provide service to the general public and make at least some of their books available for borrowing, so that readers may use them at home over a period of days or weeks. Many public libraries also serve as community organizations that provide free services and events to the public, particularly children.
    - **Special libraries:** All other libraries fall into this category by default. Many private businesses and public organizations, including hospitals, museums, research laboratories, law firms, and many government departments and agencies, maintain their own libraries for the use of their employees in doing specialized research related to their work. Special libraries may or may not be accessible to some identified part of the general public.

Also, the governments of most major countries support national libraries. Two noteworthy examples are the U.S. Library of Congress and the British Library.

Description

Libraries invariably contain long aisles with rows and rows of books.Libraries usually have nearly every item they own arranged in a specified order according to a library classification system, so that particular items may be located quickly and collections may be browsed efficiently. Some libraries have additional galleries beyond the public ones, where reference materials are stored. These reference stacks may be open to selected members of the public; others require patrons to submit a "stack request," which is a request for a librarian to find the material in the stacks and bring it out to the patron. Libraries often feature a professional librarian working from a reference desk or other central location to help users find what they are looking for.

Library Use

Many potential library patrons nevertheless do not know how to use a library effectively. This can be due to lack of early exposure, shyness, or are anxious and fearful of displaying ignorance. These problems drove the emergence of the library instruction movement, which advocates library user education. Library instruction has been practiced in the U.S. since the 19th century; one of the early leaders was John Cotton Dana, and an example of a more recent leader is Michael Lorenzen. Library instruction is closely related to the study of information literacy.

Libraries must inform the public of what materials are available in their collections, and the public must know how to access that information. Before the computer age, this was accomplished by the card catalog—a piece of wooden or metal furniture containing many drawers, each filled with standard-sized index cards identifying books and other materials. In a large library, the card catalog units often filled a large room, or else lined most hallways in the building. The emergence of the Internet, however, has led to the adoption of electronic catalog databases (often referred to as "webcats" or as OPACs, for "online public access catalog"), which allow users to search the library's holdings from any location with Internet access. This style of catalog maintenance is compatible with new types of libraries, such as digital libraries and distributed libraries, as well as older libraries that have been retrofitted.

2

UNIT

Operating Systems and Programming

1.1 BASIC CONCEPT OF AN OPERATING SYSTEM (OS)

An operating system (sometimes abbreviated as "OS") is the program that, after being initially loaded into the computer by a bootstrap program, manages all the other programs in a computer. The other programs are called applications. The applications make use of the operating system by making requests for services through a defined application program interface (API). In addition, users can interact directly with the operating system through an interface such as a command language.

All computers need some sort of Operating System. The majority of modern home computers use some form of Microsoft's operating systems. The original Microsoft operating system was called DOS (Disk Operating System) though most computers use Windows. Windows comes in various versions beginning with version 3.x then 95, 98, ME and currently XP. A few computers use IBM's O/S2. Apple's Mac use their own operating system beginning with OS 1 though most modern Macs use version 8.x or 9.x. Apple's latest version is OS 10.1.x. In the past large companies and institutions would have an operating system design exclusively for them but as the commercial operating systems become more sophisticated the benefits of this practice is becoming less apparent. Some computer professionals, Internet Service Providers (ISP) and mainframe computer users use an operating system such as UNIX (or a variant such as Linux), Windows NT or 2000 (Win2k) or one of the other network or server based operating systems.

There are many smaller operating systems. The problem is that software is currently being developed only for the main operating systems and only the newest versions of these OS. Many older computers with unique operating systems have lots of software already developed for them but there is very little new software being developed for the older computers. The older proprietary operating systems are less likely to offer technical support than the more modern operating systems.

The operating system controls the input and output or directs the flow of information to and from the CPU. Much of this is done automatically by the system but it is possible to modify and control your system if you need to.

When you turn your computer on it first needs to load the operating system sometimes referred to a booting up. Basically the computer starts from scratch every time you turn the power on. It checks all its components and will usually display a message if there is a problem. Loading the system is usually automatic.

Once the system is loaded the user can start the application or program that they are going to use.

Most computer users will run Microsoft Windows, Mac OS or Linux as their operating system. These OS are Graphic User Interface (GUI) which allows the user to control or run the computer using a Mouse and Icons. The user simply moves the mouse on a flat surface, rolls the trackball, or moves their hand over the touchpad to control a pointer. They then choose the option they want by pressing a button or touching the pad.

Without a GUI the user controls the computer using the keys on the keyboard. This is referred to as a Command Line Interface (CLI)

An operating system performs these services for applications:

- In multitasking operating systems where multiple programs can be running at the same time, the operating system determines which applications should run in what order and how much time should be allowed for each application before giving another application a turn.
- It manages the sharing of internal memory among multiple applications.
- It handles input and output to and from attached hardware devices, such as hard disks, printers, and dial-up ports.
- It sends messages to the applications or interactive user (or to a system operator) about the status of operation and any errors that may have occurred.

- It can offload the management of what are called batch jobs (for example, printing) so that the initiating application is freed from this work.
- On computers that can provide parallel processing, an operating system can manage how to divide the program so that it runs on more than one processor at a time.

1.2 COMMONLY USED OPERATING SYSTEMS

Different computers use different operating systems. Many IBM and IBM-compatible microcomputers use MS-DOS alone, a combination of MS-DOS and a version of Microsoft Windows, or a version of Windows alone with the DOS based command prompt built-in.

All major computer platforms (hardware and software) require and sometimes include an operating system. UNIX, Windows 95, 98, 2000, Xp, DEC's VMS, IBM's OS/2, AIX, and OS/390 are all examples of operating systems.

1.3 FUNCTIONING OF THE OS

An operating system is a program that is loaded into the computer on boot up that is responsible for running other applications and provides an interface with which to interact with other programs. This interface can be command line based like DOS or Unix or can include a Graphical User Interface(GUI) such as Windows operating systems.

Operating System (OS) is a program or rather more frequently a set of programs whose primary purpose is to provide an interface between applications programs such as word processing or table calculations and a computer hardware such as processor, disk or keyboard.

Application Software: word processing, airline reservation system, computer games
Operating System
Hardware

Fig. 2.1: Operating System as an Interface Layer Between Computer Hardware and Application Program

The objective is to build a machine that is a pleasure to work with but at the same time ensure ease of use, efficiency, good performance and high utilization of available resources which include hardware such as the central processing unit (CPU), memory unit and I/O devices as depicted in following figure.

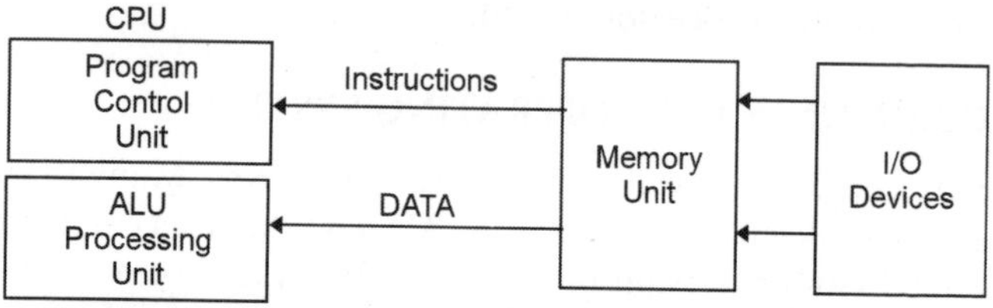

Fig. 2.2: Computer Organization: Hardware Resources

The software resources are shown in following figure.

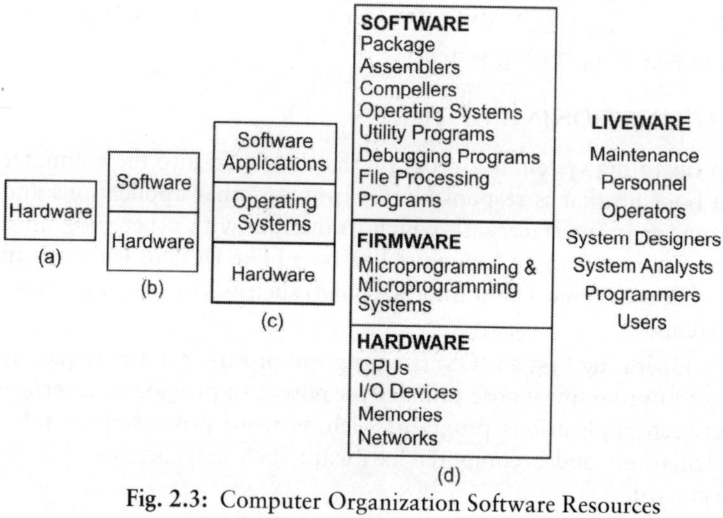

Fig. 2.3: Computer Organization Software Resources

More formally, the OS performs various supervisory and control functions over the users programs in execution which are called tasks or processes. The OS is responsible for.

1. Creating and killing processes.
2. Overseeing and controlling the progress of processes executions (*e.g.*, a detection of an infinite loop should lead to a timeout or a transfer of control to another program)
3. Scheduling, *i.e.*, defining the order in which processes are to be executed.

4. Handling of exceptions, errors or protection violations
5. Allocating hardware and software resources to processes
6. Providing access to software and data, e.g., compilers and assemblers, library programs, text files, pictures, music
7. Ensuring required quality-of-service with respect to security, real-time, dependability and others
8. Supporting communication and web services (even a browser is viewed by some as a part of the OS).

An operating system is conceptually broken into three sets of components:

- a user interface (which may consist of a graphical user interface and/or a command line interpreter or "shell"),
- low-level system utilities, and a
- kernel—which is the heart of the operating system.

As the name implies, the shell is an outer wrapper to the kernel, which in turn talks directly to the hardware.

Hardware ⟷ Kernel ⟷ Shell ⟷ Applications

In some operating systems the shell and the kernel are completely separate entities, allowing you to run varying combinations of shell and kernel (*e.g.*, UNIX), in others their separation is only conceptual.

Processor wise operating systems can be divided into 2 groups:

- Single-process
- Multi-process.

Single process operating systems are capable of working on 1 task at a time while a Multi-process OS can work on several processes at once by breaking tasks into threads. There are several terms related to multiprocessing systems that you will need to know as follows:

- **Multitasking:** This is the ability to work on several different tasks at a time. This is accomplished by switching back and forth between the tasks. There are few different types of multitasking:
- **Task Switching:** Allows for multiple applications to be run at the same time. The window that is in the foreground is the active window while the other applications run in the background. Used in Windows 1.0.
- **Cooperative Multitasking:** Applications can control the system resource until they are finished. When the hourglass is displayed

on the screen, you would be unable to perform any tasks until the system had finished the task that it was working on. If a task caused faults or other problems, it would cause the system to become unstable and force a reboot. Used in Windows 1.x.

- **Preemptive Multitasking:** Applications are allowed to run for a specified period of time depending on how important the application is to the operation of the system (priority basis). This means that even though you may see an hourglass on the screen, you can still launch or use other application at the same time. If a particular task is causing problems or faults, that application can be stopped without the system becoming unstable. Used in Windows 1.x.

Operating systems may be classified by both how many tasks they can perform 'simultaneously' and by how many users can be using the system 'simultaneously'. That is single-user or multi-user and single-task or multi-tasking. A multi-user system must clearly be multi-tasking.

Within the broad family of operating systems, there are generally four types, categorised based on the types of computers they control and the sort of applications they support. The broad categories are:

1.4 SINGLE-USER, SINGLE-TASKING OPERATING SYSTEM

As the name implies, this operating system is designed to manage the computer so that one user can effectively do one thing at a time. The Palm O.S. for Palm handheld computers is a good example of a modern single-user, single-task operating system.

We are all familiar with the concept of sitting down at a computer system and writing documents or performing some task such as writing a letter. In this instance there is one keyboard and one monitor that you interact with.

Operating systems such as Windows 95, Windows NT Workstation and Windows 2000 Professional are essentially single user operating systems. They provide you the capability to perform tasks on the computer system such as writing programs and documents, printing and accessing files.

Consider a typical home computer. There is a single keyboard and mouse that accept input commands, and a single monitor to display information output. There may also be a printer for the printing of documents and images.

In essence, a single-user operating system provides access to the computer system by a single-user at a time. If another user needs access to the computer system, they must wait till the current user finishes what they are doing and leaves.

Students in computer labs at colleges or university often experience this. You might also have experienced this at home, where you want to use the computer but someone else is currently using it. You have to wait for them to finish before you can use the computer system.

Single-user, Multi-tasking Operating System

This is the type of operating system most people use on there desktop and laptop computers today. Windows 98 and the Mac O.S. are both examples of an operating system that will let a single user has several programs in operation at the same time. For example, it's entirely possible for a Windows user to be writing a note in a word processor while downloading a file from the Internet while printing the text of an e-mail message.

2.1 MULTI-USER OPERATING SYSTEMS

A multi-user operating system allows many different users to take advantage of the computer's resources simultaneously. The operating system must make sure that the requirements of the various users are balanced, and that each of the programs they are using has sufficient and separate resources so that a problem with one user doesn't affect the entire community of users.

A multi-user operating system lets more than one user access the computer system at one time. Access to the computer system is normally provided via a network, so that users access the computer remotely using a terminal or other computer.

In the early days of large multi-user computers, multiple terminals (keyboards and associated monitors) were provided. These terminals sent their commands to the main multi-user computer for processing, and the results were then displayed on the associated terminal monitor screen. Terminals were hard-wired directly to the multi-user computer system.

Today, these terminals are generally personal computers and use a network to send and receive information to the multi-user computer system. Examples of multi-user operating systems are UNIX, Linux (a UNIX clone) and mainframes such as the IBM AS400.

Following are the famous multi-user Operating Systems.

- Linux
- UNIX
- Windows 2000

2.2 ADVANTAGES AND PROBLEMS OF MULTI-USER OPERATING SYSTEMS

The advantage of having a multi-user operating system is that normally the hardware is very expensive, and it lets a number of users share this expensive resource. This means the cost is divided amongst the users. It also makes better use of the resources. Since the resources are shared, they are more likely to be in use than sitting idle being unproductive.

One problem with multi-user computer systems is that as more users access it, the performance becomes slower and slower. Another disadvantage is the cost of hardware, as a multi-user operating system requires a lot of disk space and memory. In addition, the actual software for multi-user operating systems tend to cost more than single-user operating systems.

3.1 PROGRAMMING LANGUAGES

Programming languages are one of two types of system software. It is similar to the languages that you and I are most familiar with-Hindi, Sanskrit, Tamil, English, Spanish, French, German, Chinese, etc. Just like we speak Hindi, Sanskrit, Tamil, English, or Spanish, or French, computers, too, speak a language. The languages that computers speak are called programming languages.

A computer language is a language used by, or in association with computers. Often, the term is used synonymously with programming language, but in general a computer language need not be a programming language.

It is understood that markup languages like HTML are generally not held to be programming languages, but they are computer languages. In general, as with any other kind of language, a computer language is created wherever there is a need to communicate some information from one entity to another.

Programming languages foster the communication of programs among programmers and computers; markup languages communicate the formatting or structure of documents among humans and computers and so on.

These are the languages that programs and applications are written in so that they can be understood and carried out by the computer. Just like the natural languages that we, as humans, speak, programming languages also have grammar, vocabularies, and syntax. The major difference between natural languages and programming languages is that natural languages are ambiguous. They can be interpreted in different ways, and often have different rules. Anyway, a computer's programming language must be very specific. The rules that they have for grammar and spelling must be very precise. This is obviously not the case with natural languages. With English, for example, there are tons of rules but then, there are tons of exceptions to these rules.

3.1.1 Features of Programming Language

Each programming language can be thought of as a set of formal specifications concerning syntax, vocabulary, and meaning.

These specifications usually include:

- Data and Data Structures
- Instruction and Control Flow
- Reference Mechanisms and Re-use
- Design Philosophy

Most languages that are widely used, or have been used for a considerable period of time, have standardization bodies that meet regularly to create and publish formal definitions of the language, and discuss extending or supplementing the already extent definitions.

Data Types

Internally, all data in modern digital computers are stored simply as zeros or ones (binary). The data typically represent information in the real world such as names, bank accounts and measurements and so the low-level binary data are organised by programming languages into these high-level concepts.

The particular system by which data are organized in a program is the type system of the programming language; the design and study of type systems is known as type theory. Languages can be classified as statically typed systems, and dynamically typed languages. Statically typed languages can be further subdivided into languages with manifest types, where each variable and function declaration has its type explicitly declared, and type inferred languages. It is possible to perform type inference on programs written in a dynamically typed language, but it is entirely possible to write programs in these languages that make type inference infeasible. Sometimes dynamically typed languages are called latently typed.

With *statically typed languages*, there usually are pre-defined types for individual pieces of data (such as numbers within a certain range, strings of letters, etc.), and programmatically named values (variables) can have only one fixed type, and allow only certain operations: numbers cannot change into names and vice-versa. Most mainstream statically typed languages, such as C, C++, Java, Delphi and C#, require all types to be specified explicitly; advocates argue that this makes the program easier to understand, detractors object to the verbosity it produces. Type inference is a mechanism whereby the type specifications can often be omitted completely, if it is possible for the compiler to infer the types of values from the contexts in which they are used—for example, if a variable is assigned the value 1, a type-inferring compiler does not need to be told explicitly that the variable is an integer. Type-inferred languages can be more flexible to use, particularly when they also implement parametric polymorphism. Examples of type-inferring languages are Haskell, MUMPS and ML.

Dynamically typed languages treat all data locations interchangeably, so inappropriate operations (like adding names, or sorting numbers alphabetically) will not cause errors until run-time—although some implementations provide some form of static checking for obvious errors. Examples of these languages are Objective-C, Lisp, JavaScript, Tcl, Prolog, Python, and Ruby.

Strongly typed languages: do not permit the usage of values as different types; they are rigorous about detecting incorrect type usage, either at runtime for dynamically typed languages, or at compile time for statically typed languages. Ada, Java, ML, and Oberon are examples of strongly typed languages.

Weakly typed languages do not strictly enforce type rules or have an explicit type-violation mechanism, often allowing for undefined behavior, segmentation violations, or other unsafe behavior if types are assigned incorrectly. C, assembly language, C++, and Tcl are examples of weakly typed languages.

Note that strong vs, weak is a continuum; Java is a strongly typed language relative to C, but is weakly typed relative to ML. Use of these terms is often a matter of perspective, much in the way that an assembly language programmer would consider C to be a high-level language while a Java programmer would consider C to be a low-level language.

Note that strong and static are orthogonal concepts. Java is a strongly, statically typed language. C is a weakly, statically typed language. Python is a strongly, dynamically typed language. Tcl is a weakly, dynamically typed language. But beware that some people incorrectly use the term strongly typed to mean strongly, statically typed, or, even more confusingly, to mean simply statically typed—in the latter usage, C would be called strongly typed, despite the fact that C doesn't catch that many type errors and that it's both trivial and common to defeat its type system (even accidentally).

3.1.2 History of Programming Languages

The development of programming languages, unsurprisingly, follows closely the development of the physical and electronic processes used in today's computers.

Charles Babbage is often credited with designing the first computer-like machines, which had several programs written for them (in the equivalent of assembly language) by Ada Lovelace.

In the 1940s the first recognisably modern, electrically powered computers were created. Some military calculation needs were a driving force in early computer development, such as encryption, decryption, trajectory calculation and massive number crunching needed in the development of atomic bombs. At that time, computers were extremely large, slow and expensive: advances in electronic technology in the post-war years led to the construction of more practical electronic computers. At that time only Konrad Zuse imagined the use of a programming language (developed eventually as Plankalkl) like those of today for solving problems.

Subsequent breakthroughs in electronic technology (transistors, integrated circuits, and chips) drove the development of increasingly reliable and more usable computers. The first widely used high level programming language was Fortran, developed during 1954–57 by an IBM team led by John W. Backus. It is still widely used for numerical work, with the latest international standard released in 2003. A Computer Languages History graphic shows a timeline from FORTRAN in 1953.

Dennis Ritchie and Brian Kernighan developed the C programming language, initially for DEC PDP-11 in 1970. Later with lead of Bjarne Stroustrup the programming language C++ appeared starting from 1985 as the Object-oriented language vertically compatible with C. Sun Microsystems developed Java in 1995 which became very popular as the initial programming language taught at universities. Microsoft presented the C# programming language, bringing garbage collection, generic types, and introspection to a language that C++ programmers could learn easily, in 2001. There are other languages such as Python, Visual Basic, etc.

3.2 TYPES OF PROGRAMMING LANGUAGES

There are two types of programming languages:

1. Low-Level Programming Language
2. High-Level Programming Language

1. Low- Level Languages (Machine Language)

Machine languages are also called low-level programming languages. They are the most basic types of programming languages and can be directly understood by the computer.

They are usually written as a series of 1s and 0s—the digits in the binary system. Machine languages can be used to tell computers what computations to perform, where the data needed to carry out a certain instruction is contained in the memory, where in the memory to store the results of an operation, as well as what operation to perform next!

A low-level programming language is a language that provides little or no abstraction from a computer's microprocessor. The word "low" does not imply that the language is inferior to high-level programming languages but rather refers to the reduced amount of abstraction between the language and itself; because of this, low-level languages are sometimes described as being "closer to the hardware".

Low-level programming languages are sometimes divided into two categories: First generation, and Second generation.

(*a*) *First Generation*

The first generation programming language, or 1GL, is machine code. It is the only language a microprocessor can understand natively. Machine code cannot be written or read using a text editor, and therefore it is rarely used by a person directly.

Machine Code or Language

A system of codes directly understandable by a computer's CPU is termed this CPU's native or machine language. Although machine code may seem similar to assembly language they are in fact two different types of languages. Assembly code consists of both binary numbers and simple words whereas machine code is composed only of the two binary digits 0 and 1. Every CPU model has its own machine language, although there is considerable overlap between some. If CPU A understands the full language of CPU B it is said that A is compatible with B. CPU B may not be compatible with CPU A, as A may know a few codes that B does not.

The "words" of a machine language are called instructions; each of these causes an elementary action by the CPU, such as reading from a memory location. A program is just a long list of instructions that are executed by a CPU. Older processors executed instructions one after the other, but newer superscalar processors are capable of executing several instructions at once. Program flow may be influenced by special jump instructions that transfer execution to an instruction other than the following one. Conditional jumps are taken (execution continues at another address) or not (execution continues at the next instruction) depending on some condition.

Instructions are simply a pattern of bits—different patterns correspond to different commands to the machine. Humans use mnemonic codes to refer to the useful bit-patterns: this more readable rendition of the machine language is called assembly language. For example, on the Z80 processor, the machine code 00000101 causes the CPU to decrement the B register. In assembly language we write this DEC B.

Some languages give all their instructions the same number of bits, while the instruction length differs in others. How the patterns are organised depends largely on the specific language. Common to most is the division of an instruction into fields, of which one or more specify the exact operation (for example "add"). Other fields may give the type of the operands, their location, or their value directly (operands contained in an instruction are called immediate).

Computer software (or simply software) refers to one or more computer programs held in the storage of a computer for some purpose. Program software performs the function of the program it implements, either by directly providing instructions to the computer hardware or by serving as input to another piece of software.

(*b*) *Second Generation*

The second generation programming language, or 2GL, is Assembly Language. It is considered a second generation language because while it is not a microprocessor's native language, an assembly language programmer must still understand the microprocessor's unique architecture (such as its registers and instructions).

2. High-level Programming Languages

These types of programming languages are more advanced than the low-level programming languages (it doesn't really take a rocket scientist to figure this out). This type of language is more "English-like." In other words, the statements and sentences that comprise of most high-level languages are very similar to human languages. They aren't written in the 1s and 0s that the machine languages are written in.

High-level programming languages are often used to write up very complex programs and applications. After all, it is much easier written programs in human-like languages rather than as a strand of 1s and 0s! It is also much quicker to write applications in high-level languages.

High-level versus low-level is a relative property; a Java programmer might consider C to be a low-level language, while an assembly language programmer would consider C to be a high-level programming language. A Python programmer might even consider Java a low-level language, (though unlikely, as object-oriented languages such as Java are rarely considered low-level, even by those using much higher-level languages).

3.2.1 BASIC

BASIC is an easy-to-use and easy-to-learn high-level programming language. As a matter of fact, it is even often offered as a high school course! It is often taught to beginner programmers because while it is easy to use and understand, it is also composed of some of the concepts used by more advanced languages like PASCAL and C.

BASIC is an acronym for Beginners' All-purpose Symbolic Instruction Code. It was developed in the early 1960s by John Kemeny and Thomas Kurtz at Dartmouth College. In the years after it's development, BASIC has increasingly become more and more popular, especially with microcomputers. For a while, it was the most popular programming language. However, BASIC is not limited to microcomputers. As a matter of fact, it can also be used with personal computers as well.

BASIC has never really been standardized, and currently it comes in many different versions. For example, there are the versions used by Apple and IBM, as well as versions used by larger computers and businesses. All of the versions of the BASIC language operate in pretty much the same way.

In a BASIC program, every single line has a number. First, the line with the lowest number is executed, and then the next lowest, and so on. Often, programmers will number their lines in multiples of ten (*i.e.* 10, 20, 30, etc.). This is so additional lines could be added later without causing major problems.

Variables and constants in the BASIC language are both represented by a letter of the alphabet. No special characters, like @, !, and % are allowed. Variable names can be comprised of up to 255 letters or numerical digits. However, if you had two variables who have the same first two letters in their names, then the computer would read them as the same variable. For example, if you had two variables named "DOG" and "DOWN", the computer would think of them as the same variable.

Finally, there are some basic control statements that BASIC uses. Usually, BASIC carries out instructions by doing the lowest line number first, then the second lowest, etc. By using control statements, it is possible to alter the order in which lines are carried out. For example, if you type in "IF THEN ELSE" in conjunction with "GOTO" and a certain line number, the program will go to the specified line number. If you use "IF THEN ELSE" with "GOSUB," the program will go to a certain subroutine (a subroutine is a group of statements that perform certain actions).

As you can see, BASIC is pretty straightforward. What we have talked about here is only a minute portion of this programming language though. If you'd like to learn more, I'd suggest going to your library, or asking your school if they have a course on BASIC!

3.2.2 C++

C++ is a high-level programming language. It is based upon the programming language C. However, C++ has additional features and more advanced components.

In C++, there are expressions and statements. To put it simply, an expression is just an action that needs to be carried out, while a statement is an expression that has been ended by a semicolon. There are three types of statements in the C++ programming language:

- Declaration Statement
- Assignment Statement
- Output Statement

In C++, statements can be grouped together into functions. Every C++ program must contain the function main(). This is because when any C++ program is first opened, it must know what instructions to first carry out. C++ programs always begin execution with the first statement in the main() function. Meanwhile, a C++ program always ends with the last statement in the main() function.

In the C++ language, there can be things called comments. Comments are placed within the characters /* and */. They are used to help clarify things to humans that are reading the language. For example, they could be used to clarify parts of the code or to identify a variable.

There are five major data types in C++ :

1. **Char:** Char is short for the word character. Char data types represent individual characters, like alphabetical letters and numerical digits. Characters are a single byte, or 8 bits. One bit is one binary digit-a 1 or a 0. The binary system is the basic language of all computers.
2. **Int:** Int is short for integer. In computer language, an integer value is a machine word. A word is equivalent to either 16 or 32 bits.
3. **Float:** Float data types are floating point single precision values. This type of data is one word in length.
4. **Double:** Double data types are floating point double precision values. These are comprised of two words.

5. **Long Double:** Long double data types are like the data types float and double. It represents floating point extended precision values, and is comprised of three or four words.

C++ is a very advanced programming language. It would be impossible to briefly cover all aspects of this language. If you would like to learn more about C++, I would suggest going to your local library! There are tons of primers and tutorials about how to write in the C++ programming language!

A high-level programming language is a programming language that is more user-friendly, to some extent platform-independent, and abstract from low-level computer processor operations such as memory accesses. See programming language for a detailed discussion.

The word "high" does not imply that the language is superior to low-level programming languages but rather refers to the higher level of abstraction from machine language. For example: the difference between the programming language Java and assembly language is that Java abstracts programming functionality that assembly does not, for example, strings.

In general, "*high-level*" languages make complex programming simpler, while low-level languages produce more efficient code. In a high-level language, complex elements can be broken up into simpler, though still fairly complex, elements for which the language provides abstractions, keeping programmers from having to "reinvent the wheel". The cost of this convenience is often less efficient code overall. For this reason, code which needs to run particularly quickly and efficiently may be written in a lower-level language, even if a higher-level language would make the coding easier.

Note that the terms "*high-level*" and "*low-level*" are inherently relative. Originally, assembly language was considered low-level and COBOL, C, etc. were considered high-level. Many programmers today might refer to these latter languages as low-level. For more on this distinction, see the external link below.

3.3 CLASSIFICATION OF COMPUTER LANGUAGES

Computer languages can be classified into several kinds, including but not limited to the following:

- Programming languages
- Specification languages

- Query language (*e.g.*, SQL, XQuery)
- Markup languages (typically used for producing documents)
- Transformation languages (*e.g.*, XSLT)
- Communication protocols for networks

3.4 ALGORITHM

An algorithm (the word is derived from the name of the Persian mathematician Al-Khwarizmi), is a finite set of well-defined instructions for accomplishing some task which, given an initial state, will terminate in a corresponding recognizable end-state (contrast with heuristic). Algorithms can be implemented by computer programs, although often in restricted forms; an error in the design of an algorithm for solving a problem can lead to failures in the implementing program.

The concept of an algorithm is often illustrated by the example of a recipe, although many algorithms are much more complex; algorithms often have steps that repeat (iterate) or require decisions (such as logic or comparison) until the task is completed. Correctly performing an algorithm will not solve a problem if the algorithm is flawed or not appropriate to the problem. For example, a hypothetical algorithm for making a potato salad will fail if there are no potatoes present, even if all the motions of preparing the salad are performed as if the potatoes were there.

Different algorithms may complete the same task with a different set of instructions in more or less time, space, or effort than others. For example, given two different recipes for making potato salad, one may have peel the potato before boil the potato while the other presents the steps in the reverse order, yet they both call for these steps to be repeated for all potatoes and end when the potato salad is ready to be eaten.

Algorithms are essential to the way computers process information, because a computer program is essentially an algorithm that tells the computer what specific steps to perform (in what specific order) in order to carry out a specified task, such as calculating employees' paychecks or printing students' report cards.

For any such computational process, the algorithm must be rigorously defined: specified in the way it applies in all possible circumstances that could arise. That is, any conditional steps must be systematically dealt with, case-by-case; the criteria for each case must be clear (and computable).

Because an algorithm is a precise list of precise steps, the order of computation will almost always be critical to the functioning of the algorithm. Instructions are usually assumed to be listed explicitly, and are described as starting 'from the top' and going 'down to the bottom', an idea that is described more formally by flow of control.

Example:

One of the simplest algorithms is to find the highest in an unsorted list of numbers. Upon first thinking about the solution, you will realise that you must look at every number in the list, but only once each. Taking this into account, here is the simple algorithm :

Look at each item in the list. If it's larger than any you've seen so far, make a note of it.

The latest one you noted is the largest in the list.

And here is a more formal coding of the algorithm in a pseudocode:

```
Given: a list "List"
highest = nothing
List.each_item {highest = this_item if
                           this_item > highest}
Show highest
```

Notes on notation:

"**=**" is a loose shorthand for "**changes to**". For instance, with "**highest = this_item**", it means that the highest so far changes to this item. Note that it has a subtly different meaning in actual programming languages.

As it happens, most people who implement algorithms want to know how much of a particular resource (such as time or storage) a given algorithm requires. Methods have been developed for the analysis of algorithms to obtain such quantitative answers; for example, the algorithm above has a time requirement of O(n), using the big O notation with n as the length of the list.

4.1 FLOWCHARTING

Flowcharts are maps or graphical representations of a *process*. Steps in a process are shown with symbolic shapes, and the flow of the process is indicated with arrows connecting the symbols. Computer programmers popularized flowcharts in the 1960's, using them to map the logic of programs. In quality improvement work, flowcharts are particularly useful for displaying how a process currently functions or could ideally function.

Flowcharts can help you see whether the steps of a process are logical, uncover problems or miscommunications, define the boundaries of a process, and develop a common base of knowledge about a process. Flowcharting a process often brings to light redundancies, delays, dead ends, and indirect paths that would otherwise remain unnoticed or ignored. But flowcharts don't work if they aren't accurate, if team members are afraid to describe what actually happens, or if the team is too far removed from the actual workings of the process.

A flowchart (also spelled flow-chart and flow chart) is a schematic representation of a process. They are commonly used in programming, business/economic presentations to help the audience visualize the content better, or to find flaws in the process.

Generally the start point, end points, inputs, outputs, possible paths and the decisions that lead to these possible paths are included.

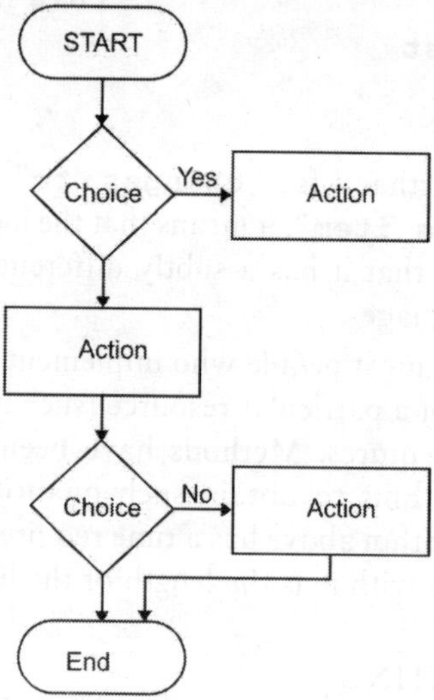

Fig. 2.4: Flowchart Basic

Flow-charts can be created by hand or manually in most office software, but lately specialized diagram drawing software has emerged that can also be used for the purpose, such as Visio,

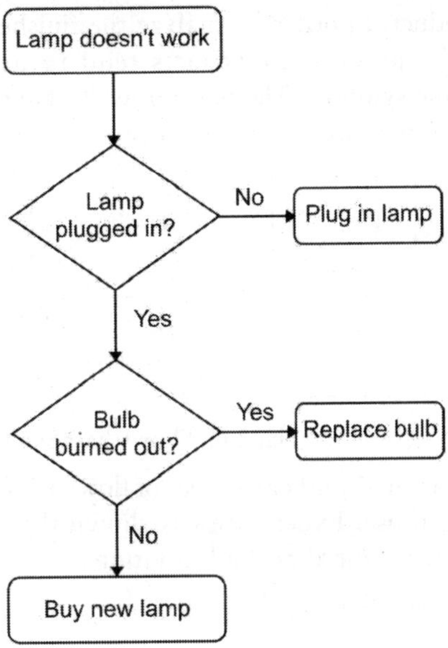

Fig. 2.5: Flowchart for Lamp Control

OpenOffice.org Draw and Dia.

Programs have been written to create flowcharts directly from computer program source.

4.2 TYPES OF FLOWCHARTS

Flowcharts are used specifically for a process. Four particular types of flowcharts have proven useful when dealing with a process analysis:

- Top-down flowchart,
- Detailed flowchart,
- Work flow diagrams, and a
- Deployment chart.

Each of the different types of flow charts tend to provide a different aspect to a process or a task. Flowcharts provide an excellent form of documentation for a process, and quite often are useful when examining how various steps in a process work together.

When dealing with a process flowchart, two separate stages of the process should be considered: the finished product and the making of the product. In order to analyze the finished product or how to operate the process, flowcharts tend to use simple and easily recognizable symbols. The basic flowchart symbols below are used when analyzing how to operate a process.

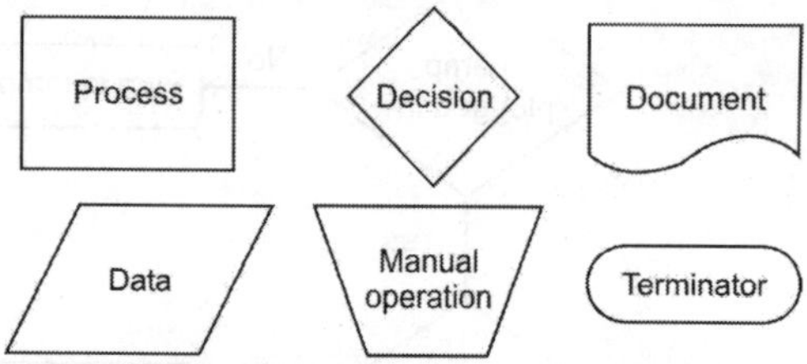

Fig. 2.6: Flow Chart Symbols

There are many other varieties of flowcharts and scores of symbols that you can use. Experience has shown that there are three main types that work for almost all situations:

- High Level Flow chart
- Detailed Flow Chart
- Deployment Flow Chart

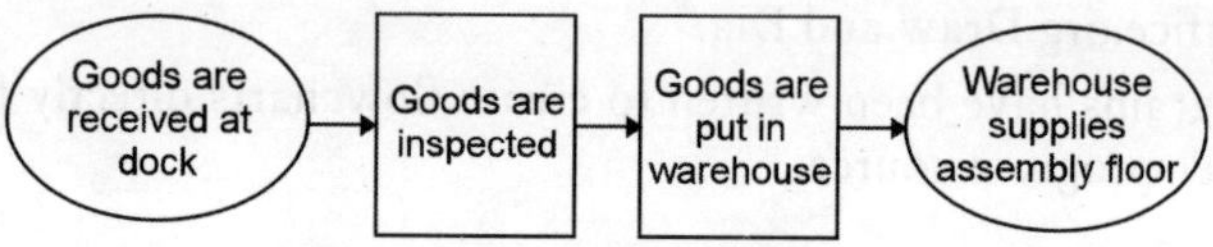

Fig. 2.7: Example of High-Level Flow Chart

High-level flowcharts map only the major steps in a process for a good overview.

Detailed flowcharts show a step-by-step mapping of all events and decisions in a process.

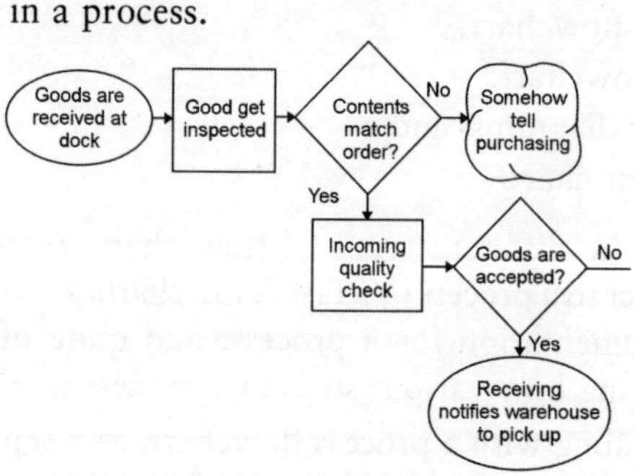

Fig. 2.8: Detailed Flow Chart

Deployment flowcharts which organize the flowchart by columns, with each column representing a person or department involved in a process.

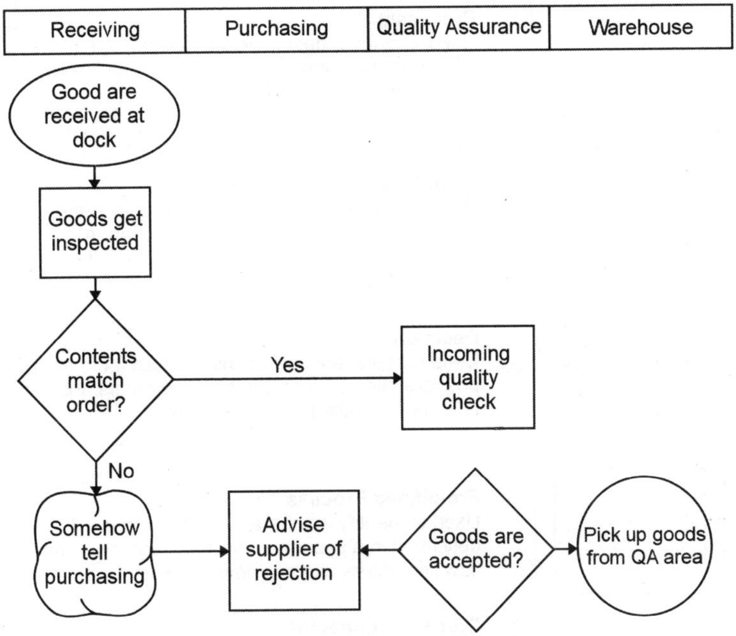

Fig. 2.9: Deployment Flowcharts

The trouble spots in a process usually begin to appear as a team constructs a detailed flowchart.

Although there are many symbols that can be used in flowcharts to represent different kinds of steps, accurate flowcharts can be created using very few (*e.g.* oval, rectangle, diamond, delay, cloud).

4.3 DRAWING FLOWCHARTS

A flowchart has different shaped boxes for different steps in the program:

1. **Terminator** (○): The beginning and end of the program have round boxes, called terminal symbols.
2. **Rectangle** (▭): Insturctions telling the computer to do something are shown in rectangular boxes. It indicates a process or arthmetical operation. By process we mean straight

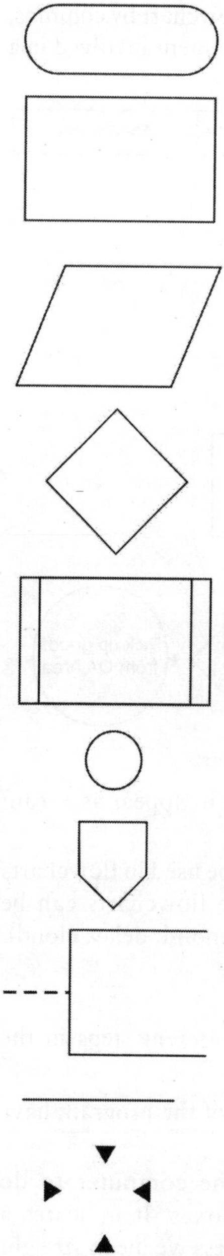

Terminal
Used to indicate the start and end of a flowchart single flowline. Flow begins or terminates here.

Processing
Used whenever data is being manipulated, most often with arithmetic operations. A single flowline enters and a single flow line exits.

Input/Output
Used whenever information is entered into the flowchart or displayed from the flowchart. A single flow line enters and a single line exits.

Decision
Used to represent operations in which there are two possible alternatives. One flow line enters and two flow lines (labelled yes and no) exit.

Pre-difined Process
Used to identify an operation that is more fully described in a separate flowchart segment. One flow line enters and one flow line exits.

On-page Connector
Used to connect remote flowchart portions on the same page. One flowline enters or exits.

Off-page Connector
Used to connect remote flowchart portions on different pages. One flowline enters or exits.

Comment
Used to add comments or clarification.

Flow Line

Flow Arrowheads

Fig 2.10: Flow Chart Symbols

through operations of word processing, filing activities, counting or incrementing. It also does addition, subtraction, multiplication, division, and logical operations of comparing data It is a rectangle.

3. **Decision boxes** (◇): The box where the computer can make decisions like finding out which numbers is bigger or smaller or equal than another number, depending on the information it receives. These are diamond-shaped boxes. The computer can decide with the help of decisions taken by the diamond box, which route is to be followed further.
4. **Connector:** A circle like symbol can be used for connecting two program segments, on different pages.
5. **Route Indicator** ↔ ↔: The lines show the possible routes the computer can follow. And arrows indicate the direction. Arrows must be used when there is any likelhood of a confusion.

   The following diagram spells the various symbols used in constructing the flowchart.

   Remember that flowcharts are made to structure the program logic roughly. The flow of diagram is from top to bottom and left to right.

6. **Showing Decisions on a Flowchart:** Flowcharts have a special symbol for asking a question. It is called a decision box. The descision box is shaped like a diamond.

A decision box always has more than one arrow coming from it. There are a YES arrow and a NO arrow.

This is DIVYAS flowchart of how she sends a letter by mail.

The YES arrow points to the next step to be done if the answer is yes. The NO arrow points to the next step to be done if the answer is no.

Arrows in a flowchart can go in any direction. Sometimes an arrow takes you all the way back to the first step. The change in direction usually comes from a decision box.

Remember that in a flowchart you must keep following the arrows until you get to STOP or END. Your answer to the question in the descision box may lead you closer to STOP, send you to an extra step, and/or lead you back to an earlier step.

After working out all the details of the program you can translate it into BASIC and test it on the computer. The program will probably not work straight away though, as there will probably be some mistakes (called bugs) in it. These may be typing mistakes made when you types the program into the computer, or errors of logic in your program.

Before you can get the program to work you have to find all the mistakes (also called bugs) and correct them.

4.4 CONSTRUCTING AN EFFECTIVE FLOWCHART

- Define the process boundaries with starting and ending points.
- Complete the big picture before filling in the details.
- Clearly define each step in the process. Be accurate and honest.
- Identify time lags and non-value-adding steps.

Circulate the flowchart to other people involved in the process to get their comments.

Flowcharts don't work if they're not accurate or if the team is too far removed from the process itself. Team members should be true participants in the process and feel free to describe what really happens. A thorough flowchart should provide a clear view of how a process works. With a completed flowchart, you can.

- Identify time lags and non-value-adding steps.
- Identify responsibility for each step.
- Brainstorm for problems in the process.
- Determine major and minor inputs into the process with a cause & effect diagram.
- Choose the most likely trouble spots with the consensus builder.

4.5 THE LIMITATIONS OF FLOWCHARTS

Flowcharts were originally introduced as aids to a systematic process of analysing problems and developing suitable computer-based solutions. In recent years flowcharts have been heavily criticised as being cumbersome and inefficient tools for the job. Newer alternatives have been introduced and widely adopted.

Particular limitations of flowcharts are:

1. Different levels of detail can easily become confused, *e.g.* details of a particular implementation can be inadvertently introduced too early in the design.
2. There are no obvious mechanisms for progressing from one level of design to the next, *e.g.,* from systems flowchart to outline program flowchart.

3. The essentials of what is done can easily be lost in the technical details of how it is done.
4. Program flowcharts, although easy to follow, are not such a natural way of expressing procedures as writing in English, nor are they easily translated into programming languages.

4.6 CREATING FLOWCHARTS ON A COMPUTER

There are various packages for creating flowcharts, according to different standards. The most common is UML, for which there are abundant packages for various platforms. The creation of simple flowcharts on a computer is fairly easy with any vector-based drawing program, but Microsoft Word (versions 97 through 2003) and Openoffice.org (Draw app) both have specialized tools for making consistent charts. For Mac OS X OmniGraffle [1] is an excellent application.

When in Word, enable the drawing toolbar and click **`Autoshapes`** then Flowcharts and finally on the appropriate shape you would like. Right-click on a **`shape`** and then click **`Add Text`** to do so. The Arrow or Line tool is used to manually draw links.

When in Draw, enable the Flowchart palette and click a shape to add it in. Double-clicking a shape will add text to it within appropriate boundaries. Connections can be automatically made between shapes using Connectors and Glue Points - click on the Connector arrow to see a selection of them before dragging from a Glue Point on a shape to another. Draw will maintain the link and automatically redraw the connector if you resize or move any shape.

3
UNIT
Networking

1.0 Networking

Historically, most communication systems have started with point-to-point links which directly connect together the users wishing to communicate using a dedicated communications circuit. As the distance between users increases beyond the length of the cable (*e.g.* the length of a telegraph wire), the connection between the users was formed by a number of sections which were connected end-to-end in series to form the circuit. The connection between the users (A and D) in the figure below (*i.e.* A and D) is represented by a series of links (AB, BC, and CD) each link connects two entities known as nodes. For a point-to-point circuit, (also known as a permanent circuit) the nodes are patch panels which provide a simple connection between the two links (*i.e.* the two transmission circuits).

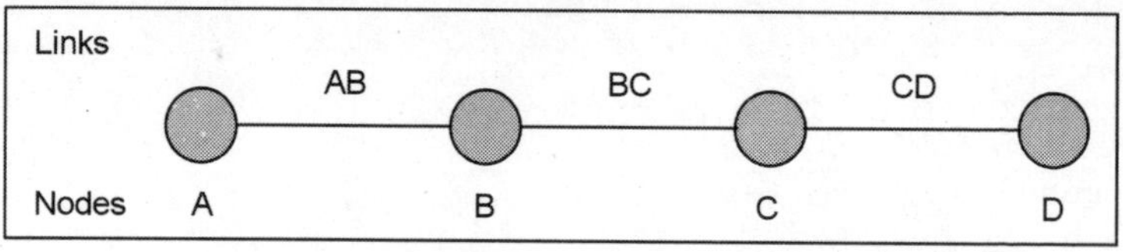

Fig. 3.1: A Connection between two Systems A and D Formed from 3 Links

As the number of connected users increased, it has become infeasible to provide a circuit which connects every user to every other user, and some sharing of the transmission circuits (known as "switching") has become necessary. To accomplish this goal, the data communication network has evolved. A network is a set of nodes that are interconnected to permit the exchange of information. Three switching techniques have been proposed for building networks:

- Circuit switching
- Message switching
- Packet switching (both virtual circuit and datagram)

Each allows sharing communication facilities among multiple users (end systems), and each uses equipment located at the nodes (intermediate systems) to replace the patch-panels used in a point-to-point connection. Packet switching is most often used for data communication. Most networks consist of many links (see the figure on next page) which allow more than one path through the network between nodes. A data communication network must be able to select an appropriate path for each required connection.

Any of the three approaches (circuit switching, message switching, and packet switching) could yield minimum delay in a particular situation, though with.The Interuations where message switching yields minimum delay are rare. The relative performance of circuit switching and packet switching depends strongly on the speed and "cost" of establishing a connection.

A network is nothing more than two or more computers connected together by a cable so that they can exchange information. With a computer network, all the computers in office are hooked up together with cables, install a special network adapter card in each computer so you'll have a place to plug in the cable, set up and configure special network software to make the network work, and viola, you have a working network. That's all there is to it.

Computer networks are collections of computers and intelligent peripheral equipments (*e.g.*, automatic teller machines and point-of-sale terminals equipped with microprocessors) that are interconnected by telephone lines, microwave relays, and other high-speed communication links for the purpose of exchanging data and sharing equipment.

Following figure shows a typical network with four PCs. You can see here that all four computers are connected together with a network cable. You can also see that the server computer has a fancy laser printer attached to it. Because of the network, computer workstation No. 1 to No. 3 can also use this laser printer.

Every computer connected to the network is said to be on the network. The technical term for a computer that is on the network is a node. When a computer is turned on and is able to access the network, the computer is said to be on-line. When the computer is unable to access the network, it is off-line. A computer could be off-line for several reasons. It could be turned off, it could be broken, the cable that connects it to the network could be unplugged, or there could be a wad of gum jammed into the disk drive.

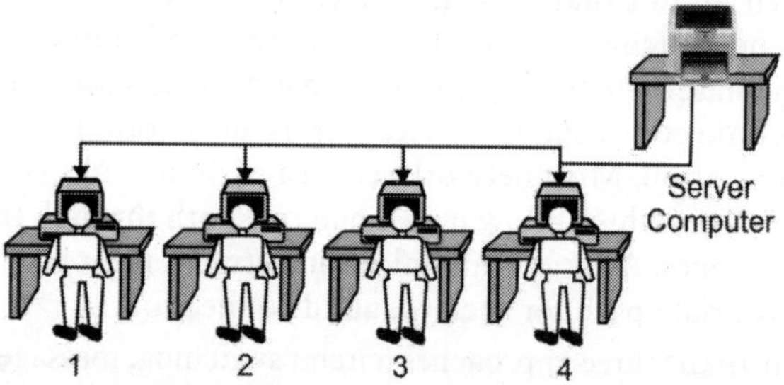

Fig. 3.2: Shared Resources through Networking.

In general terms a computer network is an interconnection of various computer systems located at the same/different places. The interconnection is done through a communication link also known as the physical layer of the network and this link is such that it is totally transparent to the users of the network, the transparency of the communication link is brought about by interfacing software known as protocols which enables a user in one location to freely access a computer system/data base/process/software in another location.

However, if the computers in the network operate together as a single unit which to the user appears as a single computer, albeit physically dispersed, then the complete system is more accurately described as a distributed system. Therefore, although, any interconnected set of computers is often conveniently referred to as a "computer network" the use of the term often implies an interconnected set of independent computers and not a distributed system. However, it may be useful when considering a distributed system to be able to recognize the particular type of network on which it is based.

1.1 TYPES OF NETWORKS

Networking has been developed on all levels, from local to international, in diverse sectors of society. Examples are networks used by government organizations for rapid retrieval of information from databases at central locations and those used by banks and retail merchants for the transfer of funds or credit verification. If telephone lines are adopted for a computer network, each computer in the network must be connected to a telephone line via a modem (modulator/demodulator). This device converts the digital pulses from a computer into analog signals to be transmitted over the telephone line and vice-versa.

1.1.1 Local Area Network (LAN)

A system of networked computers and other hardware, like printers, that are in relatively close proximity to one another. This is described in detail in paragraphs to follow.

1.1.2 Wide Area Network (WAN)

WAN is a geographically dispersed communications network, with many owners, linking computers for the purpose of communicating with each other, such as a national network for airline reservations or the Internet. The largest wide-area network is the Internet, a collection of networks and gateways linking millions of computer users on every continent.

A wide area network or WAN is a computer network covering a wide geographical area, involving a vast array of computers. This is different from personal area networks (PANs), local area networks (LANs) or metropolitan area networks (MANs) that are usually limited to a room, building or campus and big city respectively. The best example of a WAN is the Internet.

WANs are used to connect local area networks (LANs) together, so that users and computers in one location can communicate with users and computers in other locations. Many WANs are built for one particular organization and are private. Others, built by Internet service providers, provide connections from an organization's LAN to the Internet.

WANs are most often built using leased lines. At each end of the leased line, a router connects to the LAN on one side and a hub within the WAN on the other. Network protocols including TCP/IP deliver transport and addressing functions. Protocols including Packet over SONET/SDH, MPLS, ATM and Frame relay are often used by service providers to deliver the links that are used in WANs. X.25 was an important early WAN protocol, and is often considered to be the "grandfather" of Frame Relay as many of the underlying protocols and functions of X.25 are still in use today (with upgrades) by Frame Relay.

1.1.3 Metropolitan Area Network (MAN)

MAN is short for Metropolitan Area Network, a data network designed for a town or city. In terms of geographic breadth, MANs are larger than local-area networks (LANs), but smaller than wide-area networks (WANs). MANs are usually characterized by very high-speed connections using fiber optical cable or other digital media.

It might cover a group of nearby corporate offices or a city and might be either private or public. A MAN can support both data and voice, and might even be related to the local cable television network.

MAN system incorporates a large computer networks usually spanning a campus or a city. They typically use wireless infrastructure or optical fiber connections to link their sites.

For instance, a university or college may have a MAN that joins together many of their local area networks (LANs) situated around site of a fraction of a square kilometer. Then from their MAN they could have several wide area network (WAN) links to other universities or the Internet.

Some technologies used for this purpose are ATM, FDDI and SMDS. These older technologies are in the process of being displaced by Ethernet-based MANs (*e.g.* Metro Ethernet) in most areas. MAN links between LANs have been built without cables using either microwave, radio, or infrared free-space optical communication links.

DQDB, Distributed Queue Dual Bus, is the Metropolitan Area Network standard for data communication. It specified in the IEEE 802.6 standard. Using DQDB, networks can be up to 30 miles long and operate at speeds of 34 to 155 Mbit/s.

Several notable networks started as MANs, such as the Internet peering points MAE-West and MAE-East and the Sohonet media network.

1.1.4 Value-added Network (VAN)

A communications channel leased from a telephone company to offer customers with modems access to network services through a local or toll-free number.

Backbone—A high-speed network that connects several LANs in a location, such as a university campus, to provide services like e-mail and Internet access beyond the LAN environment.

When a machine in a network communicates with another computer, both follow a predetermined protocol (*i.e.,* a set of procedures), according to which data are exchanged.

1.2 LAN

A local area network (LAN) is a communication network consisting of many computers (mostly personal computers and workstations) that are placed within a local area, such as a single building or company. Since the mid-1980s, LANs have become widespread because personal computers and workstations are used extensively in offices, enabling individual users to run their specific programs easily and to have their own files at hand on floppy disks. Individual users can also share data or files on a LAN as if the data or files reside on their respective computers; computers that are designed or used for this purpose are called servers. Laser printers and other peripheral equipment also can be connected to a network for common use. Coaxial cables or fibre-optic cables are popular communication lines because they provide fast data transmission and are easy to install within buildings. In the case of LANs, there are a few different protocols, such as Ethernet and Token Ring, developed by Xerox and IBM, respectively.

A LAN is made up of network interface cards that fit inside the connected computers, cable (or an equivalent wireless connection) to connect these computers together, protocol software to move data from computer to computer, user interface software to connect the user and the network, and operating system software to actually service users' needs for resources like files and printers. Finally, a LAN runs end-user applications.

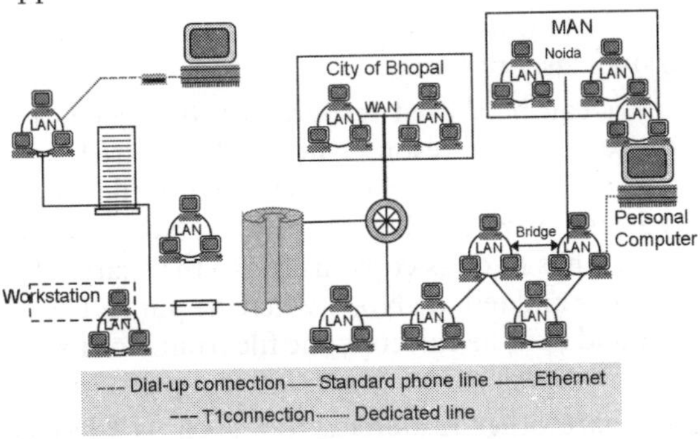

Fig. 3.3: Typical Composite Network

Once a network spans more than a few miles, such as a campus environment, it can be referred to as a metropolitan area network (MAN). A MAN spans a region such as a city. A wide area network (WAN) brings companies into the sphere of networking by connecting computers in the entire enterprise, which may span over several cities, states, and countries.

1.3 ADVANTAGES OF NETWORKING

Frankly, computer networks are a bit of a pain to set up. So why bother? Because the benefits of having a network make the pain of setting one up bearable. You don't have to be a Ph.D. to understand the benefits of networking. In fact, you learned everything you need to know in kindergarten: Networks are about sharing three things: files, resources, and programs. Sharing Files:

Networks let you share information with other computers on the network. Depending on how you set your network, you can do this in one of two ways. The most direct way is to send the file from your computer directly to your friend's computer. The second way is to send your file to an intermediate resting place, where your friend can pick it up later, kind of like dropping a bag full of ransom money at a phone booth. A third way is to permanently store the file at that intermediate place, where both of you can get at the file whenever you want. One way or the other, the data travels to your friend's computer over the network cable.

Sharing Resources

This means that you can set up certain computer resources—like a disk drive or a printer—so that all of the computers on the network can access them. For example, the laser printer attached to server computer is a shared resource. That means that anyone on the network can use it. Without the network Computer user No 1 to 3 would have to buy their own laser printers.

Disk drives can be shared resources, too. In fact, a disk drive must be set up as a shared resource in order to share files with other users. Suppose Computer No. 3 wants to share a file with the No. 4, and a shared disk drive has been set up on No. 2's computer. All Computer No. 3 has to do is copy his file to the shared disk drive in No. 2's Computer and tell the No. 4 where he put it. Then, when the No. 4 gets around to it, he can copy the file from No. 1's computer to his own.

You can share other resources, too, such as CD-ROM drives (those useful for large clip art libraries and encyclopedias) or modems (which let you access other computers that aren't on your network).

Sharing Programs

Sometimes, it's best to put programs that everyone uses on a shared disk, rather than keep separate copies of the programs on each person's computer. For example, if you have ten computer users who all use

WordPerfect, you could store ten copies of WordPerfect—one on each computer—or you could store just one copy of WordPerfect on a shared disk.

Backups

As all data is stored on the server, baking up critical data becomes a simple process.

Communication

The biggest benefit however, comes in the form of better communication-electronic mail and groupware applications. Through electronic mail or E-mail, members of a network can send messages and ensure safe delivery of data to other members, even when they are not there. Groupware applications allow users to work collectively on the same document.

1.4 DISADVANTAGES OF A NETWORK

Crahes: The biggest disadvantage is on a server-based network. When the server crashes, work gets disrupted as all network resources and its benefits are lost. Unless proper precautions are taken to ensure regular backups, the crash may result in the loss of days' and even months' of critical data and time.

Data Security: As all the data resources are all pooled together, it is possible for unauthorised personnel to access classified information if network security is weak or poorly implemented.

Privacy: A network may also mean loss of privacy as anyone, especially your boss, with the right network privileges may be in a position to read your private e-mail.

There are advantages and disadvantages to sharing a program such as WordPerfect. On the plus side, it makes it easier to keep WordPerfect up-to-date. If you decide to upgrade to an new version of WordPerfect, you have to do the upgrade only once to work with printers that are on the network. And it's cheaper, because a network version of WordPerfect that allows up to one users is less expensive than ten single-user copies of WordPerfect.

On the minus side, a shared version of WordPerfect can be harder to configure properly and might run just a bit slower because the WordPerfect program files have to travel over the network.

1.5 NETWORK COMPONENTS

The key (tele) Communications network components are and alternatives are detailed in Table 3.1.

Table 3.1: Key (tele) Communications Network Components

Network	*Examples of alternatives available component*
Media	Twisted pair wire, coaxial cable, Fibreoptics, Microwave, satellites, cellular radio, transreceivers
Processors	Modems, multiplexers, concentrators, routers, bridges gateways, Front-end processors, client and server computers
Software communication software	Network operating system Netware, Slip or PPP, POP, SMTP, MSN internet explorer or Netscape Navigator, middleware
Channels	Analog/digital, switched/non-switched, circuit/message/packet switching, simplex/ duplex, synchronous/ asynchronous, speed and baud rate
Topology	Point-to-Point, multidrop, star, ring, mesh, Ethernet
Architecture	OSI, IEEE, ISDN PSTN

1.5.1 OSI Reference Model

The International Standards Organisation (ISO) developed what they called the Open Systems Interconnection (OSI) Reference Model. This Model proposed that all network protocols should consist of seven distinct layers. The principles that were applied to reach these seven layers were:

Each layer should perform a well defined function of set of related tasks.

- The function of each layer should be designed to allow the widest set of network operations. That is, the function for a layer should not be such that only one network is possible.
- The layer boundaries should be chosen so that there is only a small amount of information flow across the boundary.

Fig. 3.4: Practical view of Several Layers of OSI

The OSI model is a concept that describes how data communications should take place. It divides the process into seven groups, called layers. Into these layers are fitted the protocol standards developed by the ISO and other standards bodies, including the Institute of Electrical and Electronic Engineers (IEEE), American National Standards Institute (ANSI), and the International Telecommunications Union (ITU), formerly known as the CCITT (Commit Consultatif International_Tlphonique et Tlgraphique).

All networks are based on a theoretical model called the Open System Interconnection (OSI) model with defines how network devices interact with each other. It is an ISO standard for worldwide communications that defines a framework for implementing protocols in seven layers (See adjoining figure).

Table 3.2 illustrates the Open System Interconnection (OSI), developed by ISO. This protocol reference model is composed of seven layers. The figure describes the functions that protocols and networks must handle at each of the seven levels in order for data to be transmitted from one user to another.

Table 3.2: The Open System Interconnection (OSI) Model.

Layer	*Description*
1. Physical	Handles voltages, electrical pulses, connectors, and switches so that data can be transmitted over network media.
2. Data link	Controls grouping data into blocks (message packets) and transferring blocks from one point in the network to another.
3. Network	Controls the routing of data and messages through the channels of a network.
4. Transport	Controls data transfer for the complete transmission path, from sending point to receiving point.
5. Session	Establishes and terminates communications links between computers.
6. Presentation	Formats data for transfer between different systems.
7. Application	Provides network services to users and user applications, including file transfer. Provides user interfaces with the systems.

2.1 CATEGORISATION OF NETWORKS

Networks are usually classified using three properties:

- Topology
- Protocol
- Architecture

Topology specifies the geometric arrangement of the network. Common topologies are a bus, ring, and star. Protocol specifies a common set of rules and signals the computers on the network use to communicate. Most networks use Ethernet, but some networks may use IBM's Token Ring protocol.

Architecture refers to one of the two major types of network architecture:

- Peer-to-peer
- Client/server.

In a Peer-to-Peer networking configuration, there is no server, and computers simply connect with each other in a workgroup to share files, printers, and Internet access. This is most commonly found in home configurations, and is only practical for workgroups of a dozen or less computers. In a client/server network, there is usually an NT Domain Controller, which all of the computers log on to. This server can provide various services, including centrally routed Internet Access, mail (including e-mail), file sharing, and printer access, as well as ensuring security across the network. This is most commonly found in corporate configurations, where network security is essential.

2.2 NETWORK TOPOLOGIES

A network topology is the geometric arrangement of nodes and cable links in a LAN, and is used in two general configurations: bus and star. These two topologies define how nodes are connected to one another. A node is an active device connected to the network, such as a computer or a printer. A node can also be a piece of networking equipment such as a hub, switch or a router. A bus topology consists of nodes linked together in a series with each node connected to a long cable or bus. Many nodes can tap into the bus and begin communication with all other nodes on that cable segment. A break anywhere in the cable will usually cause the entire segment to be inoperable until the break is repaired.

The term topology refers to the physical pattern to which a network conforms; that is, the physical layout of the network's computers, terminals, and links. The four topologies observed most often are the hierarchical, star, bus, and ring networks. While most industrial-strength networks incorporate complex interconnections that don't conform neatly to any of these four simple topologies, they are the fundamental structures on which most networks are built.

2.3 LAN TOPOLOGIES

LAN topologies define the manner in which network devices are organized. Four common LAN topologies exist:

- Bus
- Ring
- Star
- Tree

The topology of a network is the shape of the communications link between the stations, *i.e.* how the stations on the network are arranged in relation to each other.

Understanding the topology of LAN technologies can tell you a lot about your alternatives when installing or expanding a LAN. At its basic level, the topology of a network refers to the way in which all of its pieces have been connected. That is, it refers to the layout of the computers, printers, and other equipment hooked to the network in your building.

Because cable connects these scattered computing resources together into a network, your network's topology is also a function of the way in which the cabling is organized, whether it is arrayed in a bus, ring, or star, which are the three basic physical topologies available to LAN designers (see Figure below). Although recent technological advances have blurred the distinctions between the physical and logical arrangements, the topology you select (or are forced to select) may also dictate the media-access control method (that is, Ethernet or Token Ring) under which your network will operate.

A network's logical layout may differ from its physical layout. The logical topology defines the electrical path; the physical path defines how the cables, concentrators, and nodes are arranged. For example, Ethernet must be a logical bus network; however, it can be physically configured as a bus or star. Token Ring is a logical ring, but is physically configured as a star. FDDI, a logical ring, is physically configured either as a ring or a star.

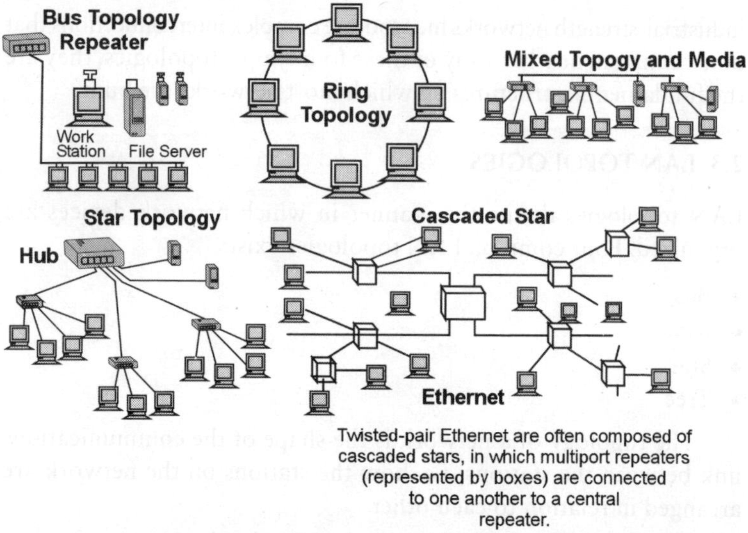

Fig. 3.5: Network Topologies

Network topology defines the way in which resources are logically or physically connected to form a network. Three topologies are most commonly used for computer networks: a bus, a ring and a star.

2.4 STAR LAN

A network architecture that consists of a central hub to which all computers connect. Star topologies are often used with Local Area Networks (*e.g.*, twisted pair Ethernet).

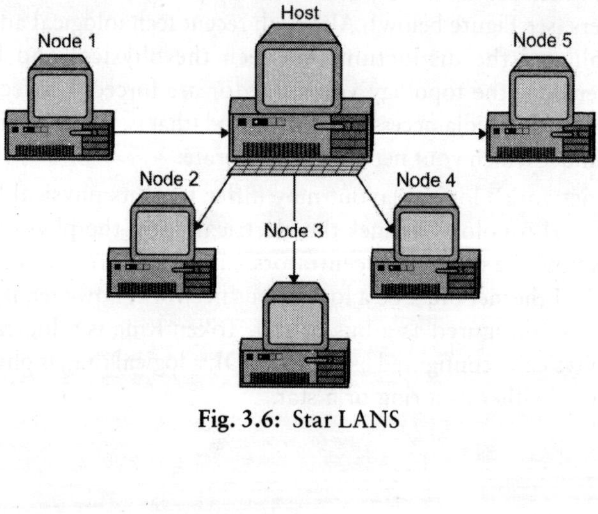

Fig. 3.6: Star LANS

An example of a Star topology. In star topologies in the network are connected to a central host. If the host fails, the networks fails.

In a Star network several devices are connected to a central device. Thus, a network in which several terminals or workstations are serviced by a host computer is an example of a Star network. Yet another example is the PBX, which will be discussed later. Micro-to-mainframe linkages also form Sstar networks in most instances.

For devices to communicate with one another in a Star network, their messages have to be routed through the central "hub" device. The processor located at the hub does most of the processing and us usually the site where the database is located. All terminal-to-terminal communications must go through the hub. If the hub fails, the entire network is disabled.

A Star topology is a LAN architecture in which the endpoints on a network are connected to a common central hub, or switch, by dedicated links. The simple star diagram of Figure 5.6 illustrates the star topology, with each station connected to a central switch by a dedicated physical link. The switch provides a path between any two devices wishing to communicate, either physically in a circuit switch or logically in a packet switch. The main advantages of a star LAN are that access to the network, *i.e.* the decision on when a station can or cannot transmit, is under central control. Star LANs tend to imply simple attachment hardware at each station, although the switches themselves can be complex and therefore expensive. Speeds are generally limited and the central switch is an obvious potential source of catastrophic failure.

Until recently, the Star topology has been found mostly in minicomputer and mainframe environments. These typically consist of a system of terminals or PCs, each wired to a central processor. It is also used by AT&T in both its Star LAN network and its Private Branch Exchange (PBX) based network. The star topology is ideal for Wide Area Network (WAN) applications in which outlying offices must communicate with a central office.

A principle advantage of the Star topology is that it not only allows centralizing key networking resources—concentrators or line conditioning equipment—but also gives the network administrator a focal point for network management. When something goes wrong with the network, the administrator can troubleshoot it from one place, usually a wiring closet, but possibly from a remote management console.

The Star-based network requires a substantial investment in cable, however. Each workstation is connected to the central concentrator by its own dedicated line. In some Star-based network technologies (ARCnet, for example) this line is coax cable that runs from an active hub to a workstation. (ARCnet can also operate as a bus.)

The 10Base-T Ethernet standard permits operating traditionally bus-based Ethernet in a Star-wired configuration using unshielded twisted-pair (or high grade telephone) wiring.

Cascaded Star

The use of a modular multiport repeater (also known as a hub or concentrator) with Ethernet allows creating large networks made of what can be called cascaded stars. In this arrangement, one centralized multiport repeater serves as the focal point for many other multiport repeaters, in effect creating a series of Star-based Ethernets.

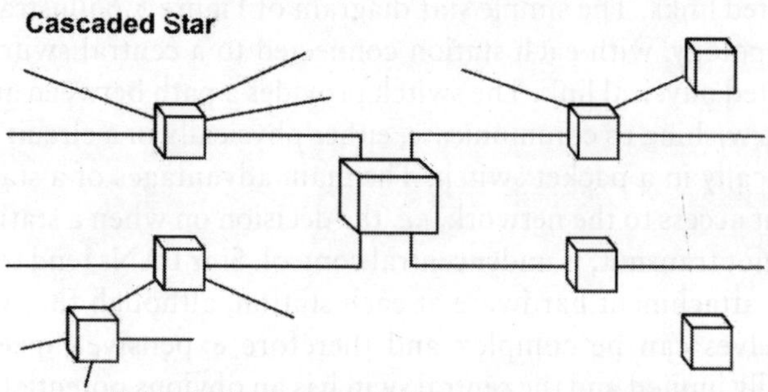

Twisted-pair Ethernet are often composed of cascaded Stars, in which multiport repeaters (represented by boxes) are connected to one another and to a central repeater.

Fig. 3.7: Cascaded Star

Using modular multiport repeaters also permits mixing star- and bus-based Ethernet workgroups into a single large network. In this instance, the modular repeater must only be able to accept modules that support the many Ethernet-compatible cable types.

2.5 RING LANS: TOKEN RING

In ring topology type LAN architecture a series of devices are connected to one another by unidirectional transmission links to form a single closed loop. Both Token Ring/IEEE 805.5 and FDDI networks implement a ring topology. Following Figure depicts a logical ring topology.

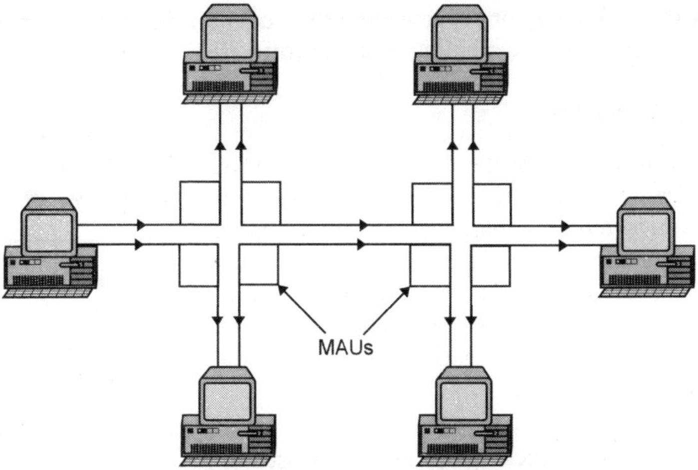

Although a physical star, IBM's Token Ring Implementation is logically or electronically a ring. The arrows represent a data packet as it moves from one station to another around the ring

Fig. 3.8: Cascaded Star

IBM popularized the ring topology with its Token Ring technology. Like the bus, a Token Ring network uses a single cable. Unlike the bus, the cable's ends are looped to form a complete logical circle or ring. Physically, Token Ring is a star-wired network. Each workstation is connected directly to a central device called a Media Access Unit (MAU). Logically (or electronically), however, the Token Ring remains a true ring.

Unlike the bus, Token Ring uses a deterministic, rather than a contention-based, access method. In the Token Ring access method, an electronic signal called a token is passed from station to station on the ring, with each station regenerating the token as it passes by.

When a station wishes to transmit data over the network, it must wait until the token is passed to it by its neighboring station. It takes control of the station and then places a data packet on the network. Only after the data packet has made a full circuit of the ring, returning to its originator, does the station release the token for the next workstation.

Token Ring can also be expanded by linking multiple rings together, just like Ethernet. In these arrangements, one Token Ring, usually a 16 Mbits/sec ring, is dedicated as an inter-network loop, with work group or departmental rings connecting to the company-wide ring via a PC running bridging software.

2.6 BUS LANS

Bus networks are the most common LANs. They have no switches and, in their simplest form, no repeaters, but simply share a common, linear communication medium, as shown below. Each station requires a tap (hardware for attachment to the medium), which must be capable of delivering the signal to all the stations on the bus.

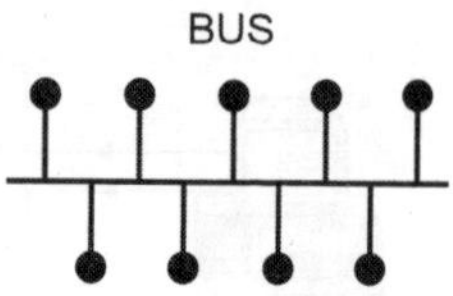

Fig. 3.9: Bus LAN Topology

The data is sent in packets, and each station 'hears' all the transmissions, picking up those addressed to it. Most bus networks have the advantage of being passive, *i.e.* all the active components are in the stations, and a failure affects only that one station. They can be somewhat limited in distance, however, and since usually only one station at a time can transmit there has to be a complex access mechanism at each station.

The most significant bus network to date is Ethernet, invented by Xerox in the mid 1970s and now adopted as an international standard called IEEE 805.5. This standard is enunciated by the Institute of Electrical and Electronics Engineers committee number 802 and subcommittee number 3 and hence the reference IEEE 805.5. The principle of operation of IEEE 805.3 networks using CSMA/CD protocol is explained in the next section.

These topologies are logical architectures, but the actual devices need not be physically organized in these configurations. Logical bus and ring topologies, for example, are commonly organized physically as a star. A bus topology is a linear LAN architecture in which transmissions from network stations propagate the length of the medium and are received by all other stations. Of the three most widely used LAN implementations, Ethernet/IEEE 805.3 networks, including 100BaseT, implement a bus topology.

In a bus topology, all workstations on the network are attached to a single cable. Ethernet, AppleTalk, and IBM's PC Network are examples of bus-based networks. This sharing of the transmission media (or cable) has several important ramifications. Most importantly, it means that the cable can carry only one message at a time, and each workstation on the network must be capable of knowing when it can and cannot transmit using this shared medium.

Ethernet employs what is called a Carrier Sense Multiple Access/ Collision Detection (CSMA/CD) access method to arbitrate use of the cable and to maximize its throughput. In this method, each station on the bus is always listening on the cable for transmission for other stations. It only transmits when the cable is not busy with another transmission. It is able to sense the collision that occurs when it and another station on the bus transmit at the same moment. Having sensed that a collision has occurred—and that the transmission has miscarried—each workstation waits a random time period (usually several microseconds) before retransmitting.

Naturally, frequent retransmission can slow down an Ethernet LAN; this limits the number of workstations that can be placed on any network segment. Fortunately, network managers have access to numerous devices, such as switches, bridges, and routers, that divide Ethernets in small segments to mitigate this problem.

3.1 LAN IMPLEMENTATION

This section addresses the methods and devices used in three major type systems:

- Ethernet/IEEE 802.3
- Token Ring/IEEE 802.5, and
- Fiber Distributed Data Interface (FDDI).

ATM is another system on which LANs can be designed.

There are a number of technologies available for connecting devices together in networks or in a simple data-sharing arrangement. Some of these technologies are listed here and covered in more detail under the appropriate headings.

- **Token Ring An** IBM developed ring network topology that operates at 4 Mbits/sec or 16 Mbits/sec.
- **Ethernet:** A traditional shared-access local area network with a data rate of 10 Mbits/sec.
- **Fast Ethernet Ethernet:** at 100 Mbits/sec over Category 5 unshielded twisted-pair cable.
- 100VG—AnyLAN Another 100-Mbit/sec Ethernet standard that uses a priority access method.

- **Gigabit Ethernet:** A gigabit-speed connection method that requires fiber-optic cabling, although copper cabling may become part of the specification in the future.
- **FDDI:** A traditional 100-Mbit/sec networking technology. Typically used for backbones and to interconnect servers. Costly to deploy to the desktop.
- **ATM (Asynchronous Transfer Mode):** A high-speed cell-based backbone network technology used by the carriers and in private network environments.

3.2 ETHERNET

In 1985, the Institute of Electrical and Electronic Engineers (IEEE) in the United States of America produced a series of standards for Local-Area Networks (LANs) called the IEEE 802 standards. These have found widespread acceptability and now form the core of most LANs. One of the IEEE 802 standards, IEEE 802.3, is a standard known as "Ethernet". This is the most widely used LAN technology in the world today. Ethernet was developed by the Xerox Corporation's Palo Alto Research Centre (known colloquially as Xerox PARC) in 1972 and was probably the first true LAN to be introduced. Although IEEE 802.3 differs somewhat from the original standard (the "blue book" defined in September, 1980) it is very similar, and both sets of standards may be used with the same LAN.

The IEEE standards have been adopted the International Standards Organisation (ISO), and is standardised in a series of standards known as ISO 8802-3.

Ethernet is a local-area network (LAN) protocol. A protocol is a set of rules followed by all of the computers in a given network so that data being transferred do not collide into each other.

Ever since its introduction to the world, Ethernet has become increasingly popular. It is now one of the most common LAN protocols used throughout the world.

As you know, networks can connect all sorts of computers and other devices. To connect these devices (called nodes), Ethernets use some sort of medium. The most common medium used to connect devices on a network is a twisted pair of fiber optic cabling. The nodes communicate in what is called a frame. Frames are short messages, and can be compared to sentences in the human language.

With all the nodes connected to a network, that network has quite a job regulating the messages and making sure they don't collide with another. How Ethernet does this can be described in the acronym CSMA/CD. CSMA/CD stands for Carrier Sense Multiple Access with Collision Detection.

First, say a computer connected to the network wants to transfer information to another computer. Before it does so, it checks to see if there is any other data or information being transferred. If not, it will wait until the medium is free. If so, the computer will transmit the data. When data is transmitted, all of the devices connected to the computer can hear the transmission.

Despite the precautions that are taken to make sure that no two computers are trying to transfer data at the same time, sometimes collisions still occur. Collisions are when two devices on a network try to transmit information at the same time. When this happens, the information being transmitted is sent back to the device it came from. Then, that device will wait for a random amount of time, before sending the information again.

Recently, newer versions of Ethernet have been developed. They are much faster than the original Ethernets. One of the newest versions, Fast Ethernet, can transfer data up to 100 megabits (100 million bits) per second! There is also the Gigabit Ethernet, which transfers 1 gigabit (1 billion bits) of data per second!

The simplest form of Ethernet uses a passive bus operated at 10 Mbps. The bus is formed from a 50 Ohm coaxial cable which connects all the computers in the LAN. A single LAN may have up to 1024 attached systems, although in practice most LANs have far fewer. One or more pieces of coaxial cable are joined end to end to create the bus, known as an "Ethernet Cable Segment". Each segment is terminated at both ends by 50 Ohm resistors (to prevent reflections from the discontinuity at the end of the cable) and is also normally earthed at one end (for electrical safety). Computers may attach to the cable using transceivers and network interface cards.

Ethernet LANs may be implemented using a variety of media (not just the coaxial cable described above). The types of media segments supported by Ethernet are:

- 10B5 Low loss coaxial cable (also known as "thick" Ethernet)
- 10B2 Low cost coaxial cable (also known as "thin" Ethernet)

- 10BT Low cost twisted pair copper cable (also known as Unshielded Twisted Pair (UTP))
- 10BF Fibre optic cable

The network design rules for using these types of media are summarised below:

Segment Type	*Max Number of systems per cable*	*Segment*	*Max Distance of a cable segment*
10B5	(Thick Coax)	100	500 m
10B2	(Thin Coax)	30	185 m
10BT	(Twisted Pair)	2	100 m
10BFL	(Fibre Optic)	2	2000 m

There is also a version of Ethernet which operates using twisted pair cabling or fibre optic links at 100 Mbps and at 1 Gbps. 100 Mbps networks may operate full duplex (using a Fast Ethernet Switch) or half duplex (using a Fast Ethernet Hub). 1 Gbps networks usually operate between a pair of Ethernet Switches. Many LANs combine the various speeds of operation using dual-speed switches which allow the same switch to connect some ports to one speed of network, and other ports at another speed. The higher speed ports are usually used to connect switches to one another.

3.3 LAN MEDIA-ACCESS METHODS

Media contention occurs when two or more network devices have data to send at the same time. Because multiple devices cannot talk on the network simultaneously, some type of method must be used to allow one device access to the network media at a time. This is done in two main ways:

Carrier Sense Multiple Access collision detect (CSMA/CD) and token passing.

In networks using CSMA/CD technology such as Ethernet, network devices contend for the network media. When a device has data to send, it first listens to see if any other device is currently using the network. If not, it starts sending its data. After finishing its transmission, it listens again to see if a collision occurred. A collision occurs when two devices send data simultaneously. When a collision happens, each device waits a random length of time before resending its data. In most cases, a collision will not occur again between the two devices. Because of this type of network contention, the busier a network becomes, the more collisions occur. This is why performance of Ethernet degrades rapidly as the number of devices on a single network increases.

In token-passing networks such as *Token Ring* and *FDDI*, a special network frame called a token is passed around the network from device to device. When a device has data to send, it must wait until it has the token and then sends its data. When the data transmission is complete, the token is released so that other devices may use the network media. The main advantage of token-passing networks is that they are deterministic. In other words, it is easy to calculate the maximum time that will pass before a device has the opportunity to send data. This explains the popularity of token-passing networks in some real-time environments such as factories, where machinery must be capable of communicating at a determinable interval.

3.4 FIBRE DISTRIBUTED DATA INTERFACE (FDDI)

FDDI is a standard for data transmission on fiber optic lines in a local area network that can extend in range up to 200 km (124 miles). The FDDI protocol is based on the token ring protocol. In addition to being large geographically, an FDDI local area network can support thousands of users.

The previous networks were based on twisted pairs or coax media. Fiber optic media can supply an enhanced speed. In fact, Token Ring networks over fiber optic cables were introduced. FFDI, running at 100 Mbps over larger areas and many stations can serve both as a high-speed LAN or as a backbone to connect other LANs.

An FDDI network contains two token rings, one for possible backup in case the primary ring fails. The primary ring offers up to 100 Mbps capacity. If the secondary ring is not needed for backup, it can also carry data, extending capacity to 200 Mbps. The single ring can extend the maximum distance; a dual ring can extend 100 km (62 miles).

FDDI is a product of American National Standards Committee X3-T9 and conforms to the open system interconnect (OSI) model of functional layering. It can be used to interconnect LANs using other protocols. FDDI-II is a version of FDDI that adds the capability to add circuit-switched service to the network so that voice signals can also be handled. Work is underway to connect FDDI networks to the developing Synchronous Optical Network SONET which in turn is part of broadband ISDN.

A FDDI backbone network example:

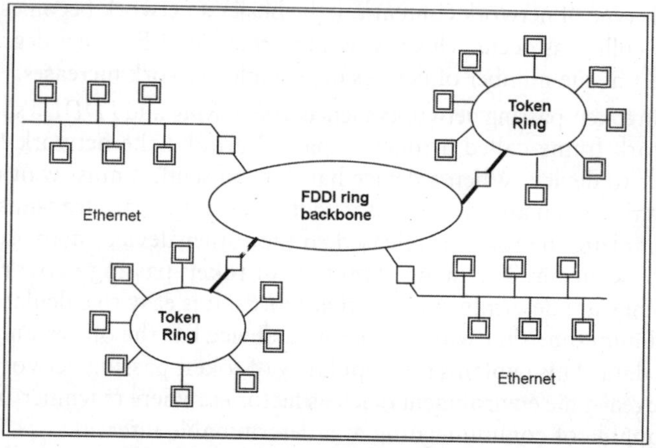

Fig. 3.10: FDDI

FDDI has been accepted as a standard since 1990, yet high cost has meant that companies have not implemented it widely, of FDDI adapters has caused many companies to look at a twisted-pair wire alternative to FDDI: Copper Distributed Data Interface (CDDI).

There are a broad range of FDDI products currently available from a variety of vendors including Cisco, 3Com, Hewlett-Packard (HP) and DEC.

FDDI Network Architecture: FDDI operates at a basic data transfer rate of 100 Mbits/sec. The maximum size of the ring is 50 km. The latency of communication varies with the ring size, up to about 1 msec.

FDDI uses a dual ring topology, which is to say that it is comprised of two counter-rotating rings. A dual-attached rooted station on the network is attached to both of these rings.

A dual-attached station on the ring has at least two ports - an A port, where the primary ring comes in and the secondary ring goes out, and a B port where the secondary ring comes in, and the primary goes out. A station may also have a number of M ports, which are attachments for single-attached stations. Stations with at least one M port are called concentrators.

The sequence in which stations gain access to the medium is predetermined. A station generates a special signaling sequence called a Token that controls the right to transmit. This Token is continually passed around the network from one node to the next. When a station has something to send, it captures the Token, sends the information in well formatted FDDI frames, then releases the token. The header of these frames includes the address of the station(s) that will copy the frame. All nodes read the frame as it is passed around the ring to determine if they are the recipient of the frame. If they are, they extract the data, retransmitting the frame to the next station on the ring. When the frame returns to the originating station, the originating station strips the frame. The token-access control scheme thus allows all stations to share the network bandwidth in an orderly and efficient manner.

3.5 A SYNCHRONOUS TRANSFER MODE (ATM)

ATM is a high-performance switching and multiplexing technology that utilizes fixed-length packets to carry different types of traffic. Information is formatted into fixed-length cells consisting of 48 bytes (8 bits per byte) of payload and 5 bytes of cell header. The fixed cell size guarantees that time-critical information (*e.g.*, voice or video) is not adversely affected by long data frames or packets. Of course, if the cells were longer in length the system would be more efficient, because the header would take up a smaller percentage of the total cell.

Multiple streams of traffic can be multiplexed on each physical facility and can be managed so as to send the streams to many different destinations. This enables cost savings through a reduction in the number of interfaces and facilities required to construct a network.

4.1 METHODS OF DATA-TRANSMISSION: COMMUNICATION MEDIA

A variety of media link devices in a communication network. The communication channels that these provide can carry either analog or digital signals. All communications media fall into one of two broad classes: physical lines or wireless media. These media are shown in following figure:

An example of communication media

The communication links are established using Telephone lines

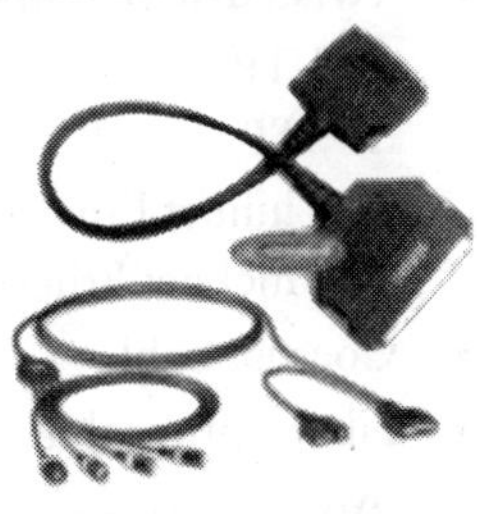

Fig. 3.11: Media Cable

- Paired wires and cables
- Co-axial cables
- Optical Fibre Cables
- Wireless

4.2 TYPES OF CABLE MEDIA

These types are compared in Table 3.2 and described in the following section

Table 3.2: Characteristics of Cable Media

Cable *Type*	Cost	Max. Dist *meters*	Max Number *of Nodes*	Common Usage
UTP	Very low	100	1024	Star network 2 to 155 Mbps
STP	Fairly expensive more	100	270	IBM Tokenring than UTP or Thinnet 4 to 16 mBPS
Thinnet	Relatively inexpensive	185	30	per Thinnet Ethernet less than STP or UTP segment 2 to 10 Mbps
Thicknet	Fairly expensive more	500	100	per Thicknet-ethernet than Thinnet, STP, or UTP; segment 2 to 10 Mbps less than fiber-optic
Fiber	Most expensive, but costs	20km	No limit	FDDI Optic are dropping 100 Mbps in some cases
Wireless	Very High	No limitation		

4.3 PHYSICAL LINES

The three principal types of physical lines that dominate today's telecommunication systems are:

- Twisted-wire pairs
  - ❐ UTP
  - ❐ STP
  - ❐ Thinnet Ethernet
  - ❐ Thicknet Ethernet
- Coaxial cable
- Fiber optic cable.

While both twisted-pair and coaxial cable are still commonly used, fiber optic cable is the wave of the future.

4.3.1 Twisted-wire Pairs

Twisted pair is the ordinary copper wire that connects home and many business computers to the telephone company. To reduce cross-talk or electromagnetic induction between pairs of wires, two insulated copper wires are twisted around each other. Each sets or desktop locations require multiple connections, twisted pair is sometimes installed in two or more pairs, all within a single cable. Shielded twisted pair is often used in business installations. The more common kind of wire that is installed to your home is unshielded twisted pair.

Twisted-wire pairs are the least expensive type of communications medium. For years, they have been used by the telephone companies to develop telephone networks. Small wires are twisted together in pairs, each of which can handle a single phone conversation, and the pairs are bound together into large cables. Twisted-wire pairs were developed primarily as a voice-oriented, telephone communications medium; their capacity is typically lower than the others discussed next. However, the capacity of twisted-wire pairs can be expanded by proper shielding and communication equipment. Because of this, local area network developers are seeing new interest in twisted-wire pairs.

Unscreened (or Unshielded) Twisted Pair UTP

A type of cable widely used for constructing local area networks based on the ethernet standard, and for other types of communication including voice telephone. In fact, a UTP cable often consists of a number of twisted pairs of wires, and is terminated at each end using telephone-type plugs.

Unshielded of Twisted Pair (UTP)

UTP has the advantages of

- A high installed base
- Cheap to install
- Easy to terminate

Its disadvantages are:

- Very noisy
- Limited in distance
- Suffers from interference

The popularity of UTP mostly stem from its low cost. Even good-quality UTP cable is still considerably cheaper than the coaxial alternative, but the bandwidth over long distances is much lower.

The current working standard for twisted pair data cables is known as Category 5. In UTP based ethernet networks, the system requires one or more hubs, into which all the cables are connected.

To reduce crosstalk or electromagnetic induction between pairs of wires, two insulated copper wires are twisted around each other. Each signal on twisted pair requires both wires. Since some telephone sets or desktop locations require multiple connections, twisted pair is sometimes installed in two or more pairs, all within a single cable. For some business locations, twisted pair is enclosed in a shield that functions as a ground. This is known as shielded twisted pair (STP).

Although twisted pair is often associated with home use, a higher grade of twisted pair is often used for horizontal wiring in LAN installations because it is less expensive than coaxial cable.

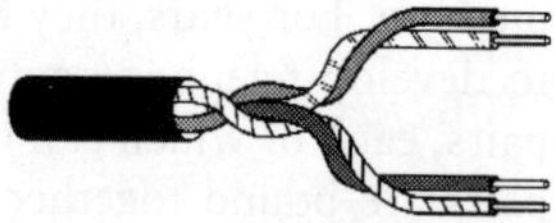

Fig. 3.12: Twisted Pair UTP

Twisted pair used in Local Area Networks has several ratings. Category 3 has a speed rating of 10 million bits per second (the speed of ethernet), whereas category 5 has a speed rating of 100 million bits per second. CIT is cabled using Category 5 UTP cable. UTP (Unshielded twisted pair) is cable which has no ground shield. Cables are often provided with a ground shield which helps to reduce signal interference from external sources, thus making the signal travelling down the cable less prone to alteration. Twisted pair cable is provided in two forms, UTP and STP. The adjoining diagram shows UTP Category 5 cable.

STP (Shielded Twisted Pair)

Shielded twisted pair is a special kind of copper telephone wiring used in some business installations. An outer covering or shield is added to the ordinary twisted pair telephone wires; the shield functions as a ground.

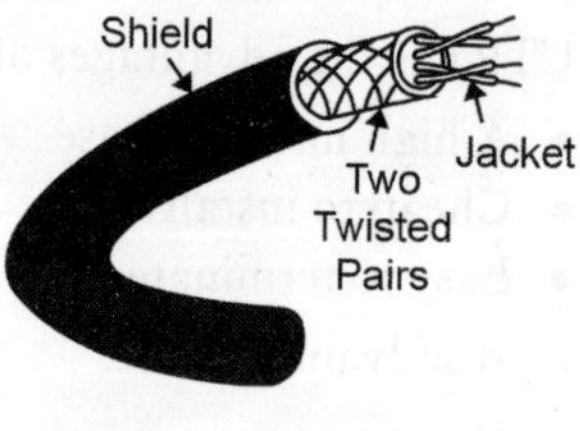

Fig. 3.13: STP

The only difference between STP and UTP is that STP cable has a shield between the outer jacket or casing and the wires. The shield which is usually made of aluminum/polyester, makes STP less vulnerable to EMI because the shield is electrically grounded; however, STP is not much susceptible to attenuation than UTP. The diagram adjoining shows category 3 STP.

Thin Ethernet (10B2 / IEEE 804.3a)

One form of cabling supported by Ethernet is low cost 50 Ohm coaxial cable as shown in the Figure below. This type of cable is cheap to purchase and is flexible and easy to install. It has become known in the communications industry as "Thin Ethernet", so called to differentiate it to "Thick Ethernet" which was originally used in Ethernet LANs. The official name for this cable is 10 Baseband2 (10B2), indicating that it is specified for baseband communications (*i.e.* not modulated) at 10 Mbps over distances up to approximately 200m.

The connection of a network interface card in a computer to 10B2 cabling is usually via a BNC connector as shown in the figure below. Some buildings use a special type of connector in place of the more common BNC to try to prevent problems arising when users disconnect or reconnect their equipment. (This may otherwise cause a break in the transmission medium inhibiting communication between other users). The end of each segment of cable must be terminated with a 50 Ohm BNC terminator.

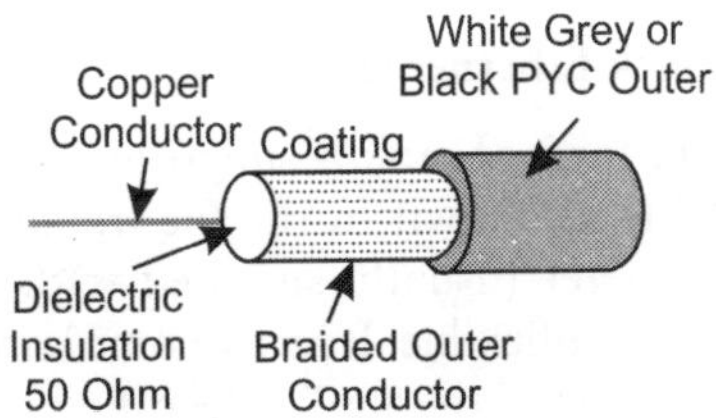

Fig. 3.14: Connection of Transceiver on to a 10B2 Ethernet Network

10B2 cabling is flexible and the network may be constructed in any convenient way by simply connecting all devices to the cable (see below):

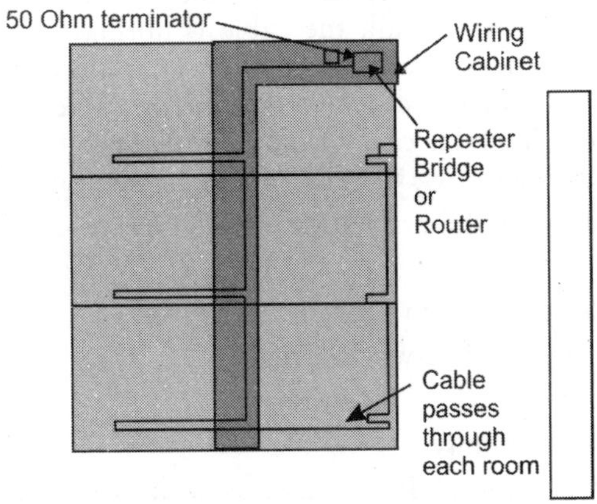

Fig. 3.15: A typical Cabling Scheme for 10B2 Cabling in an Office

A summary of the properties of this type of cabling is given below:

- Segment length < 185 m and > 0.5 m
- Up to 30 attached nodes
- Cable flexible and cheap
- Integrated or external transceiver connected via a BNC 'T' connector
- Used mainly for workgroups
- Difficult to manage (*i.e.* breaks in cable difficult to locate)

Thick Ethernet (10B5)

One form of cabling supported by Ethernet is low-loss 50 Ohm coaxial cable as shown in the Figure adjoining. This type of cable is 0.5" diameter (usually supplied with a yellow outer PVC coating) and rather inflexible. It has become known in the communications industry as "Thick Ethernet". The official name for this cable is 10 Baseband5 (10B5), indicating that it is specified for baseband communications (*i.e.* not modulated) at 10 Mbps over distances up to 500 m.

The 10B5 cabling system uses thick (low-loss) coaxial cable which forms a shared bus. Upto 100 transceivers may be used to connect computers to the bus. The cable is difficult to install (due to its weight, large 0.5" diameter, and constraints on minimum curvature). Its key advantage is the extended transmission distance and the fact that it has good noise immunity. Modern 10B5 cabling is used for backbone connections and it is now fairly uncommon to find this type of cabling using to connect user's workstations. Since this type of cabling is relatively difficult to install, the cable is normally installed in a communications duct or along a corridor in an office building (see below). An AUI drop cable is then run from the cable to an AUI port on the network interface card of each computer to be connected. A typical example of the use of 10B5 cabling in an office environment.

A summary of the properties of this type of cabling is given below:

- Segment length > 500m
- Up to 100 attached network devices
- Cable run needs careful attention
- Good noise immunity
- Vampire or In-Line (N-Type connector) with an external transceiver
- Used mainly for building backbones

Co-axial Cable

Co-axial or co-ax literally means "having the same axis". Coaxial cable is called "coaxial" because it includes one physical channel that carries the signal surrounded (after a layer of insulation) by another concentric physical channel, both running along the same axis. The outer channel serves as a ground. Many of these cables or pairs of coaxial tubes can be placed in a single outer sheathing and, with repeaters, can carry information for a great distance.

Co-axial cable has a center conductor, an insulator, a solid or braided shield around this insulator with a tough jacket on the outside. The inner insulation provides a constant distance between the center conductor and the shielding, providing a superior quality signal over longer distances, which gives higher bandwidth and better immunity to external interference than simple twisted pair cable provides. However, coaxial cable is more expensive, is more difficult to connect because it requires special tools to install connectors on the cable, and is very stiff making physical installation more difficult as well. It can not bend sharply which means routing to some locations can be difficult. The high bandwidth it offers offsets the installation problems when high speed and/or error-free transmission is necessary.

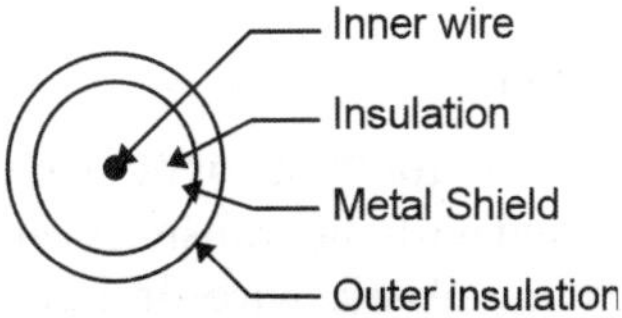

Fig. 3.16: Formation of Co-axial Cable

Co-axial cable is also used by cable TV companies between the community antenna and user homes and businesses. Coaxial cable is sometimes used by telephone companies from their central office to the telephone poles near users. It is also widely installed for use in business and corporation Ethernet and other types of local area network.

Co-axial cable permits communications at much higher rate than conventional wires (even twisted pair wires) because of its much better high-frequency transmission characteristics. Figure shows a twisted wire pair and a coaxial cable. In the coaxial cable, a central wire which carries the data signal is surrounded by a conductive sleeve to form an electrical shield around the central wire. Coaxial cable is used in implementing a local area network called the Ethernet. For longer distance and still higher speed data transfers, transmission media other than coaxial cable are used . These assure that the data arrive at their destination quickly and without error.

Coaxial cable has the advantages of

- Cheap to install
- Conforms to standards
- Widely used

Its disadvantages are:

- Limited in distance
- Limited in number of connections
- Terminations and connectors must be done properly

Optical Fibre

Optical fiber (or "fiber optic") refers to the medium and the technology associated with the transmission of information as light pulses along a glass or plastic wire or fiber. Optical fiber carries much more information than conventional copper wire and is in general not subject to electromagnetic interference and the need to retransmit signals. Most telephone company long distance lines are now of optical fiber.

The chief advantage of optical fiber over copper wire is that fiber supports significantly higher bandwidths than copper. It is also immune to electrical interference, such as that caused by lightning, electrical motors and other sources of strong magnetic fields. It is also significantly more expensive than copper and much more difficult to connect or repair when broken.

It is relatively new among the media technologies consists of thousands of hair-thin strands of glass that are bound together as a cable. Data are sent along the cable as light waves, from a high-powered, laser beam source.

Compared to coaxial cable, fiber optic cable has greater channel capacity and is lighter, faster, and compacter. Transmission speeds in excess of 5 gigabits (5 billion bits) per second are reported. Such rates would make it possible to transmit the entire contents of a 30-volume encyclopedia from coast-to-coast in less than a second. Also, because it is virtually impossible to cut the cable to send data to unauthorized points, these cables are highly resistant to being tapped. Fiber optic cable will be the primary media of future communications systems.

Fiber optic cable transmits light signals rather than electrical signals. Each fiber has an inner core of glass or plastic that conducts light. A layer of glass that reflects the light back into the core, called cladding, surrounds the inner core. A plastic sheath surrounds each fiber. The sheath can either be tight or loose.(Fig. 3.18)

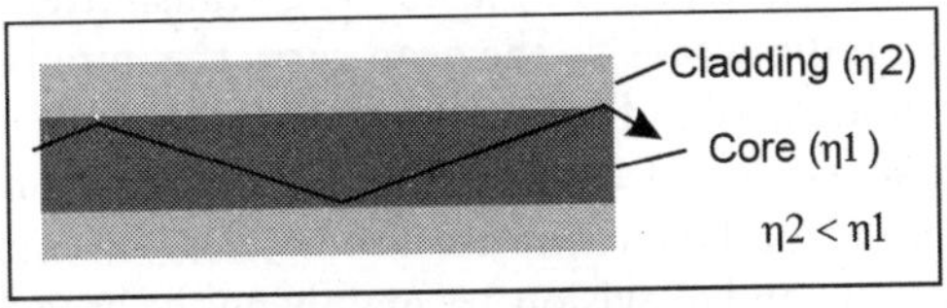

Fig. 3.17: Fibre Optic Cable

Fiber optic cable is enormously more efficient than the other network cable media. It has much lower attenuation than copper wires, mainly because the light is not radiated out in the way that electricity is radiated from copper cables. Current fiber optic technologies allow data rates from 100 Mbps to 2 Gbps. Two disadvantages of fiber optic cable are that it is more expensive than the other types of cable media and that it is more difficult to install.

An integrated services digital network (ISDN) is linking nations with interconnecting digital webs capable of handing the huge communications demands of the future.

The use of fiber optic lines is increasing, replacing coaxial cable and conventional telephone lines. Each fiber optic cable is approximately the diameter of a human hair and made of clear glass strands. Data is transmitted in the form of light beams, generated by laser devices. This technology has made possible transmission speeds as high as 1 billion bps. Fiber optic cables are replacing conventional cables because they are lighter, more durable, occupy less space, and offer 10 times the datacarrying capacity of standard coaxial cable.

Fibre optic cable has the advantages of:

- High capacity
- Immune to interference
- Can go long distances

Its disadvantages are:

- Costly
- Difficult to join

4.4 Wireless Communications

Various telecommunication systems that use radio waves to carry signals and messages across distances. Wireless communications systems use devices called transmitters to generate radio waves. A microphone or other mechanism converts messages, like sounds or other data, into electronic impulses. The transmitters change, or modulate, the radio waves so they can carry the impulses, and then transmit the modulated radio signals across distances. Radio receivers pick up these signals and decode them back into the original messages. Commercial radio and television are also wireless telecommunication systems, but radio and television are mainly public broadcast services rather than personal communication systems.

Wireless communication allow people greater flexibility while communicating, because they do not need to remain at a fixed location, such as a home or office. Wireless technologies make communication services more readily available than traditional wire-based services (such as ordinary telephones), which require the installation of wires. This is useful in places where only temporary communication services are needed, such as at outdoor festivals or large sporting events.

These technologies are also useful for communicating in remote locations, such as mountains, jungles, or deserts, where telephone service might not exist. Wireless services allow people to communicate while in a car, airplane, or other moving vehicle. Police, fire, and other emergency departments use two-way radio to communicate information between vehicles that are already responding to emergency calls, which saves valuable time. Construction and utility workers frequently use hand-held radios for short-range communication and coordination. Many business people use wireless communications, particularly cellular radio telephones, to stay in contact with colleagues and clients while traveling.

All wireless communication devices use radio waves to transmit and receive signals. These devices operate on different radio frequencies so that signals from one device will not overlap and interfere with nearby transmissions from other devices. The number of companies offering wireless communication services has grown steadily in recent years. In 1988 about 500 companies offered cellular radio telephone (cell phone) services. By 1995 that number had grown to over 1500 companies serving millions of subscribers. Wireless communication is becoming increasingly popular because of the convenience and mobility it affords, the expanded availability of radio frequencies for transmitting, and improvements in technology.

4.4.1 Modes of Wireless Communication

Wireless communications systems have grown and changed as technology has improved. Several different systems are used today, all of which operate on different radio frequencies. New technologies are being developed to provide greater service and reliability. Following modes are available for wireless communication:

- Radio Waves (Short Wave)
- Microwaves
    - Terrestrial microwave
    - Satellite microwave
- Hand-Held Radio Transceivers (police wireless)
- Cellular Radio Telephones (So called mobile)
- Sea and Air Transceivers

4.4.1 Radio Waves

Radio waves have frequencies between 10 kilohertz (KHz) and I gigahertz (GHz). The range of the electromagnetic spectrum between 10KHz and IGHz is called radio frequency (RF). Radio waves fall into three categories:

- Low-power, single frequency transceivers operate at only one frequency.
- High-power, single frequency transmissions are similar to low-power, but can cover larger distances and go through and around objects. High power rates improve the signal's resistance to attenuation, and repeaters can be used to extend the signal range.
- Spread-spectrum broadcasts signals over a range of frequencies. The signal is coded with a technique called chips, which gives the technology both security and redundancy. The available frequencies are divided into channels or hops. The adapters tune into a specific frequency for a predetermined length of time and then switch to a different frequency.

The two most common in use today are micro wave and satellite communications.

4.4.2 Microwaves

Micro wave systems do not use cable as a transmission medium, rather they use the air. Using very high frequency signals, they support thousands of telephone channels and several television channels on the one circuit. Transmitters and receivers must be located within sight of each other, about 30 kilometers apart.

Microwave is currently the most popular long-distance transmission method in the United States. It uses line-of-sight communication. Microwave systems consist of two radio transceivers: one to transmit and one to receive. These antennas are often installed on towers to give them more range and raise them above anything that may block their signal.

Fig. 3.18: Microwave Transceiver

Microwave is a radio system which uses very high frequencies to send and receive data. Because of the high frequencies involved, stations are located about 30 kilometers apart and in line of sight (visible to each other). Microwave systems have sufficient bandwidth capacity to support a large number of voice channels and one or two TV channels.

Microwave systems have the advantage of

- Medium capacity
- Medium cost
- Can go long distances

Its disadvantages are:

- Noise interference
- Geographical problems due to line of sight requirements
- Becoming outdated

Microwave communication comes in two types:

- Terrestrial microwave
- Satellite microwave

Microwave signals are normally telecommunicated by microwave stations satellite. Both communication satellites and terrestrial microwave stations are especially designed for sending large quantities of data one way at a time. Both ways can efficiently transmit large quantities of data at the highest possible speeds. Land-based microwave transmission stations are generally 25 to 30 miles apart and must have no major obstructions between them. Signals received from one microwave station are amplified and passed on to the next station. Microwave stations are usually built on top of all buildings and mountain peaks for the best transmission capability. Because of the long transmission distances involved, satellite systems may not be suitable for some applications that require fast-response, two-way interactive communication.

Terrestrial Microwave

Terrestrial microwave stations are ground-based communication stations that are designed to communicate with each other or with satellite. Ground stations that communicate with each other must be within line of sight. Even in flat terrains, the curvature of the earth limits the distance between stations to 30 miles. Terrestrial microwave uses earth-based transmitters and receivers in the low gigahertz range of frequencies. Communications are line-or-sight and cannot go around corners or through buildings.

Sometimes, they must be placed on top of tall buildings or mountains in order to ensure an obstacle-free communication path. As with any telecommunication media, the transmitted signal weakens (attenuates) as it travels from sender to receiver. To compensate for signal attenuation, a receiving microwave station amplifies the signal before passing it along to another station. Unlike other media, heavy rain can affect microwave transmission.

Since it is impractical to dot the entire surface of the earth with microwave stations, especially over the ocean, communication satellites are important.

The high-speed transmission of data is not limited to wire service. Microwave transmissions are another way to telecommunicated data at high speeds, within a wideband channel. Microwaves are high-frequency radio signals which can handle computer-generated data, as well as sound and video images such as those seen on television. Once this data is converted to microwave impulses, it can be transmitted throughout the world.

Satellites

Beyond certain distance it is not possible to use communications through ground methods due to two main reasons, one due to fading of signals which can not be amplified without noise and second due to geographical reasons say sea, deserts ,mountains etc. In such cases the data may be beamed to a communication satellite that acts as a reflector by accepting signals from one point on earth and returning the same signals to some other point on earth. The microwave signals because it's precisely positioned 22,300 miles above the equator with an orbit speed that matches the earth's rotation. Dozens of satellites are now in orbit to handle international and domestic dat, voice, and video communications.

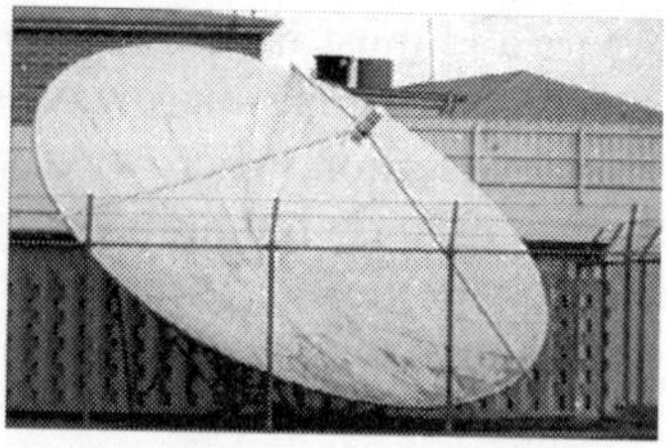

Fig. 3.19: Satellite

Both the microwaves and lower-frequency radio waves can be used to create long-distance satellite transmission. At present, nearly 100 communication satellites are in stable "parking orbits" approximately 22,300 miles above the earth. These satellites are operated by private companies, such as American Satellite Company, At & T, COMSAT, Intellsat, RCA, Satellite Business Systems, and Western Union, that sell communication services to customers. The earth station operated by these private carries are located in major U.S. cities and project radio waves to the satellites. These signals are then reflected to yet other earth stations operated by these private carriers are located in major U.S. cities and project radiowaves to the satellites. These signals are then reflected to yet other earth stations for distribution to computer users. Like microwaves, satellite transmission based on radiowaves is a long-distances line-of-sight communication medium.

Satellite systems support tens of thousands of speech channels and tens of television channels. The cost is very high per circuit (about $4 million per transponder). Signals are received using a special dish.

Satellite ground stations with large dishes communicate with a communication satellite in geostationary orbit around the earth. Each channel is managed by a transponder, which can support thousands of speech channels and about 4 TV channels simultaneously. The cost of satellite links is still very expensive (about $ 4M per transponder). It is primarily used for intercontinental links.

Satellite systems are comprised of ground-based transmitter and receiver dishes, with an orbital satellite circuit (called a transponder). Signals are transmitted to the orbiting satellite, which relays it back to another ground station. The footprint coverage of a single satellite system is very large, covering thousands of square kilometers (entire countries).

Satellite systems have the advantage of

- Low cost per user (for PAY TV)
- High capacity
- Very large coverage

Its disadvantages are:

- High install cost in launching a satellite
- Receive dishes and decoders required

4.5 CELLULAR RADIO

The first commercially available radio and telephone system, known as improved mobile telephone service (IMTS), was put into service in 1946. This system was quite unsophisticated, but then there was no solid state electronics available.

With IMTS, a tall transmitter tower was erected near the center of a metropolitan area. Several assigned channels were transmitted and received from the antenna atop this tower. Any vehicle within range could attempt to seize one of those channels and complete a call. Unfortunately, the number of channels made available did not come even close to satisfying the need. To make matters worse, as the metropolitan area grew, more power was applied to the transmitter or receiver, the reach was made greater, and still more erstwhile subscribers were unable to get dial tone.

The solution to this problem was cellular radio. Metropolitan areas were divided into cells of no more than a few miles in diameter, each cell operating on a set of frequencies (send and receive) that differed from the frequencies of the adjacent cells. Because the power of the transmitter in a particular cell was kept at a level just high enough to serve that cell, these same sets of frequencies could be used at several places within the metropolitan area. Beginning in 1983, two companies, one called a wireline company and the other called a non-wireline carrier, were given a franchise to operate in each major territory.

Two characteristics of cellular systems were important to their usefulness. First, the systems controlled handoff. As subscribers drove out of one cell and into another, their automobile radios, in conjunction with sophisticated electronic equipment at the cell sites (also known as base stations) and the telephone switching offices (also known as mobile telephone switching office [MTSO]), transferred from one frequency set to another with no audible pause. Second, systems were also designed to locate particular subscribers by paging them in each of the cells. When the vehicle in which a paged subscriber was riding was located, the equipment assigned sets of frequencies to it, and conversation could begin.

In a Cellular System, Vehicle-to-Cell-Site Comunications take place using FDMA, TDMA, CDMA, GSM, etc.; communications between the cell Site and the MTSO utilize conventional techniques

The initial transmission technology used between the vehicle and the cell site was analog in nature. It is known as advanced mobile phone service (AMPS). The analog scheme used was called frequency division multiple access (FDMA).

But the age of digital transmission was upon us, and many companies operating in this arena concluded that a digital transmission scheme would be preferred. The result was time division multiple access (TDMA). In Europe, the selected scheme was an adaptation of the TDMA used in the United States, and it was called group special mobile. Since then, the name has been changed to global system for mobile communications (GSM).

As if that was not enough, a third group of companies determined that a special spread-spectrum or frequency-hopping scheme would be even better, and this also was developed and trialed. This is called code division multiple access (CDMA). Thus, there are at least four schemes that may be used for communication between a vehicle and the cell site. Communications between the cell site and the MTSO utilized more conventional techniques, such as microwave, copper pairs, or fiber optics.

The continuing growth of cellular communications (there are presently about 20,000 new subscribers signing on each day), led government and industry in the United States to search for additional ways to satisfy the obvious need not only for ordinary telephone service but also for special services and features, smaller telephones, and cellular phone use. This search led to the PCS industry. Additional frequency bands were allocated for their use, and rather than assign them to the first comers or by way of a lottery, the FCC auctioned them off through a sophisticated bidding contest that brought the U.S. treasury billions of dollars.

5.0 LOCAL AREA NETWORKING HARDWARE

In addition to computers, the hardware components needed to create LAN include the

- Network Cable
- Cable Interface Unit
- Network Interface Cards for each computer.
- LAN Devices

5.1 NETWORK CABLE

The network cable can be twisted-pair cable, coaxial cable, or fiber-optic cable:

- Twisted-pair cable costs the least but is the slowest. It is also vulnerable to interference from large machines, such as air conditioners. This interference can destroy data.
- The two kinds of coaxial cable are baseband and broadband. Baseband coaxial cable can carry only one signal at a time, but it is fast (up to 10 million bits per second). Broadband coaxial cable can carry more than one signal at a time. Cable TV companies use broadband coaxial cable.
- Fiber optic cable are made from tiny strands of glasslike material. These strands transmit light pulses with very high efficiency. Each pulse represents one bit of data. Fiber optic cables can carry massive amounts of data (up to billions of bits per second). Fiberoptic cables are unaffected by magnetic or electrical interference; they are, however, expensive and hard to install.

5.2 NETWORK INTERFACE CARD [NIC]

All LANs that use wires as the physical transmission media require the client machine to have a network interface card [NIC] installed to provide the circuitry and connection to the wires. These NICs must be compatible with both the computer's own internal bus architecture and the type of network architecture that is used. NICs used to attach to networks using IEEE 802.3 Ethernet standard are commonly called Ethernet Cards. Likewise, computers attached to computers using the IEEE 802.5 are called Token Ring Adaptors. NICs are also available for laptop computers; they are sometimes called pocket adapters because of their small size.

A network interface card is used to connect a computer to an Ethernet network. The card (shown in the figure below) provides an interface to the media. This may be either using an external transceiver (as shown) or through an internal integrated transceiver mounted on the network interface card PCB. The card usually also contains the

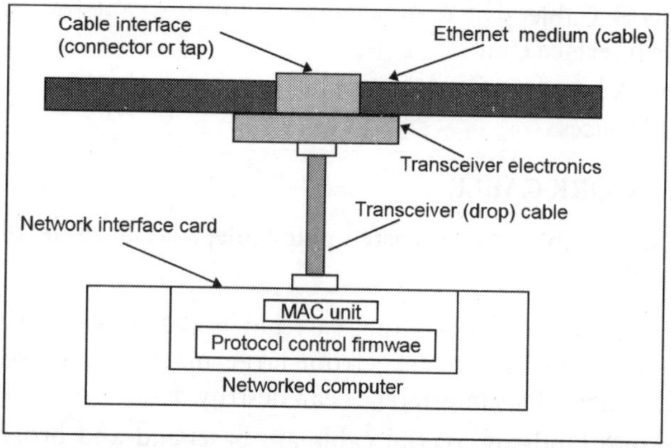

Fig. 3.20: Network Operating System

protocol control firmware and Ethernet Controller needed to support the Medium Access Control (MAC) data link protocol used by Ethernet.

It sits in a slot in the PC with one end of the network cable plugged in to the back. The other end of the network cable generally plugs in to a wall-jack connection, although it can plug directly in to the hub or switch for a small network. Another cable runs from there in to the network hub, which completes the connection.

The network interface cards convert the data on the computer's internal bus to conform to the protocol standards of the network architecture. NICs may be equipped with their own CPUs and other electronic devices to provide better data handling and throughput. Each NIC is provided with its own unique device address so that messages can be sent to it. These addresses are different than the ones you are familiar with in sending e-mail and NIC addresses are seldom entered directly by the end user. NICs, like other devices attached to the computer's internal bus via an expansion slot or a PCMCIA slot, require a device driver to operate. These NIC drivers are software, supplied with the network operating system [NOS] or by the manufacturer of the NIC.

5.3 CABLE INTERFACE UNIT (HUB)

The cable interface unit, sometimes called a hub, sends and receives signals on the network cable. This unit is a box outside the computer.

The network interface card is inserted into an expansion slot inside the computer. The interface card sends and receives messages to and from the LAN. The card is connected to the cable interface unit by wire.

Hubs are needed to connect one incoming port to form several similar outgoing ports. These can be of three types:

- Standalone Hubs
- Stackable Hubs
- Modular Hubs

Standalone hubs: Standalone hubs are single products with a number of ports. Standalone hubs usually include some method of linking them to other standalone hubs for network expansion. Standalone hubs are usually the least expensive type of hub, and are best suited for small, independent workgroups, departments, or offices, typically with fewer than 12 users per LAN.

Stackable Hubs: Stackable hubs work just like standalone hubs, except that several of them can be "stacked" (connected) together, usually by short lengths of cable. When they are connected together, they act like a modular hub, because they can be managed as a single unit. These hubs are ideal if you want to start with a minimal investment, but realize that your LAN will grow.

Fig. 3.21: Hub Stackable

Modular Hubs: Modular hubs are popular in networks because they are easily expanded and always have a management option. A modular hub is purchased as a chassis, or card cage, with multiple card slots, each of which accepts a communications card, or module. Each module acts like a standalone hub, and usually has 12 twisted pair ports. Modules supporting different types of network cabling, like coaxial or token ring, can also be purchased.

5.4 SWITCHES

Switches are another type of device used to link several LANs and route packets between them. A switch has multiple ports, each of which can support either a single station or an entire Ethernet or

Token Ring LAN. With a different LAN connected to each of the ports, it can switch packets between LANs as needed.

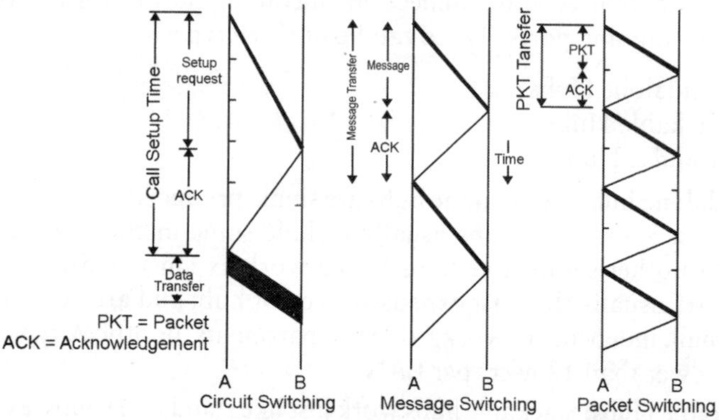

Fig. 3.22: Switches

Sample Network Layout With a LAN Switch, Courtesy of Bay Networks

There are maximums on distances between workstations and hubs, hubs and hubs, and stations connected to a single LAN. You can exceed these maximums by linking two LAN segments (groups of users/devices) together using a Bridge or Router.

6.1 GATEWAYS

Workstations in one network often need access to computer resources in another network or another part of a WAN. For example, an office manager using a local area network might want to access an information service that is offered by a VAN over the public phone system. In order to accommodate this type of need, gateways and bridges are often necessary.

A gateway is a collection of hardware and software facilities that enables devices on one network to communicate with devices on another, dissimilar network.

Gateway is a server computer that performs protocol conversion between different types of networks or applications. For example, a gateway can convert a TCP/IP packet to a NetWare IPX packet and vice-versa or from AppleTalk to DECnet, from SNA to AppleTalk and so on.

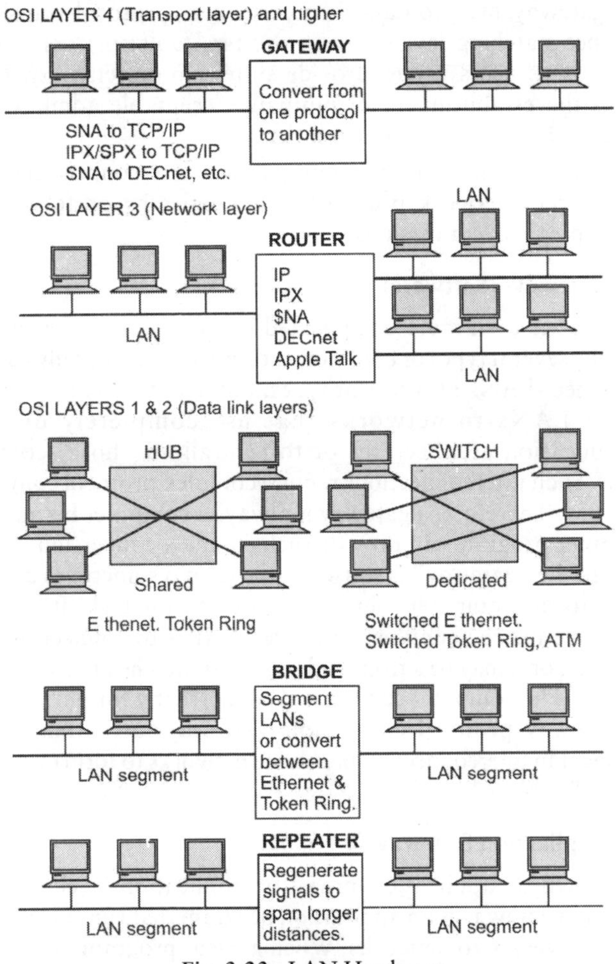

Fig. 3.23: LAN Hardware

They perform complete conversion from one protocol to another rather than simply support one protocol from within another, such as IP tunneling. Sometimes routers can implement gateway functions.

6.1.1 Interior Gateway Protocols

There are different types of routers for local-area networks and wide-area networks. Interior Gateway Protocols (IGP) are the protocols that manage routers within a small LAN, for example, a network in a small business.

A gateway may contain devices such as protocol translators, impedance matching devices, rate converters, fault isolators, or signal translators as necessary to provide system interpretability. It also requires the establishment of mutually acceptable administrative procedures between the two networks.

A protocol translation/mapping gateway interconnects networks with different network protocol technologies by performing the required protocol conversions.

6.1.2 Gateways (Software)

Gateways are even more complicated interconnection devices. It is one of the several types of communication servers. Not only can they interconnect LANs of different types, but also they can be used to connect LANs to networks that use completely different communication architectures or to centralized "host" computer systems. Such interconnections require complex protocol conversion operations. Therefore, a gateway is always a computer because only a computer program can provide such complex conversion services. Although the "gateway" is software, the term is sometimes used also to refer to the computer that is dedicated to the task. It may be a stand alone microcomputer with several NICs to connect different LAN types, or it may be a front-end processor attached to a mainframe computer. The computer can be used concurrently for purpose other than gateway protocol conversion. Gateways are often a major component in interconnecting backbone networks to form even larger, regional WANs.

6.1.3 Application Gateway

Also known as application proxy or application-level proxy, an application gateway is an application program that runs on a firewall system between two networks. When a client program establishes a connection to a destination service, it connects to an application gateway, or proxy. The client then negotiates with the proxy server in order to communicate with the destination service. In effect, the proxy establishes the connection with the destination behind the firewall and acts on behalf of the client, hiding and protecting individual computers on the network behind the firewall. This creates two connections: one between the client and the proxy server and between the proxy server and the destination. Once connected, the proxy makes all packet-forwarding decisions. Since all communication is conducted through the proxy server, computers behind the firewall are protected. While this is considered a highly secure method of firewall protection, application gateways require great memory and processor resources compared to other firewall technologies, such as stateful inspection.

6.1.4 Network Gateway

A network gateway is an internetworking system, a system that joins two networks together. A network gateway can be implemented completely in software, completely in hardware, or as a combination of the two. Depending on their implementation, network gateways can operate at any level of the OSI model from application protocols to low-level signaling.

Because a network gateway by definition appears at the edge of a network, related functionality like firewalling tends to be installed on the network gateway.

6.2 BRIDGES

Bridges are simple devices that connect LANs using exactly same network architecture. Bridges are similar to network hubs, but have one important difference. Hubs pass on all data messages; hubs never filter out any messages. Bridges, on the other hand, only pass on messages that are for nodes on other LANs. It does this "filtering" by maintaining a list of device addresses that are assigned to the network interface cards attached to the network. In that way, bridges can help reduce unnecessary network traffic.

In order to traverse a LAN, every packet is encapsulated in a LAN frame. Because bridges direct traffic by the MAC address, it doesn't mater what type of packet is carried in the data field of the LAN frame. Packets based on any number of different communication protocols can be passed transparently through a bridge. End stations, such as workstations, file servers, and printers, communicate directly and transparently across bridges, as if they are on a single LAN. The LAN frame that is generated by the originating station is automatically passed to the destination station regardless of whether there are bridges in between. The bridges dynamically learn the addresses of the end stations so that, unlike repeaters, they forward only those frames that are not addressed to a station on the same network segment.

Bridges have the same general characteristic as gateways, but connect networks that employ similar protocols and topologies.

Routers are similar to bridges in that they connect two similar networks. However, unlike bridges, this connection is done at level 3 of the OSI model the network level. Bridges establish connections at levels 1 and 2 and are less complex than routers.

A bridge enables users to partition a single local area network into multiple sub networks or to connect two or more networks to form one single logical LAN. When the number of stations on a LAN increases, the performance of that LAN decreases. This is specially true of LAN's with a CSMA/CD media access control. In such situations, it can be beneficial to partition the network with a bridge. By using a bridge, the performance of each LAN is increased, while each user still has access to the servers on both sub-networks.

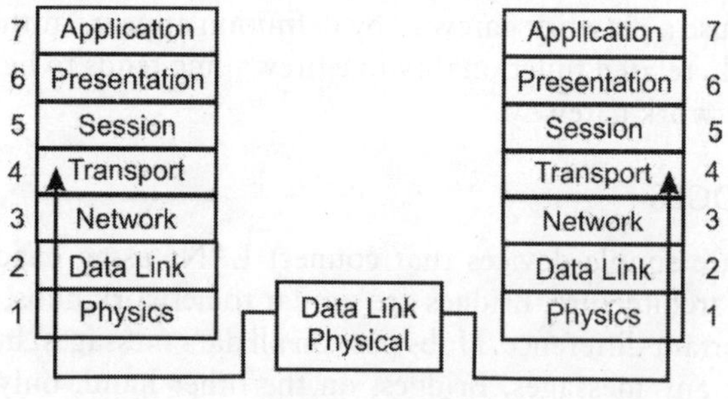

Fig. 3.24: Bridge

A bridge connects the LAN's at the Data Link Layer (OSI level 2). The unit of data at this level is called a frame, and the bridge uses the MAC addresses in the frame to decide which frames to forward to the other LAN. The bridge reads all the frames on the originating LAN, matches the address in the frame against a table stored in the bridge, and forwards the frame to the destination LAN. Any frame that is addressed to the originating LAN is ignored. The bridge works in both directions.

For a bridge to function, the LAN's connected must have the same Data Link Layer, *e.g.* IEEE 802.2. Bridges are often used to connect networks of different departments on to a corporate high speed backbone. Such a bridge arrangement has three advantages.

- It keeps local departmental traffic off the corporate backbone.
- It enables each station to see any other station that is allowed to see.
- It reduces the size of each network to increase performance and manageability.

Bridges are simpler and less expensive then routers. Bridges make a simple do/don't decision on which packets to send across two segments they connect. Filtering is done based on the destination address of the packet. If a packet's destination is a station on the same segment where it originated, it is not forwarded. If it is destined for a station on another LAN, it is connected to a different bridge port and forwarded to that port.

6.2.1 Types of Bridges

Bridges can be grouped into categories based on various product characteristics. Using one popular classification scheme, bridges are either local or remote. Local bridges provide a direct connection between multiple LAN segments in the same area. Remote bridges connect multiple LAN segments in different areas, usually over telecommunications lines. Following figure illustrates these two configurations.

Local and remote bridges connect LAN segments in specific areas.

6.3 BRIDGING AND ROUTING TECHNOLOGY

Bridges and routers are internetworking devices that connect Local Area Networks (LANs) to each other and to Wide Area Networks (WANs). The purpose of interconnecting LANs and WANs is to enable users on one network to share information and resources with users on other networks. By using the right combination of internetworking devices, it is possible to build large, unified networks that are enterprise-wide, or even global in scope. Although bridges and routers perform virtually the same function, they work in different ways. Bridges forward data from one network to another with no alteration of the original LAN frame. Because bridges do not process networking protocols, they are easy to install, can accommodate all protocols, and offer excellent throughput.

6.3.1 Routing

Routing is the process of sending messages through switches such that they reach their destinations in an efficient way. Routing is one of the major design considerations in wide-area switching networks. When a message sets out from, say, London to Glasgow, it should follow the shortest, fastest route. This will usually be the one that involves the fewest switching steps, unless it is necessary to avoid

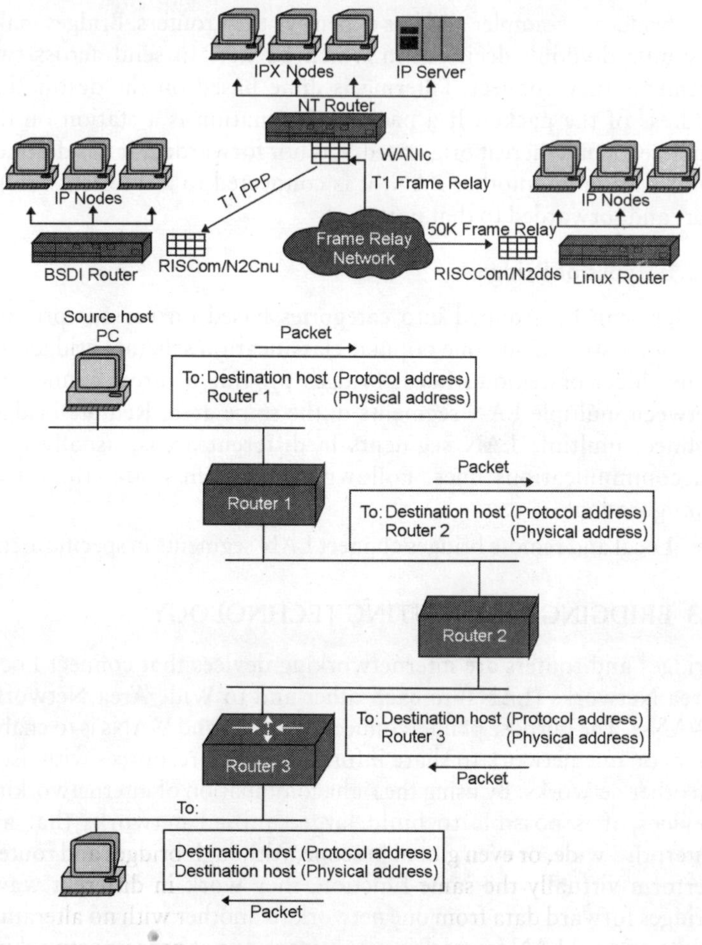

Fig. 3.25: The Routing

congestion in some part of the network. The decision on which route to follow is similar to the one that a car driver must make when planning a long journey. Some parts of the country are covered by motorways (high bandwidth) and some by smaller roads (lower bandwidth).

If one can avoid going through towns this will probably be an advantage, but it may be necessary to do so to avoid congestion on the highways. If the network has a static structure, and the load on various lines is reasonably constant, then routing can be fixed. Each message that is delivered from node A to node B will always follow the same route. In practice, networks change in structure and the loads at various points change from minute to minute. This means that adaptive routing is often to be preferred. This means that routing information is updated from time to time to take account of changing conditions on the network.

By comparison to bridges, routers direct traffic by the network address for each packet. When routers are used for internetworking, the ends stations communicate indirectly. A router receives only the packets that are addressed to it by an end station or another router. End stations communicate with a router, which communicates in turn with either the destination station or another router. This continues until the destination station is reached. Since it uses the network layer of each protocol to direct traffic, a router must have the appropriate routing software for every protocol it needs to route.

6.3.2 Routers

Routers are the devices that make the Internet possible. Routers are connected to two or more networks at crossing points. A router is a device that enables users to link together enterprise-wide networks. When they receive a packet of information, they decide which way they should send the packet to ensure that it gets to its destination quickly and efficiently. For example, say you want to send an e-mail to your friend. The e-mail message will first be broken up into small packets. Each packet will contain the IP Address of the destination computer. Then, your computer will ship all the packets to the nearest router, which will be connected to many networks. The router then decides which network it should send the packet to in order to best reach your friend's computer!

A router provides much more services than a bridge, because it operates at network layer 3 of the OSI model. This enables routers to intelligently manage network traffic. Routers maintain a map of the inter-network and are aware of conditions on the network, so that they can send packets along the most efficient path. They can also make use of redundant paths, which enables them to reroute traffic when some links go down. The provides the network with some fault tolerance.

Routers are more complex and more expensive than bridges. They use information within each packet to route it from one LAN to another, and communicate with each other and share information that allows them to determine the best route through a complex network of many LANs.

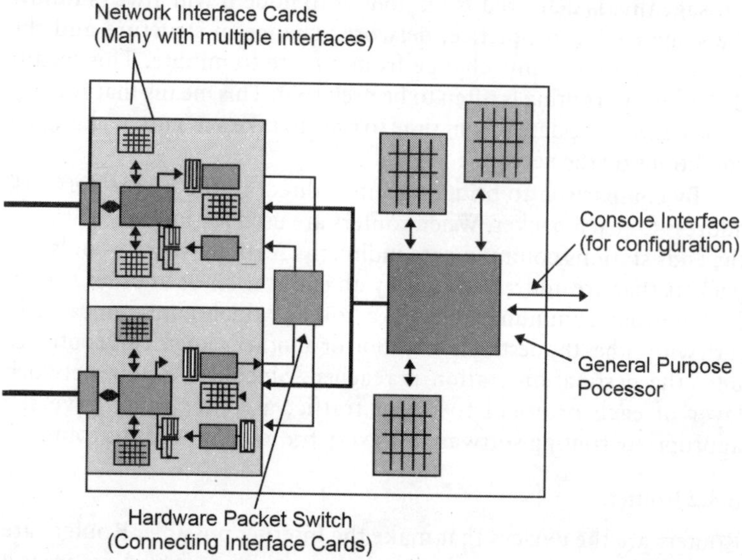

Fig. 3.26: Routers

Routers are dependent on the network layer protocol, therefor all networks connected to the routers must have the same network layer. A token ring network can be connected to an ethernet network, as long as they use the same network layer protocol like *e.g.* IPX or IP. There are however routers available that support multiple protocols.

Block diagram of a complete router showing the interfaces Interfaces to the network (over which packets are received or transmitted) are shown left, and the control interface (used to set up and reconfigure the router) is shown right.

Routers are used to connect LANs together that may be either the same or a different type of LAN architecture. Like bridges, they forward only messages that need to go to other networks. But, unlike bridges, they perform additional tasks. Routers may choose the best route to send a message by when several possible routes exist in a complex network topology. It does this by maintaining a routing table of to identify the optimal path to use in traversing the interconnected network. The routing table contains addresses. These addresses differ from the device addresses used by bridges that hook LANs together. They are called internetwork addresses and are written according to the protocols of the communication architecture such as TCP/IP, the Internet's communication architecture.

6.3.3 Advantages

The relative advantages of bridging and routing are outlined briefly below:

Advantages of Bridging

- Provides simpler installation and configuration
- Automatically locates users via 'self-learning' capabilities
- Handles all protocols
- Provides more flexibility
- Protects investment as needs change
- Does not require software upgrades for new releases of each protocol
- Uses automatic path optimisation
- IEEE 802.1 standard spanning Tree Protocol
- One path control algorithm for all protocols
- Provides automatic re-routing
- Results in lower cost of ownership
- Fewer staff needed
- Fewer software upgrades required

Advantages of Routing

- Provides simpler administration of network security
- Helps comply with TCP/IP address restrictions if using the Internet Network
- Uses WAN links more efficiently because they do not carry LANframing overhead
- Isolates broadcast traffic to individual LANs
- May reduce total overhead
- Useful for lower-speed WAN links
- Complements bridging in large networks

6.4 MODEM

The term stands for modulate and demodulate which changes the signal from digital, which computers use, to analog, which telephones use and then back again. A high speed connection also requires a modem but because the information is transferred digitally it isn't required to change the signal from digital to analog but is used to create the connection between your computer and the computer you are connecting with.

A modem is a device that interprets digital and analog signals, enabling data to be transmitted over voice-grade telephone lines. At the source, digital signals are converted to a form suitable for transmission over analog communication facilities. At the destination, these analog signals are returned to their digital form. Following figure illustrates a simple modem-to-modem connection through a WAN:

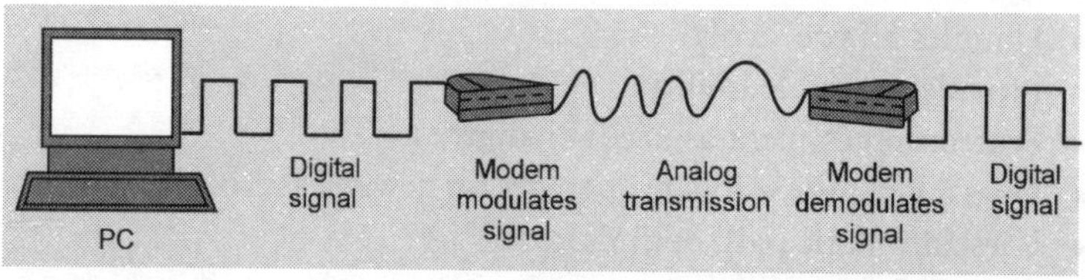

Fig. 3.27: A Modem Connection through a WAN Handles Analog and Digital Signals.

Modems also use Error Correction which corrects for transmission errors by constantly checking whether the information was received properly or not and compression which allows for faster data transfer rates. Information is transferred in packets. Each packet is checked for errors and is resent if there is an error.

Anyone who has used the Internet has noticed that at times the information travels at different speeds. Depending on the amount of information that is being transferred the information will arrive it's destination at different times. The amount of information that can travel through a line is limited. This limit is called bandwidth.

6.4.1 Concept of Modulation

Data is transmitted over communication channels in either analog or digital form. Analog transmission is used on traditional telephone lines: Messages are sent and received over telephone wires in the form of continuous electronic waves that follow the path of the wire. Digital transmission, by contrast, carries data as distinct "on" and "off" states, just as the computer does.

Analog transmission requires that data be converted from the digital form in which it is stored in the computer to analog form before transmission, in process called modulation. At the receiving end, data must then be demodulated converted from wave form to digital form—before the data is entered into the computer. Both modulation and demodulation are accomplished by devices called modems, a term derived from the words modulation and demodulation. Digital transmission requires no data conversion, so it reduces the time required to send messages and creates fewer transmission errors than analog transmission.

There are three basic types of modulation:

- Amplitude Modulation (AM) modifies the amplitude of the basic carrier,
- Frequency Modulation (FM) modifies the frequency of the basic carrier.
- Phase Modulation (PM) modifies the phase of the basic carrier frequency.

Because communication is usually a two-way process, modulation and its opposite, demodulation, can usually be carried out in the same piece of equipment, called a modem.

Thus modulation is process of changing the properties of some carrier signal, usually of approximately constant frequency, to represent a data stream. Modulation is a means by which data is transmitted from one place to another. There are many different ways in which this can be done.

6.4.2 Connecting the Modem

Three types of modems are used for preparing data for transmission over telephone lines:

- Internal
- External Direct connect
- Cable Modem

An internal modem installs like a sound card or video card, inside your PC. An external modem plugs in like a mouse or keyboard, into a "serial port".

A direct connect modem is attached directly to the computer and to the telephone lines. The user connects with another computer by typing the access numbers on the computer keyboard.

An acoustic coupler modem differs from a direct-connect modem is that it must be used with a standard contoured telephone receiver. The receiver of the telephone is placed on two cups built into the modem, thus linking the acoustic coupler with the telephone. An internal modem is built into or direct-connect modem.

Fig. 3.28: Three types of Modem

Direct connect modems and acoustic couplers can be used to link personal computers. Data are transmitted as a serial stream of bits using the ASCII code. An RS-232C serial interface connection is generally used to link the personal computers to the modems. In addition to the data bits, each character transmitted may also be accompanied by a "star" bit and a "stop" bit.

For PCs, modems are available as internal or external units. An internal modem is plugged into one of the PC's bus slots, while an external one plugs into a serial port.

Even if the modem is internal, it will usually still communicate with the PC using a serial process; this means it will have to be allocated a serial port number. Good choices are 3 and 4.

If you have a mouse connected to serial port 1 (often labelled 'COM1') then you must use a modem on port 4; if you have a mouse on COM2, then you must use a modem on COM3. This is to avoid the dreaded interrupt clash. If you want to want to use of modem for Internet access, you will need to make appropriate arrangements with an Internet service provider. The service provider will probably be able to offer advice about the type of modem to buy and how to set it up.

6.4.3 Modem Speeds

A modem's speed is measured in bits per second (bps), the number of 0s and 1s it can send or receive in one second. This is sometimes referred to as the modem's baud rate; for high-speed modems, baud rate and bps are essentially the same. You may also see measurements in kilobits per second (Kbps), where a kilobit is 1000 bits. The table below lists common modem speeds, and will help you pick the proper modem speed for you.

Modems are measured by the speed that the information is transferred. The measuring tool is called the baud rate. Originally modems worked at speeds below 2400 baud but today analog speeds of 56,000 are common. Cable, wireless or digital subscriber lines (DSL) modems can transfer information much faster with rates of 300,000 baud and up.

6.4.4 Internal or External Type Modems

An internal modem rests on a card which resides inside the computer's case; an external modem resides outside the case and is plugged into a serial port on the back of the computer case.

Speeds are essentially similar whether the modem lives inside or outside the case. Which type should you get? Well, if you're buying a new computer system, the decision is easy. Your new system will almost certainly include an internal modem at 28.8 or faster. Simply make sure the computer contains a modem that is fast enough to meet your needs. If you're installing a new modem in an existing system, there are advantages and disadvantages to each type. Internal modems are generally about $ 20 cheaper than their external counterparts, and they don't take up extra space on your desk. On the other hand, external modems don't require you to open your computer case (a scary operation for novices), and they are usually easier to configure than internal modems.

If opening the computer case scares you, get an external modem and don't worry about it. If you aren't afraid of opening the case... get an external modem anyway. I also recommend that you spend a few extra dollars on a well-known name brand rather than getting the cheap brand that happens to be in the bargain bin this week. Internal modems and cheap modems are more likely to have configuration problems (difficulties with COM and IRQ settings, which are not fun to resolve), even with Windows 95's "Plug and Play" interface. A name-brand external modem is most likely to be problem-free.

6.4.5 Cable Modems

A relatively new development is a device that provides high-speed Internet access via a cable TV network. With speeds of up to 36 Mbps, cable modems can download data in seconds that might take much longer with a dial-up connection. Because it works with your TV cable, it doesn't tie up a telephone line. Best of all, it's always on, so there is no need to connect—no more busy signals! This service is now available in some cities in the United States and Europe.

With most homes already wired by cable operators, cable modem is possibly the best (and the cheapest) solution for Internet access from home. The cable operator has to acquire a dedicated Internet connection from any of the Internet Service Providers (ISPs) and pump it down the cable through a Cable router. Cable operators can provide both upstream and downstream bandwidths up to 30 Mbps. All that a subscriber has to do is to plug in a cable modem at the other end of the pipe.

The download times in the table above are relative and are meant to give you a general idea of how long it would take to download different sized files at different connection speeds, under the best of circumstances. Many things can interfere with the speed of your file transfer. These can range from excessive line noise on your telephone line and the speed of the web server from which you are downloading files, to the number of other people who are simultaneously trying to access the same file or other files in the same directory.

But cable modem is not without its drawbacks. As cable TV was initially designed as a broadcast medium, it is set-up in a 'tree and branch topology. You have to share that 30 Mbps bandwidth with other cable modem subscribers. So if there are 10 other subscribers on your Internet cable network, your actual connect speed could be about 3 Mbps (depending upon what other subscribers are doing). Another cable modem disadvantage is that you have to do one thing at a time, watch cable TV or connect to the Internet.

6.4.6 ADSL Modem

ADSL or Asymmetric Digital Subscriber Line modem allows you high-speed connections over ordinary copper telephone wires by utilising vacant frequencies in the range of 0- to 2-MHz in the copper wires. With a down-load speed of 9 Mbps and an upload speed of 3.5 Mbps, ADSL modems are slower than cable modems. Connect speed is high as the bandwidth is not shared with other subscribers.

To be able to hook up to the Internet with ADSL, you will need an ADSL modem at your end while your ISP will need to have ADSL switch for you to connect to. However, one disadvantage of ADSL is that the further you are located from the ADSL switch, the weaker the modem's performance. And if you are beyond 3 km from the switch, you might as well look for other options to connect to the Net.

7.1 THE NETWORK PROTOCOLS

Your computer sends data to, or receives data from, a host computer over the Internet. A program such as Telnet breaks up the data into pakcets. Protocols, which are standards on which the computing community has agreed, specify how packets should be layered, or packaged, into even smaller packets. Different layers of packets address a variety of software and hardware needs in order to send information over different networks and communication links.

Many of the addressing, error-checking, retransmission, and acknowledgment services most commonly associated with networking take place at the Network and Transport OSI layers. Protocol suites are often referred to by the suite's Transport and Network protocols. In TCP/IP, for instance, TCP is a Transport layer protocol and IP is a Network layer protocol. (Note, however, that TCP/IP predates OSI and diverges from OSI in a number of ways.)

IPX/SPX is another protocol suite known by its Transport and Network layer protocols, but the order of the protocols is backward from the way the protocols are listed in TCP/IP. IPX is the Network layer protocol; SPX is the Transport layer protocol.

The lower Data Link and Physical layers provide a hardware specific foundation, addressing items such as the network adapter driver, the media access method, and the transmission medium. Transport and Network layer protocols such as TCP/IP and IPX/SPX rest on that Physical and Data Link layer foundation, and, with the help of the NDIS and ODI standards, multiple protocol stacks can operate simultaneously through a single network adapter.

Upper-level protocols provide compatibility with a particular networking environment. For instance, the so-called NetBIOS over TCP/IP stack provides Microsoft clients with TCP/IP.

There are a wide variety of actual protocols and protocol suites, such as TCP/IP, IPX/SPX, NetBEUI, AppleTalk, and DLC.

7.1.1 Common Internet Protocols

In addition to TCP/IP and IPX/SPX, some of the common Transport and Network layer protocols are the following:

- NWLink. Microsoft's version of the IPX/SPX protocol essentially spans the Transport and Network layers.
- NetBEUI. Designed for Microsoft networks, NetBEUI includes functions at the Network and Transport layers. NetBEUI isn't routable and therefore doesn't make full use of Network layer capabilities.
- AppleTalk Transaction Protocol (ATP) and Name Binding Protocol (NBP). ATP and NBP are AppleTalk.

Transport Layer Protocols

Datagram Delivery Protocol (DDP). DDP is the AppleTalk Network layer protocol.

- NWLink is Microsoft's version of IPX/SPX.
- Windows NT (like other Microsoft operating systems such as Windows for Workgroups and Windows 95) services client requests by using the Server Message Block (SMB) protocol. SMB is an Application layer protocol.

Three stages must take place before a protocol is operational:

- A model describes the general function of the protocol.
- The protocol is defined in complete detail.
- The protocol must be realized by software and hardware designers in real products.

Consider the process of designing a building. The architect first produces sketches that describe the general nature of the building. Then the architect, possibly working with a specialist in particular building trades, develops blueprints that describe every detail of the building. Finally, an actual building is constructed.

7.2 Networks and Native Interface Concepts

By default following network interfaces had been used worldwide:

- IPX/SPX for DOS based workstations
- TCP-IP for Unix and Window based workstations
- NETWARE for Clint-Server applications

IPX

IPX, which stands for the internetwork packet exchange, is the underlying protocol used by NetWare's file-redirection modules. It is a router with IPX routing can interconnect LANs, so that Novell NetWare clients and servers can communicate.

IPX is an adaptation of a protocol developed by Xerox Corporation called XNS-Xerox Network Standard. IPX corresponds to the Network layer of the OSI model; IPX performs addressing, routing, and switching to deliver a message (packet) to its destination. This protocol is speedler than the session-oriented SPX protocol. Although delivery is not guaranteed, Novel indicates that IPX packets are correctly received about 95 per cent of the time.

Netware's file-redirection modules use the IPX protocol (not the session-oriented SPX protocol) to send and receive file-service packets to and from the file server. This method is safe and reliable because every such request from a workstation requires a response from the file server. The file-redirection modules never assume that a file-service packet (to write data to a file, for example) has been processed by the file server until the proper acknowledgement response is returned.

If you use NetWare, you obviously already have IPX. Depending on the version of NetWare you have, you may have SPX as well.

Internet Connections Protocols: The TCP/IP

Computers on the Internet "talk" to each other by using protocols. The two protocols used by the Internet are Transmission Control Protocol (TCP) and Internet Protocol (IP), referred to as TCP/IP.

It is the general protocol suite of the Internet, encompassing protocols for network activities such as datagram delivery and acknowledgement, and protocols for user activity such as remote login (Telnet) and file transfer (FTP). A given implementation of TCP/IP typically contains a dozen or more protocols and facilities essential to the use of TCP/IP applications in a TCP/IP environment. It is a two layer predominant transport protocol, that provides a reliable connection-oriented, byte stream service.

The computer *sends data* to, or *receives data from*, a host computer over the Internet. A program such as Telnet breaks up the data into pakcets. Protocols, which are standards on which the computing community has agreed, specify how packets should be layered, or packaged, into even smaller packets. Different layers of packets address a variety of software and hardware needs in order to send information over different networks and communication links.

TCP breaks the application data from the end user down into TCP packets. Each packet has a header with the address of the host, information for putting the data back together, and information for making sure the packets don't get corrupted.

IP breaks TCP packets down even further. An IP packet has a header with address information and carries TCP information and data. IP packets are not very reliable, but the TCP level just keeps resending packets until the correct IP packets get through.This further uses Ethernet or FDDI link.

The subnets can break IP packets down and add their own specialized address information. An IP packet may go through many subnets before the packet reaches the host. If the end user and the host are not connected by subnets, one further protocol may come into play. SLIP (outgoing) and PPP (in coming) transmit pakcets over telephone lines, allowing dialup access to the Internet. The physical level transmits the actual signal along the various network. If you were able to look at the signal, all you'd see would be a series of pulses.

Once the data arrives at the host, each level gets unpacked, allowing the TCP level to reassemble the data into a format the host computer can use.

The Internet protocol suite (also commonly called the TCP/IP protocol suite) was originally developed by the United States Department of Defense (DoD) to provide robust service on large internet works that incorporate a variety of computer types. In recent years, the Internet protocols constitute the most popular network protocols currently in use.

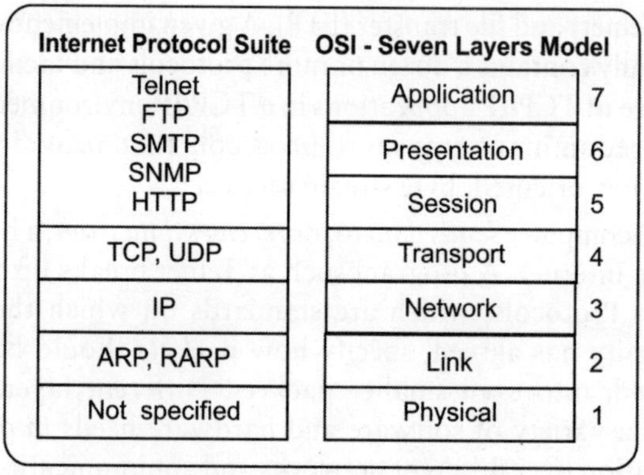

Fig. 3.29: Relationship of the Protocols in the Internet suite to the Layers of the OSI Reference Model

One reason for the popularity of TCP/IP is that no one vendor owns it, unlike the IPX/SPX, DNA, SNA, AppleTalk protocol suites, all of which are controlled by specific companies. TCP/IP evolved in response to input from a wide variety of industry sources. Consequently, TCP/IP is the most open of the protocol suites and is supported by the widest variety of vendors. Virtually every brand of computing equipment now supports TCP/IP.

Much of the popularity of the TCP/IP protocols comes from their early availability on Unix. The protocols were built into the Berkeley Standard Distribution (BSD) Unix implementation. Since then, TCP/IP has achieved universal acceptance in the Unix community and is a standard feature on all versions of Unix.

Following figure illustrates the relationship of the protocols in the Internet suite to the layers of the OSI reference model. Notice that the suite doesn't include protocols for the Data Link or Physical layers. TCP/IP was designed to work over established standards such as Ethernet. Over time, TCP/IP has been interfaced to the majority of Data Link and Physical layer technologies.

The Internet protocols do not map cleanly to the OSI reference model. The DoD model was, after all, developed long before the OSI model was defined. The model for the Internet protocol suite has four layers (refer to Figure). From this model, you can see the approximate relationships of the layers. The DoD model's layers function as follows:

- The Network Access layer corresponds to the bottom two layers of the OSI model. This correspondence enables the DoD protocols to coexist with existing Data Link and Physical layer standards.
- The Internet layer corresponds roughly to the OSI Network layer.
- Protocols at this layer move data between devices on networks.
- The Host-to-Host layer can be compared to the OSI Transport layer. Host-to-Host protocols enable peer communication between hosts on the internetwork. (At the time these protocols were designed, personal computers and workstations didn't exist, and all network computers were host computers. As a result, devices on TCP/IP networks are typically referred to as hosts. The concept of a client/server relationship didn't exist, and all communicating hosts were assumed to be peers.)
- The Process/Application layer embraces functions of the OSI Session, Presentation, and Application layers. Protocols at this layer provide network services.

One huge advantage of TCP/IP is that TCP/IP is required for communication over the Internet. One disadvantage is that the size of the protocol stack makes TCP/IP difficult to implement on some older machines. (Present-day PC models should have no problem running TCP/IP.) TCP/IP has traditionally been considered slower than other protocol stacks, but again, the power of the newer machines overcomes much of this difficulty.

A large number of protocols are associated with TCP/IP. Several of these are discussed briefly in the following sections.

TCP/IP is like netBIOS, IPX and SPX in several ways. TCP/IP stands for Transmission Control Protocol/Internet Protocol. The Department of Defense designed TCP/IP for ARPAMERT, a geographically large network (not a LAN) that connects the various sites of the US DoD advanced Research projects agency. TCP/IP is a layer of protocols and not a LAN operating system. IP provides datagram communications between nodes on a network (like Novell's IPX).

TCP is like Net BIOS in that it provides point-to-point, guaranteed-delivery communications between nodes. A set of fairly standard utilities exists for transferring files (FTP), doing simple remote program execution (TELNET), and sending electronic mail (SMTP) over TCP/IP networks. These utilities do not perform file redirection. You need a product such as Sun Microsystems' NFS (Network file system) or Locus Computing's PC Interface (sold by IBM under the name AIX for DOS users) to do file redirection on a TCP/IP-based network.

The Internet Protocol (IP) is a connection less protocol that provides datagram service, and IP packets are most commonly referred to as IP datagrams. IP is a packet-switching protocol that performs addressing and route selection. An IP header is appended to packets, which are transmitted as frames by lower-level protocols. IP routes packets through internet works by utilizing dynamic routing tables that are referenced at each hop. Routing determinations are made by consulting logical and physical network device information, as provided by the Address Resolution Protocol (ARP).

In the network layer protocol of TCP/IP, IP provides the functions needed "to deliver a 'package of bits' (Internet datagram) from source to destination over an interconnected system of networks" (*e.g.*, the Internet). Performance improvements are considerable. While routers typically handle 500,000 to 1 million packets per second, IP switching techniques can provide up to 10 times that amount.

IP provides all of Internet's data transport services. Every other Internet protocol is ultimately either layered atop IP, or used to support IP from below.

IP performs packet disassembly and reassembly as required by packet size limitations defined for the Data Link and Physical layers being implemented. IP also performs error checking on the header data using a checksum, although data from upper layers is not error-checked.

- **Addressing:** IP headers contain 32-bit addresses which identify the sending and receiving hosts. These addresses are used by intermediate routers to select a path through the network for the packet.
- **Fragmentation:** IP packets may be split, or fragmented, into smaller packets. This permits a large packet to travel across a network which can only handle smaller packets. IP fragments and reassembles packets transparently.
- **Packet timeouts:** Each IP packet contains a Time To Live (TTL) field, which is decremented every time a router handles the packet. If TTL reaches zero, the packet is discarded, preventing packets from running in circles forever and flooding a network.
- **Type of Service:** IP supports traffic prioritization by allowing packets to be labeled with an abstract type of service.
- **Options:** IP provides several optional features, allowing a packet's sender to set requirements on the path it takes through the network (source routing), trace the route a packet takes (record route), and label packets with security features.

All digital computers represent data as a collection of bits. A bit is the smallest possible unit of information. It can be in one of two states—off or on, 0 or 1. The meaning of the bit, which can represent almost anything, is unimportant at this point. The thing to remember is that all computer data - a text file on disk, a program in memory, a packet on a network - is ultimately a collection of bits.

If one bit has two different states, how many states do two bits have? The answer is four. Likewise, three bits have eight states. For example, if a computer display had eight colors available, and you wished to select one of these to draw a diagram in, three bits would be sufficient to represent this information. Each of the eight colors would be assigned to one of the three-bit combinations. Then, you could picking one of the colors by picking the right three-bit combination.

A common and convenient grouping of bits is the byte or octet, composed of eight bits. If two bits have four combinations, and three bits have eight combinations, how many combinations do eight bits have? If you don't want to write out all the possible byte patterns, just multiply eight two's together - one two for each bit. Two times two is four, so the number of combinations of two bits is four. Two times two times two is eight, so the number of combinations of three bits is eight. Do this eight times—or just compute two to the eighth power—and you discover that a byte has 256 possible states.

Obviously, if a byte has 256 possible states, its exact state can be represented by a number from 1 to 256. However, since zero is a very important number, a byte is more typically represented by a number from 0 to 255. This is very common, and with bit pattern 00000000 representing zero, and bit pattern 11111111 representing 255. The numbers matching these two patterns, and everything in between, can be computed by assigning a weight to each bit, multiplying each bit's value (0 or 1) by its weight, and then adding the totals. For example, here's how 217 is represented as 11011001 in binary.

The most common bit patterns in Internet engineering are those with a string of one bits, followed by a string of zero bits. Here are all such bytes, along with their decimal representation, computed just like the example using 217.

4
UNIT

Internet: Basic Features and Tools

1.1 NETWORKING

Historically the internet is made up of millions of computers linked together around the world in such a way that information can be sent from any computer to any other 24 hours a day. These computers can be in homes, schools, universities, government departments, or businesses small and large. They can be any type of computer and be single personal computers or workstations on a school or a company network. The Internet is often described as 'a network of networks' because all the smaller networks of organisations are linked together into the one giant network called the Internet. All computers are pretty much equal once connected to the Internet, the only difference will be the speed of the connection which is dependent on your Internet Service Provider and your own modem.

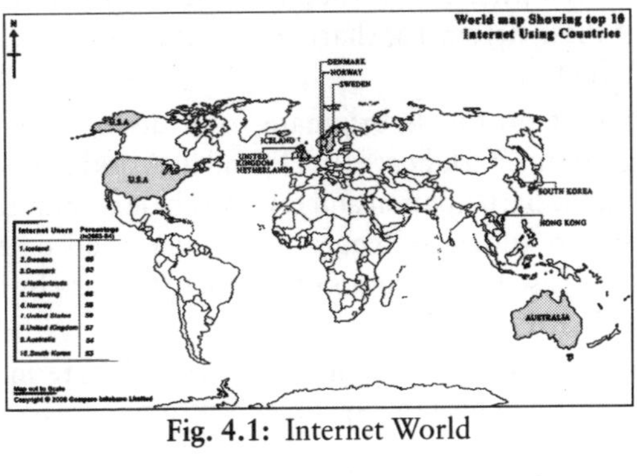

Fig. 4.1: Internet World

Thus the Internet, or simply the Net, is the publicly accessible worldwide system of interconnected computer networks that transmit data by packet switching using a standardized Internet Protocol (IP) and many other protocols. It is made up of thousands of smaller commercial, academic, domestic and government networks. It carries various information and services, such as electronic mail, online chat and the interlinked web pages and other documents of the World Wide Web.

Apart from the incredibly complex physical connections that make up its infrastructure, the Internet is held together by bilateral or multilateral commercial contracts and by technical specifications or protocols that describe how to exchange data over the network.

All computers on the Internet communicate with one another using the Transmission Control Protocol/Internet Protocol suite, abbreviated to TCP/IP. Computers on the Internet use a client/server architecture. This means that the remote server machine provides files and services to the user's local client machine. Software can be installed on a client computer to take advantage of the latest access technology.

An Internet user has access to a wide variety of services: electronic mail, file transfer, vast information resources, interest group membership, interactive collaboration, multimedia displays, real-time broadcasting, shopping opportunities, breaking news, and much more.

Unlike older communication systems, the Internet protocol suite was deliberately designed to be agnostic with regard to the underlying physical medium. Any communications network, wired or wireless, that can carry two-way digital data can carry Internet traffic. Thus, Internet packets flow through wired networks like copper wire, coaxial cable, and fiber optic; and through wireless networks like Wi-Fi. Together, all these networks, sharing the same high-level protocols, form the Internet.

The Internet protocols originate from discussions within the Internet Engineering Task Force (IETF) and its working groups, which are open to public participation and review. These committees produce documents that are known as Request for Comments documents (RFCs). Some RFCs are raised to the status of Internet Standard by the Internet Architecture Board (IAB).

"Cyberspace" is another name for Internet. This term was coined by William Gibson in his fantasy novel Neuromancer to describe the "world" of computers, and the society that gathers around them. Gibson's fantasy of a world of connected computers has moved into a present reality in the form of the Internet. In cyberspace people "exist" in the ether—you meet them electronically, in a disembodied, faceless form.

1.2 THE INTERNET AND THE WORLD WIDE WEB

Sometimes people use the words Internet and World Wide Web (WWW) synonymously but they are different. The WWW is a component of the Internet that presents information in a graphical interface. You can think of the WWW as the illustrated version of the Internet. It began in the late 1980's when physicist Dr. Berners-Lee wrote a small computer program for his own personal use. This program allowed pages, within his computer, to be linked together using keywords. It soon became possible to link documents in different computers, as long as they were connected to the Internet. The document formatting language used to link documents is called HTML (Hypertext Markup Language.)

The Web remained primarily text-based until 1992. Two events occurred that year that forever changed the way the Web looked. Marc Andreesen developed a new computer program called the NCSA Mosaic and gave it away! The NCSA Mosaic was the first Web browser.

The browser made it easier to access the different Web sites that had started to appear. Soon Web sites contained more than just text, they also had sound and video files. The development of the WWW has been the catalyst for the popularity of the Internet and is also the easiest part of the internet to use. We now have Internet Chat, Discussion Groups, Internet Phone capabilities, Video conferencing, News Groups, Interactive Multimedia, Games and so much more.

1.3 INTERNET REQUIREMENTS

To access the Internet by computer, you need a computer, a modem or other telecommunications link, and software to connect to an Internet Service Provider.

Items Needed to Connect

Windows 98 and Windows NT and XP have built-in support that enables you to connect to an ISP. You need a modem and a PPP (Point-to-Point Protocol) or SLIP (Serial Line Internet Protocol) account with your provider. The provider may also require the following:

- User name
- Password
- Local access phone number
- Host and domain name
- DNS server IP address
- Authentication technique

You will need to enter some or all of this information into your system's Dial-up Networking configuration to connect successfully. If necessary, your ISP can supply an IP address and what called an IP subnet mask. (These are sets of numbers that uniquely identify your connection to the Internet. These items usually are optional. You should only need them if your provider requires a dedicated IP address each time you connect to the Internet.)

From your with its Web and the special Internet program, TCP/IP, you get connected to your local Internet service provider. The ISP has a "point-of-presence" (POP) on the Internet with the ability to understand and contribute to Internet routing.

You also need to think about a login name and a password. Your login name will be part of your address that people type to send e-mail messages to you. It is usually anywhere from 2–12 characters long and it will have to be unique to your provider. For instance, my login name is vkjain1944. Well, I know there are lots of vkjain out there in this world, but probably there can only be one whose birth year is 1944. A lot of people use their first name initial and then their last name or maybe a nickname. Just remember, that whatever you use, it identifies you. So, you wouldn't want to use anything that characterizes you in a bad way, like stupid.

Your password is a secret code to type in that allows you to get on the Internet or to check your e-mail. This should be known only by you so no one else can get on the Internet using your name or read your private e-mail. Usually, your password is at least 6 characters long and is case-sensitive. In other words, hema1978 and Hema1978 are not the same. I think it's best to use all lower-case so you don't have to remember which characters you capitalized!

So, to get on the Internet, just call a local provider, get their software, and let them walk you through setting your system up. It's as easy as that. If you use Windows 95 or onward, you already have all the software you need. You just need to set up your system with their unique numbers and then your set!

No company can do without a proper Internet connection. Companies in India are using different modes of Internet connectivity, including dial-up, DSL and leased lines. Many companies have also taken connections from different vendors in order to maintain round-the-clock uptime and to prevent any business losses due to even a few minutes of downtime.

1.4 INTERNET SERVICE PROVIDER (ISP)

An Internet Service Provider (ISP) is a business or organization that offers users access to the Internet and related services. Most telecommunication operators are ISPs. They provide services such as Internet transit, domain name registration and hosting, dial-up access, leased line access and colocation.

Small Internet Service Providers (ISPs) hook into regional ISPs, which themselves link into major backbones that traverse the U.S. connecting major metropolitan areas. This diagram shows what a typical national backbone might look like as well as a county and state provider. While local ISPs may offer services only within their county, regional providers often span state lines.

In order to get online, you'll need to contact with an Internet service provider (ISP) to deliver an Internet connection through your modem. With a good ISP, getting online is a breeze, with customized software, helpful tech support, and immediate modem connections. On the other hand, if you hook up with a bad ISP, you can look forward to busy signals, tech support that is not available when you need it, censorship of what you read and write, and excessive fees. This guide will help you spot excellent ISPs.

An Internet Service Provider (ISP) charges users a fee (usually monthly or hourly) and, in return, provides them with a phone number and computer account which will enable them to connect to the Internet.

1.5 INTERNET CONNECTIVITY OPTIONS

Generally, an ISP charges a monthly access fee to the consumer. The consumer then has access to the Internet for an unlimited number of hours, although the speed at which this data is transferred varies widely.

Internet connection speed can generally be divided into two categories:

- Dialup and
- Broadband.

Dialup connections range from free to inexpensive and require the use of a phone line. Broadband connections can be either ISDN, wireless, cable, DSL, Satellite or Ethernet. Broadband is faster, always on, and usually more expensive.

1.6 DIALUP SERVICES

Dialed service serves a different function. It is used as a regular telephone might be. Users dial a specific telephone number to gain access to their computers. Users must for access to the computer, as only a specific number of input lines exist. Once these lines are filled, users are denied access to the system (usually a busy signal indicates all lines are in use). Users are charged message units only for the time used, with rates varying according to distance, day, time of usage, and the telephone services involved. Wide Area Telephone Service, or the WATS line, is a commonly used dialed service.

Dialed service is more flexible than leased lines, as you use a public phone to gain access to your computer. This is often an inexpensive way of accessing a computer from one's home or office. Charges are based on the length and number of calles. Transmission speeds of dialed services are not as high as leased lines. This may present some difficulties with graphics software, as higher transmission speeds (normally associated with leased lines) are needed to properly temecommunicate pictorial displays. For conventional processing activities, however, dialed services provide excellent support. Most users find that the flexibility of dialed service lines more than offsets the frustration of being occasionally denied access to the computer. For many users, it also has costs benefits.

Dial-up access is an inexpensive but relatively slow form of Internet access in which the client uses a modem to dial the internet service provider's (ISP) node, a dialup server type such as the Point-to-Point Protocol and TCP/IP protocols to establish a modem-to-modem link, which is then routed to the Internet. It is now regarded as legacy technology given the advent of widely available broadband Internet access in the Western world.

Dial-up requires no additional infrastructure on top of the telephone network. As telephone points are available throughout the world, dial-up remains useful to travellers. Dial-up is attractive in rural or remote areas where getting a broadband connection is impossible or expensive. Sometimes, dial-up access may also be an alternative to people who have limited budgets, though broadband is now increasingly available at lower prices because of competition.

Dial-up requires time to establish a telephone connection and perform handshaking before data transfers can take place, potentially a source of frustration. In locales with telephone connection charges, each connection incurs an incremental cost. If calls are time-charged, the duration of the connection incurs costs.

Dial-up access is a transient connection, because either the user or the ISP terminates the connection. Internet service providers will often set a limit on connection durations to prevent hogging of access, and will disconnect the user—requiring reconnection and the costs and delays associated with that.

Dial-up modems typically have a maximum theoretical speed of *56 kbit/*s (using the V92 protocol), although in most cases only up to *53 kbit/*s is possible due to overhead. Also, these speeds are the maximum possible; in almost all cases transfer speeds will be lower, averaging about *10 kbit/*s. Line noise further reduces achieved transfer rates.

Dial-up connections usually have high latency that can be as high as 400 ms or even more, which can make online gaming or video conferencing difficult, if not impossible.

Broadband Internet access (mostly via cable and ADSL) have been replacing dial-up connections in the last five years. The reason for this replacement is mostly because broadband connections usually have speeds which far exceed the capacity of dial-up, in some cases up to 15,360 *kbit/s*. An increasing amount of Internet content such as Macromedia Flash, online gaming and streaming media require large amounts of bandwidth. Many computer games released in 2005 (such as Battlefield 2 or Star Wars Battlefront) are not compatible for online play with dial-up modems. It is likely that this trend will continue into the future.

A significant hardware component in dialed service operations is the modem. It allows electronic signals to be converted into sound waves and vice-versa, so that data can be transmitted over telephone lines. One modem is positioned at the user's end to first convert impulses as they enter the telephone line, with a second modem at the other end of the line to reconvert data for the computer's use. The first conversion is called modulation, with the reconversion called demodulation.

Dial-up connections range from free to inexpensive and require the use of a phone line. Broadband connections can be either ISDN, wireless, cable, DSL, Satellite or Ethernet. Broadband is faster, always on, and sometimes more expensive.

Very few companies in India are using dial-up connectivity these days due to its highly unreliable nature and slow speed. However, some companies are still using dial-up connectivity mainly for Internet and e-mailing purposes. Many of them started off with it when they had less staff and limited areas of operation.

1.7 LEASED LINE CONNECTIVITY

A leased line is a (usually) symmetric telecommunications line connecting two locations together. Unlike traditional PSTN lines they do not have a telephone number, each side of the line being permanently connected to the other. They can be used for telephone, data or Internet services.

Leased or private lines are telephone lines strung between users and their online systems. The lines are dedicated to the user and are exclusively available on a 24-hour basis. Usually, a flat monthly fee is chargerd for this service. No message units are charged for the time the leased lines are used. Users can stay online as long as the processing session requires, once logged onto the system.

Leased lines give the user other advantages. They can be modified to handle data at speeds higher than possible with conventional telephone lines. Remember, the lines are used only to link users to their computer and nothing else. The upgraded line handles data at a higher rate and eliminates the need to convert data during its transmission to the computer. Leased lines are ideal for users who must have a permanent and exclusive link to their online systems.

In India, leased lines are usually available at speeds of *64k*, *128k*, *256k*, *512k*, *2Mb* and provided to the customer on X.21 presentation. Higher speeds are available on alternative interfaces.

In the US, leased lines are usually presented on a T1 bearer circuit in a number 56k or 64k timeslots. This has advantages over the UK—more than one logical connection can be provided on a single bearer, and upgrades can take place relatively easily. However, the customer must manage their own network termination equipment-Data Service Unit or Channel Service Unit (CSU/DSU).

Leased line connectivity is the most preferred mode of Internet connectivity used by Indian companies. Many have found it to be very reliable and fast. In bigger companies and corporations there is permanent demand for Internet usage and the leased line is always available and provides the required throughput.

Many companies are using Virtual Private Networks, IP telephony and running their Intranets over a leased line. Others use a leased line to effectively connect remote offices.

High volume data transfer is possible over a leased line, which in turn helps in cutting down on bandwidth costs. Besides this, many companies are using VoIP using a leased line within a legal framework. An Internet, which helps a company in maintaining connectivity with its mobile workforce, is also being used widely by companies through a leased line.

1.8 INTEGRATED SERVICES DIGITAL NETWORK (ISDN)

ISDN is a type of circuit switched telephone network system, designed to allow digital (as opposed to analog) transmission of voice and data over ordinary telephone copper wires, resulting in better quality and higher speeds, than available with analog systems. More broadly, ISDN is a set of protocols for establishing and breaking circuit switched connections, and for advanced call features for the end-user.

There are two points of view into the ISDN world. The most common viewpoint is that of the end user who wants to get a digital connection into the telephone/data network from home, whose performance would be better than an ordinary analog modem connection. The typical end-user's connection to the Internet is related to this point of view, and talk about the merits of various ISDN modems, carriers' offerings and tarriffing (features, pricing) are from this perspective. Much of the following discussion is from this point of view, but it should be noted that as a data connection service, ISDN has been mostly superseded by DSL. Because most incumbent telcos were deeply reluctant to roll out end-user digital service at reasonable prices until the cable modem prodded many of them to provide DSL.

There is however a second viewpoint: that of the telephone industry, where ISDN is not a dead issue. A telephone network can be thought of as a collection of wires strung between switching systems. The common electrical specification for the signals on these wires is T1 or E1. On a normal T1, the signalling is done with A&B bits to indicate on or off hook conditions and MF and DTMF tones to encode the destination number. ISDN is much better than this as messages can be sent much more quickly than by trying to encode numbers as long (100 ms per digit) tone sequences. This translated to much faster call set-up times which is greatly desired by carriers who have to pay for line time and also by callers who become impatient while their call hops from switch to switch.

It is also used as a smart network technology intended to add new services to the public switched telephone network (the PSTN) by giving users direct access to end-to-end circuit-switched digital services.

ISDN has never gained popularity as a telephone network in the United States and today remains a niche product. It is still commonly used in recording studios, when a voice-over actor is in one studio, but the director and producer are in a studio at another location. ISDN is used because of its "guaranteed" real-time, not-over-the-Internet service, and its superior audio fidelity as compared to POTS service.

In Japan, it became popular to some extent from around 1999 to 2001, but now that ADSL has been introduced, the number of subscribers is in decline. NTT, a dominant Japanese telephone company, provides an ISDN service with the names INS64 and INS1500, which are much less recognized than ISDN.

In the UK, British Telecom (BT) provides Home Highway and Business Highway, which are BRI ISDN services which offer connection from analog devices (such as normal phones) as well as ISDN devices (such as PCs equipped with terminal adapters). Home Highway has been bought by many home users, usually for Internet connection. Although not as fast as ADSL, it was available before ADSL, and in places where ADSL does not reach. BT also offers PRI ISDN.

In France, France Tlcom offers ISDN services under their product name Numeris (2 B+D) of which a professional Duo and home Itoo version is available. ISDN is generally known as RNIS in France and has widespread availability. The introduction of ADSL is reducing ISDN use for data transfer and Internet access, although it is still common in more rural and outlying areas.

In Germany, ISDN is very popular with an installed base of 25 mio. channels (29% of all subscriber lines in Germany as of 2003 and 20% of all ISDN channels worldwide). Due to the success of ISDN, the number of installed analog lines is decreasing. Deutsche Telekom (DTAG) offers both BRI and PRI. Competing phone companies often offer ISDN only and no analog lines.

In ISDN, there are two types of channels, B and D:

B channels are used for data: This refers both to voice and data information, and D channels are intended for signalling and control (but can also be used for data). B stands for Bearer and D stands for Delta.

There are two kinds of access to ISDN:

- **Basic Rate Interface (BRI):** Consisting of two B channels, each with bandwidth of 64 kbit/s, and one D channel with a bandwidth of 16 kbit/s. Together these three channels can be designated as 2B+D, and,
- **Primary Rate Interface (PRI):** Containing a greater number of B channels and a D channel with a bandwidth of 64 kbit/s. The number of B channels varies based on the country.

1.8.1 Types of Communications Handled

Amongst the kinds of data that can be moved over the *64 kbit/s* channels are pulse-code modulated voice calls, providing access to the traditional voice PSTN. This information can be passed between the network and the user end-point at call set-up time. In North America, ISDN is nowadays mostly used as an alternative to analog connection, most commonly for Internet access. Some of the services envisaged as being delivered over ISDN are now delivered over the Internet instead. In Europe, and in Germany in particular, ISDN has been successfully marketed as a phone with features, as opposed to a POTS phone (Plain Old Telephone Service) with little to no features. In North America, POTS phones can have all manner of advanced features (Three-Way Call, Call Forwarding, Caller ID, etc.), but in Germany, these services are usually only available to ISDN subscribers.

Where an analog connection would require a modem an ISDN connection requires a terminal adapter (TA).

1.9 CABLE MODEM (BROADBAND)

A cable modem is a special type of modem that is designed to modulate a data signal over cable television infrastructure. The term cable Internet access (or simply cable) refers to the delivery of Internet service over this infrastucture.

Cable modems should not be confused with older LAN systems such as 10base2 or 10base5 that used coaxial cables - and especially not with 10broad36, which actually utilized the same sort of cable as CATV systems.

1.9.1 Cable Modem for Broadband Internet Access

Cable modems are primarily used to deliver broadband Internet access, taking advantage of unused bandwidth on a cable television network.

Along with digital subscriber line technology, cable modems ushered in the age of broadband Internet access in developed countries. Prior to the availability of such systems, Internet access involved slow dial-up access over a public switched telephone network.

Users in a neighborhood share the available bandwidth provided by a single coaxial cable line. Therefore, connection speed can vary depending on how many people are using the service at the same time.

Often the idea of a shared line is seen as a weak point of cable Internet access. From a technical point of view, all networks, including DSL services, are sharing a fixed amount of bandwidth between a multitude of users—but because cable networks tend to be spread over larger areas than DSL services, more care must be taken to ensure good network performance.

A more significant weakness of cable networks using a shared line is the risk of loss of privacy, especially considering the availability of hacking tools for cable modems. This issue is addressed by encryption and other privacy features specified in the DOCSIS (Data Over Cable Service Interface Specification) standard used by most cable modems.

1.10 SATELLITE INTERNET

Satellite internet services are used in locations where terrestrial Internet access is not available and in locations which move frequently. Internet access via satellite is available globally, including vessels at sea. There are three types of satellite Internet service:

- one-way multicast,
- one-way with terrestrial return, and
- two-way satellite access.

Microwave

Microwaves have wavelengths approximately in the range of 30 cm (frequency = 1 GHz) to 1 mm (300 GHz). However, the boundaries between far infrared light, microwaves, and ultra-high-frequency radio waves are fairly arbitrary and are used variously between different fields of study. The existence of electromagnetic waves, of which microwaves are part of the higher frequency spectrum, was predicted by James Clerk Maxwell in 1864 from his famous Maxwell's equations. In 1888, Heinrich Hertz was the first to demonstrate the existence of electromagnetic waves by building apparatus to produce radio waves.

The microwave range include ultra-high frequency (UHF) (0.3-3 GHz), super high frequency (SHF) (3-30 GHz), and extremely high frequency (EHF) (30-300 GHz) signals.

Microwaves are used in broadcasting transmissions because microwaves pass easily through the earth's atmosphere with less interference than longer wavelengths. There is also much more bandwidth in the microwave spectrum than in the rest of the radio spectrum. Typically, microwaves are used in television news to transmit a signal from a remote location to a television station from a specially equipped van.

Radar also uses microwave radiation to detect the range, speed, and other characteristics of remote objects.

Wireless LAN protocols, such as Bluetooth and the IEEE 802.11g and b specifications, also use microwaves in the 2.4 GHz ISM band, although 802.11a uses an ISM band in the 5 GHz range. Licensed long-range (up to about 25 km) Wireless Internet Access services can be found in many countries (but not the USA) in the 3.5-4.0 GHz range.

Metropolitan Area Networks - MAN protocols, such as WiMAX (Worldwide Interoperability for Microwave Access) based in the IEEE 802.16 specification. The IEEE 802.16 specification was designed to operate between 2 to 11 GHz. The commercial implementations are in the 2.5GHz, 3.5Ghz and 5.8GHz ranges.

Cable TV and Internet access on coax cable as well as broadcast television use some of the lower microwave frequencies. Some cellphone networks also use the lower microwave frequencies.

Microwave Frequency Bands

The microwave spectrum is usually defined as electromagnetic energy ranging from approximately 1 GHz to 1000 GHz in frequency, but older usage includes lower frequencies. Most common applications are within the 1 to 40 GHz range. Microwave Frequency Bands are defined in the table below:

Table 4.1: Microwave Frequency Bands

Designation	*Frequency Range*
L band	1 to 2 GHz
S band	2 to 4 GHz
C band	4 to 8 GHz
X band	8 to 12 GHz
Ku band	12 to 18 GHz
K band	18 to 26 GHz
Ka band	26 to 40 GHz
Q band	30 to 50 GHz
U band	40 to 60 GHz
V band	50 to 75 GHz
E band	60 to 90 GHz
W band	75 to 110 GHz
F band	90 to 140 GHz
D band	110 to 170 GHz

The above table reflects Radio Society of Great Britain (RSGB) usage. The term P band is sometimes used for UHF frequencies below L-band.

1.11 DIGITAL SUBSCRIBER LINE (DSL)

Digital Subscriber Line, or DSL, is a family of technologies that provide a digital connection over the copper wires of the local telephone network. Its origin dates back to 1988, when an engineer at Bell Labs devised a way to carry a digital signal over the unused frequency spectrum. This allows an ordinary phone line to provide digital communication without blocking access to voice services. Bell's management, however, were not enthusiastic about it, since it was not as profitable as renting out a second phone line for consumers who preferred to have access to the phone when dialing out. This changed in the late 1990s when cable companies started marketing broadband Internet access. Realising that most consumers would prefer broadband Internet to a second dial out line, Bell companies rushed out the DSL technology that they had been sitting on for the past decade as an attempt to slow broadband Internet access uptake, to win market share against the cable companies.

As of 2005, DSL provides the principal competition to cable modems for providing high speed Internet access to home consumers in Europe and North America; although on average, cable is much faster than DSL in most commercial situations. Older ADSL standards could deliver 8 Mbit/s over about one mile (2 km) of copper wire. The latest standard ADSL2+ can deliver over 20 Mbit/s per user over similar distances. However many copper lines are longer than one mile (2 km) reducing the amount of bandwidth that can be transmitted. Modern cable systems, on the other hand, can provide 30 Mbit/s downstream, but this bandwidth is shared between all the users on the cable segment (which could be 100 to 200 households).

The Public Switched Telephone Network was initially designed to carry POTS calls, as the concept of data communications as we know it today did not exist. For reasons of economy, the system nominally passes audio between 300 and 3,400 Hz, which is regarded as the range required for human speech to be clearly intelligible. This is known as commercial bandwidth. Dial-up services using modems are constrained by the POTS channel's Shannon capacity, which indicates the maximum data rate which can be supported by a given amount of bandwidth.

At the Local Exchange (UK terminology) or Central Office (US terminology) the speech is generally digitised into a 64 kbit/s data stream in the form of an 8 bit signal using a sampling rate of 8,000 Hz, therefore any signal above 4,000 Hz is not passed by the phone network (and has to be blocked by a filter to prevent aliasing effects). The bandwidth between the commercial bandwidth limit and the channel limit can be utilised in a fully digital end to end connection to achieve a full 64 kbit/s on an ISDN line.

The local loop connecting the central office to most subscribers is capable of carrying frequencies well beyond the 3.5 kHz upper limit of POTS. Depending on the length and quality of the loop, the upper limit can be as high as the tens of megahertz. DSL takes advantage of this unused part of the circuit by creating 4312.5 Hz wide channels starting between 10 and 100 kHz, depending on how the system is configured. Each channel is evaluated for usability in much the same way an analog modem would on a POTS connection. Allocation of channels continues at higher and higher frequencies (up to 1.1 MHz for ADSL) until new channels are deemed unusable. More usable channels equates to more available bandwidth, which is why distance and line quality are a factor. The pool of usable channels is then split into two groups for upstream and downstream traffic based on a preconfigured ratio. Once the channel groups have been established, the individual channels are bonded into a pair of virtual circuits, one in each direction. Like analog modems, DSL transceivers constantly monitor the quality of each channel and will add or remove them from service depending on whether or not they are usable.

The commercial success of DSL and similar technologies largely reflects the fact that in recent decades, while microchips and disk drives have been getting faster and cheaper, the cost of digging holes in the ground remains very high. All flavors of DSL employ very complex digital signal processing algorithms to overcome the inherent limitations of existing POTS wires. Not long ago, the cost of such signal-processing power would have been prohibitive, but today the cost of installing a DSLAM at one end, and a DSL modem at the other end, of an existing local loop is orders of magnitude less than would be the cost of installing a fiber optic cable over its route.

Most residential and small-office DSL implementations reserve low frequencies for POTS service, so that with suitable filters and/or splitters the existing voice service continues to operate independent of the DSL service. Thus POTS-based communications, including FAX machines, can share the wires with DSL. However, in most cases only one DSL modem can use a local loop at a time; it is generally not possible for a customer to have multiple DSL connections. As of 2005, the standard way to let multiple computers share a broadband connection is to purchase an inexpensive router that establishes a local Ethernet or Wi-Fi network on the customer's premises.

Once upstream and downstream channels are established, they are used to connect the subscriber to a service such as Internet access.

Equipment

The subscriber end of the connection consists of a DSL modem. This converts data from the digital electronic pulses used by computers into a digital audio stream of a suitable frequency range for the particular DSL variant in use.

In addition the subscriber may need to install a passive electronic filter (known variously as a "filter", "micro-filter" or a "splitter") if using the POTS service on the same line. This ensures that the DSL modem and the telephone only receive the frequencies they are designed to handle. Subscribers can plug a filter into an existing telephone socket when using a "wires-only" service, or alternatively the DSL provider may install it. Some POTS devices, such as "tapeless" digital answering machines, are especially sensitive to small amounts of high-frequency signals leaking across the simple passive filters provided in the installation kit from the DSL supplier; the customer may therefore need to purchase higher-quality "active" filters from a third-party supplier and/or move some POTS devices to a room farther away from the DSL modem.

In the early days of DSL, installation required a technician to visit the premises. One splitter was installed near the point where the phone line entered the premises, from which a dedicated data line was installed. Today, many DSL vendors offer a self-install option, in which they ship equipment and instructions to the customer. In this case, since no changes are made to the cable plant on the customer premises, all the phone wires are carrying both POTS and DSL signal frequencies; therefore the customer generally needs to plug a splitter into each telephone outlet. As of 2005, establishing new cable modem or satellite broadband service generally does require a visit by a technician to the premises, even when there is existing cable television service to this customer; this constitutes one of the major competitive advantages of DSL over cable broadband service.

At the exchange a digital subscriber line access multiplexer (DSLAM) terminates the DSL circuits and aggregates them, where they are handed off onto other networking transports. It also separates out the voice component.

2.1 E-MAIL

Electronic mail, or e-mail, allows computer users locally and worldwide to exchange messages. Each user of e-mail has a mailbox address to which messages are sent. Messages sent through e-mail can arrive within a matter of seconds.

E-mail is a facility that enables computer users to send messages and files to each other via the Internet. E-mail can be used regardless of the types of computer being used, as long as each user has a modem (the hardware needed to connect to the Internet) and an account with an Internet Service Provider (ISP).

E-mail is fast becoming an integral part of business, as it is a highly effective messaging service that can quickly distribute information to many people, enabling documents, spreadsheets and other attachments to be worked with by your business associates. It has already developed its own culture ("netiquette"), and there is an art to using it efficiently in the workplace. It is often informal and is a useful additional form of communication but shouldn't be allowed to completely replace traditional and valuable forms of communication, such as telephone and face-to-face contact.

A powerful aspect of e-mail is the option to send electronic files to a person's e-mail address. Non-ASCII files, known as binary files, may be attached to e-mail messages. These files are referred to as MIME attachments MIME stands for Multimedia Internet Mail Extension, and was developed to help e-mail software handle a variety of file types. For example, a document created in Microsoft Word can be attached to an e-mail message and retrieved by the recipient with the appropriate e-mail program. Many e-mail programs, including Eudora, Netscape Messenger, and Microsoft Outlook, offer the ability to read files written in HTML, which is itself a MIME type.

2.2 SIMPLE MAIL TRANSFER PROTOCOL (SMTP)

SMTP is the de facto standard for e-mail transmission across the Internet. SMTP is a relatively simple, text-based protocol, where one or more recipients of a message are specified (and in most cases verified to exist) and then the message text is transferred. It is quite easy to test a SMTP server using the telnet program. SMTP uses TCP port 25. To determine the SMTP server for a given domain name, the MX (Mail eXchange) DNS record is used.

SMTP started becoming widely used in the early 1980s. At the time, it was a complement to UUCP which was better suited to handle e-mail transfers between machines that were intermittently connected.

SMTP, on the other hand, works best when both the sending and receiving machines are connected to the network all the time.

SMTP is a "push" protocol that does not allow one to "pull" messages from a remote server on demand. To do this a mail client must use POP3 or IMAP. Another SMTP server can trigger a delivery in SMTP using ETRN.

2.3 POST OFFICE PROTOCOL (POP)

This is the method that computers use to get mail from other computers. Remember, it's not used for sending mail, only retrieving it from mail servers. The various versions of POP are merely a result of updates to the POP technology. Don't be too surprised to see POP4 to POP10 in years to come.

POP protocol that defined how e-mail messages are to be retrieved from a mail server. POP is one of the two protocols that a PC mail program should be able to use if it is to send and receive e-mail over the Internet.

POP (Post Office Protocol) is used to provide simple remote access to mailboxes. POP3 is the latest version of POP. With POP3, once mail clients connect and download their messages, the messages are typically removed from the server and stored on the client's local machine.

This means that you have to ensure that the all the machines are backed up in order to prevent you loosing any information in e-mail messages. To resolve this problem, you could use the IMAP4 or Web Mail clients which allow messages to be manipulated on the server.

2.3.1 POP3 (Post Office Protocol) Version 3

The earlier versions of the POP protocol, POP (informally called POP1) and POP2, have been thoroughly made obsolete by POP3. In contemporary usage, the less precise term POP almost always means POP3 in the context of e-mail protocols.

POP3 are the standard rules that many e-mail clients use to handle incoming e-mail messages. E-mail clients built on standards like SMTP and POP3 can share information with each other, so we don't all need to use the same e-mail client. To set up your e-mail client, you may need to know your SMTP server address and your POP3 server address. If so, your Internet service provider (ISP) can give you this information.

POP3 is a commonly implemented method of delivering e-mail from the mail server to the client machine. Other methods include IMAP and Microsoft Exchange. The POP3 server specified in a person's e-mail program specifies where to look for e-mail that is directed to a particular e-mail address. This should never be changed unless told to by the ISP or company IT department that provides the email address.

POP3 is an application layer Internet standard protocol used to retrieve e-mail from a remote server to a local client over a TCP/IP connection. Nearly all individual Internet service provider e-mail accounts are accessed via POP3.

POP3 and its predecessors are designed to allow end-users with intermittent connections such as dial-up connections to retrieve e-mail when connected, and then to view and manipulate the retrieved messages without needing to stay connected. Although most clients have an option to leave mail on server, e-mail clients using POP3 generally connect, retrieve all messages, store them on the user's PC as new messages, delete them from the server, and then disconnect. In contrast, the newer, more capable IMAP e-mail retrieval protocol supports both connected and disconnected modes of operation. E-mail clients using IMAP generally leave messages on the server until the user explicitly deletes them. This and other facets of IMAP operation allow multiple clients to access the same mailbox. Most e-mail clients can be configured to use either POP3 or IMAP to retrieve messages; however, ISP support for IMAP is not as common.

Whether using POP3 or IMAP to retrieve messages, clients use the SMTP protocol to send messages. E-mail clients are sometimes referred to as either POP or IMAP clients, but in both cases SMTP is also used.

E-mail attachments and non-ASCII text are nearly universally conveyed in e-mail in accordance with MIME formatting rules. Neither POP3 nor SMTP require e-mail to be MIME formatted, but since essentially all internet e-mail is MIME formatted POP clients by default must also understand and use MIME. IMAP is designed to assume e-mail is MIME formatted.

Like many other older Internet protocols, POP3 originally supported only an unencrypted login mechanism. Although plain text transmission of passwords in POP3 is still common, POP3 currently supports several authentication methods to provide varying levels of protection against illegitimate access to a user's e-mail. One such method (defined in the base specification as an optional command) is APOP, which uses MD5 in an attempt to avoid replay attacks and disclosure of a shared secret; clients implementing APOP include Mozilla, Thunderbird, Eudora, and Novell Evolution. POP3 can also support IMAP authentication methods via the AUTH extension.

It is also possible to encrypt POP3 traffic using SSL. POP3 works over a TCP/IP connection using network port 110.

3.0 SENDMAIL

Sendmail was one of the first (if not the first) mail transfer agents to implement SMTP. As of 2001 there are at least 50 programs that implement SMTP as a client (sender of messages) or a server (receiver of messages). Some other popular SMTP server programs include Philip Hazel's exim, IBM's Postfix, D.J. Bernstein's and Microsoft Exchange Server.

Since this protocol started out as purely ASCII text-based, it did not deal well with binary files. Standards such as MIME were developed to encode binary files for transfer through SMTP. Today, most SMTP servers support the 8BITMIME extension, permitting binary files to be transmitted almost as easily as plain text.

Sendmail is an open source mail transfer agent (MTA): a computer program for the routing and delivery of e-mail. Its authors released the current version, Sendmail 8.13.5, on September 16, 2005. A descendant of the original ARPANET delivermail application, Sendmail is a remarka~ bly flexible program, supporting many kinds of mail transfer and delivery including the overwhelmingly popular SMTP.

Sendmail has been widely criticized as slow, overcomplicated, insecure, and difficult to maintain by comparison with other MTAs such as Exim, Postfix and Qmail. Nevertheless it remains the most popular mail transfer application on the Internet (albeit not as popular as it once was), a fact almost certainly due in part to its position as the standard MTA under most variants of the UNIX operating system. According to one study, as of November 2001 approximately 42 per cent of the publicly reachable mail servers on the Internet were running Sendmail.

Sendmail is often run as the superuser, representing a severe security threat if compromised. This is despite the recommendation since 2001 by its authors that it be run as an unprivileged user. Sendmail version 8.12 (and later) is no longer installed by default as setuid root.

4.0 Internet Protocols

Internet Protocols (set of instructions) are used to transfer files or data from one machine to the other. All computers on the Internet communicate with each other using the Transmission Control Protocol Internet Protocol (TCP/IP). Thus, data is sent from the server to the client (and vice-versa) using TCP/IP.

Mostly, the client is your browser and the server is a program running on a different computer. You use the browser on your computer (called the client machine in Internet lingo) to access the information on another computer (called the server machine). This server machine can be located thousands of miles from your workplace.

In old times (only a few years!), protocols such as Gopher, Archie etc. were used extensively. But now they have faded into oblivion; why? Thanks to the WWW.

There are many protocols available on the Internet. The World Wide Web, which is a part of the Internet, brings all these protocols under one roof. You can thus use HTTP, FTP, Telnet, e-mail etc., through your browser.

Some common protocols of the Internet are:

- **SMTP** (Simple Mail Transport Protocol): used for e-mail.
- **FTP** (File Transfer Protocol): employed for transfering files from one machine to the other.
- **Telnet Protocol:** The Telnet protocol allows you to connect to another machine. Once connected, your computer behaves like a terminal of the distant machine and you can utilize all the resources on it if you have required permissions.
- HTTP (HyperText transfer Protocol): used on the World Wide Web (WWW).

4.1 TELNET OR REMOTE LOGIN

Telnet permits your computer to log onto another computer and use it as if you were there. Telnet is a program that allows you to log into computers on the Internet and use online databases, library catalogs, chat services, and more. There are no graphics in Telnet sessions, just text. To Telnet to a computer, you must know its address. This can consist of words (locis.loc.gov) or numbers (140.147.254.3). Some services require you to connect to a specific port on the remote computer. In this case, type the port number after the Internet address. Example: telnet nri.reston.va.us 185.

Telnet is available on the World Wide Web. Probably the most common Web-based resources available through Telnet have been library catalogs, though most catalogs have since migrated to the Web. A link to a Telnet resource may look like any other link, but it will launch a Telnet session to make the connection. A Telnet program must be installed on your local computer and configured to your Web browser in order to work.

With the increasing popularity of the Web, Telnet has become less frequently used as a means of access to information on the Internet.

4.1.1 Establishing Telnet Connection

There are seven steps to a successful TELNET session:

1. Start up the TELNET program;
2. Some TELNET packages allow you to combine steps 1 and 2 into one simple step!);
3. Make a note of what the "escape character" is;
4. Log in to the remote computer;
5. Set the "terminal emulation;"
6. Play around on the remote computer; and
7. Quit.

Now let's talk about each of these steps. There are a lot of different TELNET software programs around, but each of these programs operates on the same basic principles. Or you can have session on telnet by using Run utility of windows in start menu by just typing telnet.

Step 1: Starting the TELNET program is easy. All most of you will have to do is type the word TELNET on your command line, and the program will start right up. If you are using Windows or a Macintosh (or an equivalent), double-click on the TELNET icon.

Step 2: Give the TELNET program the address of the computer that you want to access. This is where the programs start to differ. Some programs will automatically ask you to enter the address of the remote host but most will not. If your program does not automatically ask you for the address, you need to type

```
open <site address> <port number>
```

on the command line. For example, to get TELNET access to YALEINFO.YALE.EDU 7000, you would type OPEN YALEINFO.YALE.EDU 7000. (If you are using Windows or a Mac, the OPEN command may be located on a pull-down menu.) If you do not include the port number, TELNET will automatically assume that you want to connect to port 23.

Note that some TELNET programs require that you use different notations when you give an explicit port number. Examples of these notations are:

```
YALEINFO.YALE.EDU 7000 This is the most common.
YALEINFO.YALE.EDU:7000 TrmpTeL uses this.
YALEINFO.YALE.EDU#7000
YALEINFO.YALE.EDU/PORT=4000 VAX/VMS uses this notation.
"YALEINFO.YALE.EDU 7000" The quotes are required.
```

Thus, if your TELNET program appears not to accept the simple address with port number (the first example above), you might try one of these other formats or else ask your ISP for some help. By the way, the same notations that you see above may (or may not) be used in the "one step" format that we'll talk about in the next paragraph.

Step 3: Remember when I said that some nifty TELNET programs allow you to easy step? Instead of doing the two steps I just went over, you may be able to just type

```
telnet <site address> <port number>
```

on the command line (for example: TELNET YALEINFO.YALE.EDU 7000).

After you have told TELNET which computer you want it to access, and right before you gain access to the remote computer's login screen, you will see something like this:

```
telnet YALEINFO.YALE.EDU 7000 Trying
130.132.21.53 Port 7000
Connected to YALEINFO.YALE.EDU Escape character
is ..
```

This tells you that your TELNET program is trying to access the YALEINFO.YALE.EDU 7000 address, gives the IP address for YALEINFO, tells you when you are connected, and gives you the escape character. REMEMBER THE ESCAPE CHARACTER! You are going to need it in a second.

Step 4: Log in to the remote computer. Everyone should know how to log in to a computer by now. If you are accessing a public site, the "login" — the "password" that you need to access the remote computer — will probably be publicly known.

Step 5: Set the terminal emulation. All this means is that you are going to tell the remote site how data should be shown on your screen. The most common terminal emulation setting is VT100, which is the standard for terminal-based communications. If you do not have a VT100 terminal, or a terminal that can pretend it's a VT100 terminal, you may have to set your terminal emulation to either your correct terminal type or, if you do not know your correct terminal type, to a "dumb" terminal emulation.

Step 6: Fortunately, some TELNET sites automatically set the terminal emulation for you, so you do not even have to worry about it. If you end up with a screen full of gibberish, chances are you did not use the correct terminal emulation setting.

Your best bet if this happens is to disconnect from the site and try again. I think you can figure out the sixth step — play around on the remote computer— all by yourself.

Step 7: The final step is to quit. Some sites are nice and tell you how to do this but most expect you to figure out how to quit on your own. If you can figure out the command that the remote host wants you to use to quit, you should use that command to quit. This gives the remote host a better chance of "cleaning up" after your session, possibly even saving some information that you have generated. The most common commands used to quit a TELNET session a standard way to exit.

That is where the escape character comes in! Remember that once you access the remote computer, every keystroke of yours will be carried out not on your computer but on the remote computer! Typing the escape character (usually the control key and the right bracket key (]) pressed at the same time) temporarily interrupts your TELNET session and puts you into the TELNET command mode.

4.2 FILE TRANSFER PROTOCOL (FTP)

FTP stands for File Transfer Protocol. This is both a program and the method used to transfer files between computers. Anonymous FTP is an option that allows users to transfer files from thousands of host computers on the Internet to their personal computer account. FTP sites contain books, articles, software, games, images, sounds, multimedia, course work, data sets, and more. If your computer is directly connected to the Internet via an Ethernet cable, you can use one of several PC software programs, such as WS_FTP for Windows, to conduct a file transfer.

FTP transfers can be performed on the World Wide Web without the need for special software. In this case, the Web browser will suffice. Whenever you download software from a Web site to your local machine, you are using FTP.

FTP makes it possible to transfer files from one computer or the TCP/IP protocol suite. When you transfer a file from your computer to a remote computer it is called file uploading and when you transfer a file from a remote host to your computer it is referred to as file downloading. There are a number of hosts on the Internet, known as FTP servers, that allow you to take advantage of a wealth of information available from these remote hosts, also called, archive sites. Before you can upload or download a file you have to log into the remote computer.

FTP client is a program that helps you connect to an FTP server. There are two types of FTP clientstext based (or line mode clients), and GUI based. The GUI based FTP clients are easy to use as compared to line mode clients. This protocol determines how the file is broken up for transmission and reassembled at the other end, how error detection is performed etc.

There are many servers throughout the Internet that provide files via FTP a site that allows you to download or upload files is called an FTP site or an FTP server.

You may already have encountered some while browsing on the web—often you just click on a link to start file downloading you can also indicate a file transfer directly by typing in the URL for the file in the Location box in Netscape.

Hundreds of systems connected to the internet have file libraries or archives, accessible to the public. Much of this consists of free or low cost computer program for virtually every make of computer. If you want a different communications program for your IBM, or feel like playing a new game on your Amiga. You will probably be able to get it using FTP.

There are also libraries of documents as well, if you want a copy of recent US supreme court decision, copies of historical documents, song lyrics, poems and even summaries of every animaniacs episode ever made. You can also find extensive files detailing everything you would possibly want to know about the Internet.

With the Internet address or domain name of site, you can use FTP to connect to a specific location on that site (specific directory path and file name) and copy a file containing information to or from your computer for your own use. This activity is individual and private, although many FTP sights log (or register) file copy transactions for either market research or an security precautions. In the case of individuals sharing information, the file was prepared and stored specifically for you by the person distributing the information.

4.2.1 Anonymous FTP connections

There are two types of FTP connections available on the Internet:

- Anonymous
- Non-anonymous

Many FTP sites do not allow access to the general public. On these sites, a username and password will be required before being allowed to log in. There are however, FTP servers that do allow anyone to log on and view the uploaded information. These are known as "Anonymous FTP" servers and provide "Anonymous FTP" sites. The term "anonymous" is used because the person requesting the file does not need to have an account to access the publicly available files.

The most widely used type is anonymous FTP. In fact, you may have used it without even knowing it. Many web pages contain links to files that you can download. Often these links point to a file in an anonymous FTP directory. If a file is stored in an anonymous FTP directory, virtually anyone with Internet access and an FTP program of some sort, even a web browser, can download the file. Uploading, on the other hand, is not possible with anonymous FTP. Anonymous FTP, therefore, is used primarily to give the Internet public download access to a particular directory of files. Anyone can download files from the directory, but only the "owner" of directory can upload to the directory.

4.2.2 Using FTP Program

Some form of FTP is available to anyone who has access to the Internet. FTP is an information "service" available to any Internet user, just like the web and e-mail. FTP, like these other services, is independent of the software you use to access it. For example, if you use Windows, you can use both Netscape and other programs, like the Windows 95 FTP client or WS_FTP, to access internet files using FTP. If you use Unix, you would probably use the Unix FTP client. If you use the Macintosh operating system, there are likewise several FTP programs available for the Mac OS. You can use line mode FTP if you have Internet connection and for that reason in any other mode.

Simply put, FTP allows you to enter a directory on a computer connected to the Internet and transfer a file to or from that directory to another directory on another computer. Normally, you will be transferring the file to or from a large, multi-user computer and your own computer. File transfer can go in two directions. "Downloading," refers to a transfer of a file from a transfer of a file from your computer to a remote computer.

In order to upload or download a file by FTP, you need to do four things:

1. Login into a remote computer that has been configured as an FTP "server".
2. Submit a username and a password to gain access to the remote system.
3. Go into a particular directory on the remote system that contains the file you are interested in downloading or to which you wish to upload a file.
4. Transfer the file to or from the system in question.

When you connect to any FTP directory, the host system asks for your username and password before allowing you access to the directory (this process is done behind the scenes when you use a web browser to access an FTP directory). With an anonymous FTP directory, any user can gain access because the username is always "anonymous" and the password is always the user's e-mail address.

Non-anonymous FTP, on the other hand, requires you to know the unique username and password for the FTP directory in question. Normally, you will use non-anonymous FTP to get access to directories that YOU OWN on an Internet computer. For example, when you sign up with a commercial internet service provider (ISP), or get an Internet account through school or work, you normally get a "home" directory on the ISP's computer assigned to you. You can use the disk space in this home directory for any purpose you choose. To get access to that directory, however, you will normally need to use FTP. Your ISP can tell you the name of the FTP "host" computer you must connect to as well as the username and password you must use to get access to your home directory.

Tutorial 4.1 Starting Line mode FTP client (client means your computer)

1. Click start and then Run buttons from start menu in Windows 95/98 or any other advanced version of Windows. The run dialog box appears.
2. Type ftp in the text entry box against Open in this dialog box and click OK.

You can do the same thing from DOS as well.

(i) At DOS prompt type FTP and press Enter.

In both cases ftp client (software program) appears on screen as shown in following figure:

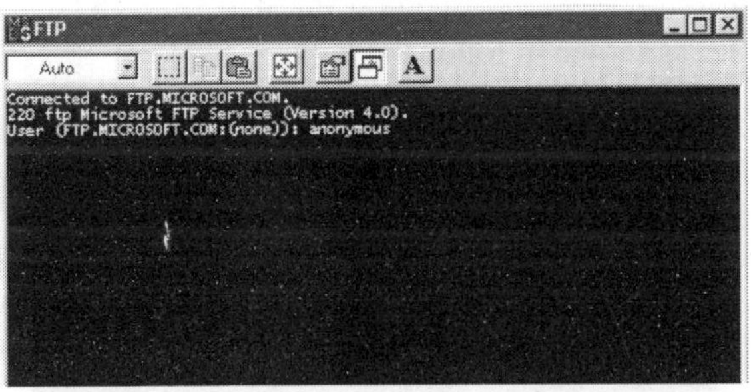

Fig. 4.2: Line Mode FTP client Program in Windows 95/984

(ii) The moment you you type ftp the words ftp.microsoft.com. If you are already connected to internet you will be connected to specified FTP site. You are now prompted to enter the username.

(iii) Type anonymous as username and press Enter. The program will ask you to enter the password. Type your e-mail password supplied by the ISP and press enter. It noteworthy that while you were typing the password it was not transparent on screen. Once you are logged on to ftp, the files and directories will be accessible to you.

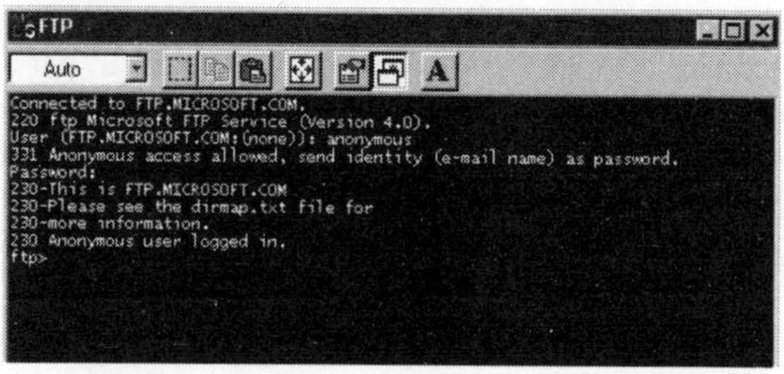

Fig. 4.3: Connection to FTP's Archived Site

(iv) You can also use open command to log on to required ftp site. The syntax of open command is

`Open hostname`

You can write

`Open ftp.microsoft.com`

or abbreviate this as

`O ftp.microsoft.com`

For getting disconnected simply type Disconnect.

(v) For listing the directory type dir and press Enter. The list of files or directories available at this particular site will be listed by ftp as shown in following Figure:

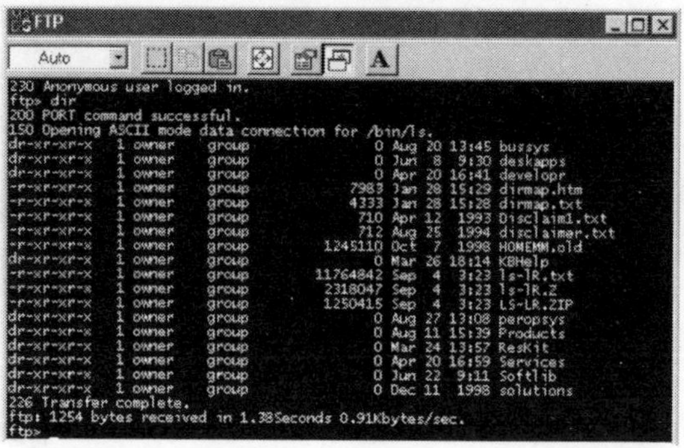

Fig. 4.4: Directory listing of FTP Site of microsoft.com

The dir command is almost identical to ls command of UNIX. Examine the listing carefully. the word dr indicates that the details given against that particular line are of a subdirectory.

(vi) You can change the directory by using cd command as in Unix/ Dos. Type cd products to reach product directory (as listed in following Figure) and press Enter. The list of files already available in this list will be shown as:

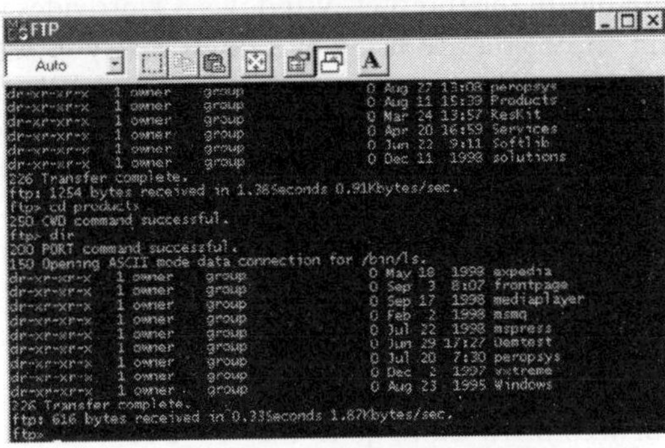

Fig. 4.5: Listing files of a Directory

(vii) For downloading files from Internet to your computer's hard disk you have to use get command. This causes the concerned file from the remote site to get copied to your default hard disk. Type get disclaimer.txt at the ftp prompt and press Enter. The file will be downloaded to your hard disk and FTP will report the size of the file as shown in following Figure:

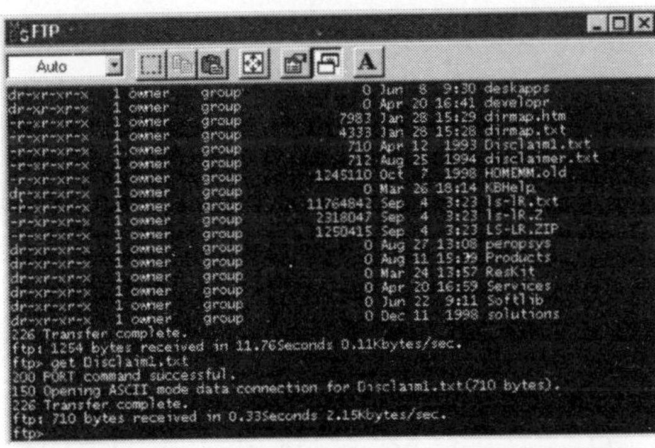

Fig. 4.6: Connecting to FTP Sites with WS_FTP

In this tutorial, we will first discuss the fundamentals of how FTP does what it does. Then we will do some hands-on downloading and uploading with WS_FTP, a popular shareware FTP program for Windows. In addition we will do some FTP downloading with Netscape, a popular world wide web browser for Windows.

While you can use the FTP program that comes with Windows 95 to get files using FTP, the Windows 95 FTP program is a text-based tool modeled after the Unix FTP client and is not very user-friendly. A much better program is the freeware program, "WS_FTP", which is available on the Internet. Windows 3.x provides no built-in FTP program so, if you are using Windows 3.x, you will have to download your own FTP program, such as WS_FTP. WS_FTP comes in a 32 bit Windows 95 version and a 16 bit Windows 3.x version.

How to Download WS_FTP

WS_FTP comes as a self-extracting, self-installing download file. You must first download the file to your hard drive. Then you run (double click) the download file to install the WS_FTP program and the associated files.

- Using your web browser, to go to the WS_FTP web site: http://members.cnx.net/reboot/ftps.htm
- If your web browser tells you that the file type is unknown and asks you what you want to do with it, just choose save to disk .
- When prompted, save the file to your disk. Write down the name of the file and the name of the directory you are saving it to.
- Start the File Manager or Explorer.
- Double click the download file to install the WS_FTP program.
- How to Run WS_FTP:

When you install WS_FTP, an icon to run it will be added to your system. If you are running Windows 95, click on Start and Programs and find the WS_FTP section. Click on the WS_FTP program to run it. If you are using Windows 3.x, find the new group called "WS_FTP", look inside of it, find the WS_FTP icon, and double click it.

Example

- Filling out Your "Session Profile"

You should start on the "General" tab, as illustrated in following figure. Click on "New" to create a new session profile.

- **Profile Name:** The name you assign to a particular FTP session profile. This is completely up to you. The name you use should identify the site clearly so that you know what the site is and why you want to connect to it.
- **Host Name:** The name of the Internet computer that hosts (contains) the FTP directory you want to connect to. Sometimes this is the

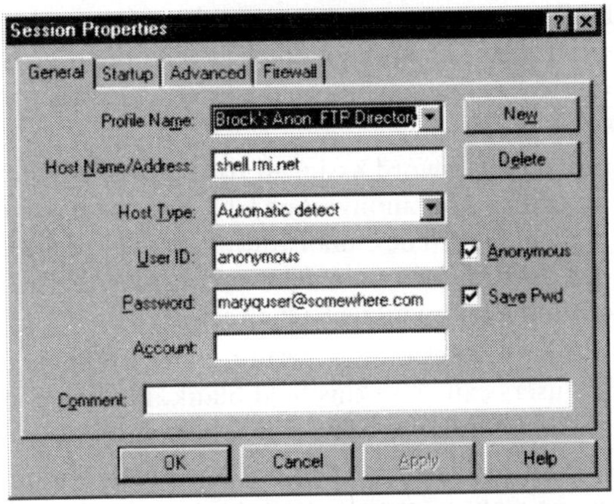

Fig. 4.7: FTP Getting WS_FTP Interface

Internet "domain" name of the FTP computer. You will have to get this name from the owner of the FTP directory or from some other source. If you are the owner of the FTP directory, just ask the technical support folks at your Internet service provider for the host name you should use to connect to this directory. The FTP dircctory we will connect to in this tutorial is stored on a computer owned by Rocky Mountain Internet, an Internet service provider used by the author of this tutorial.

- **Host Type:** The type of operating system that the FTP host is running. Most internet computers, but not all, run the powerful, multi-user, Unix operating system. The dialog box in example 1 is set to "Automatic detect", which will make WS_FTP automatically determine the type of operating system it is connecting to. You can usually leave this setting on Automatic detect.
- **Anonymous (Checkbox):** Unless you are the owner of this directory, or have been given the username and password by the owner, you will be logging in anonymously, as discussed above. If so, check this checkbox.

  Check this box so you don't have to fill in your password each time you login to this FTP site. If your computer is not in a secure environment, you might want to fill in your password each time you login. In that case, leave this checkbox unchecked. If you check this box, anyone with access to your computer can pretend to be you and login to this FTP directory.

- **User ID:** The username assigned to the person who owns this directory. If this is an anonymous FTP site, the username will be "anonymous" so that anyone on the Internet may get access to the directory.
- **Password:** The password assigned to the person who owns this directory. If this is an anonymous FTP site, the password will be your e-mail address. The e-mail address password is a courtesy to the owner of the anonymous FTP site. It tells the owner who is logging into the anonymous FTP directory.
- **Account:** This applies only to certain older FTP host computers. Most FTP users can leave this field blank. For more information on this field, see the WS_FTP on-line help.
- **Comment:** In this field, you can write in a description of the site and why you want to connect to it. Or, you can leave it blank, as illustrated above.

OK, now click on the "Startup" tab.

You will see the following window:

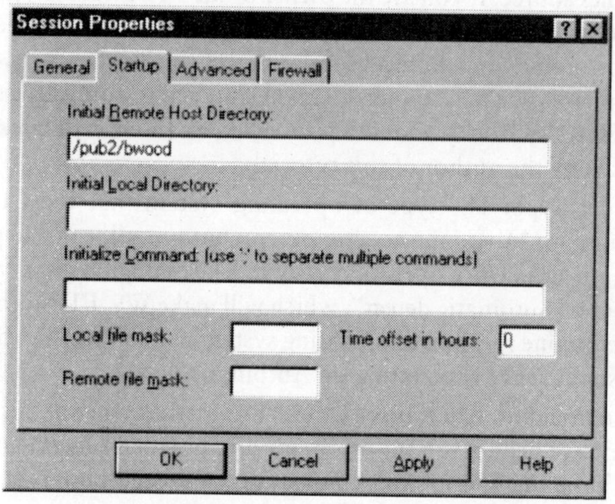

Fig. 4.8: Filling Sessions Properties Dialog Box

- **Initial Remote Host Directory:** This specifies the FTP directory on the remote host that contains the files you are interested in getting access to. Note that the slashes that separate the directories are "forward" slashes (/), as opposed to the "back" slashes (\) that DOS and Windows users are familiar with. This is because the FTP host computer in the example runs the Unix operating system.

If you leave this dialog box blank, you will login to a default directory on the host system. When in doubt, leave this box blank.

- **Initial Local Directory:** This specifies the directory on your local computer where downloaded files will go to and uploaded files will come from. You can fill in this box with the name of a directory that actually exists on your hard drive. Or, if you leave this dialog box blank (as shown in the example), a default directory (normally the WS_FTP directory) will be used.

You can leave all of the other fields blank. This will automatically be saved.

Once you connect... Note the split screen. WS_FTP conveniently splits your screen into two lists of files. On the left side of the window are your local files, the ones on your computer. On the right side you will find the files that reside on the FTP site.

4.3 HYPERTEXT TRANSPORT PROTOCOL (HTTP)

HTTP is the primary method used to convey information on the World Wide Web. The original purpose was to provide a way to publish and receive HTML pages.

Development of HTTP was coordinated by the World Wide Web Consortium and working groups of the Internet Engineering Task Force, culminating in the publication of a series of RFCs, most notably RFC 2616, which defines HTTP/1.1, the version of HTTP in common use today.

HTTP is the communications protocol used to connect to servers on the Web. Its primary function is to establish a connection with a Web server and transmit HTML pages to the client browser or any other files required by an HTTP application. Addresses of Web sites begin with an http:// prefix; however, Web browsers typically default to the HTTP protocol. For example, typing www.yahoo.com is the same as typing http://www.yahoo.com.

HTTP is a "stateless" request/response system. The connection is maintained between client and server only for the immediate request, and the connection is closed. After the HTTP client establishes a TCP connection with the server and sends it a request command, the server sends back its response and closes the connection (see cookie).

HTTP differs from other TCP-based protocols such as FTP, in that connections are usually terminated once a particular request (or related series of requests) has been completed. This design makes HTTP ideal for the World Wide Web, where pages regularly link to pages on other servers. It can occasionally pose problems for Web designers, as the lack of a persistent connection necessitates alternative methods of maintaining users' "state". Many of these methods involve the use of "cookies".

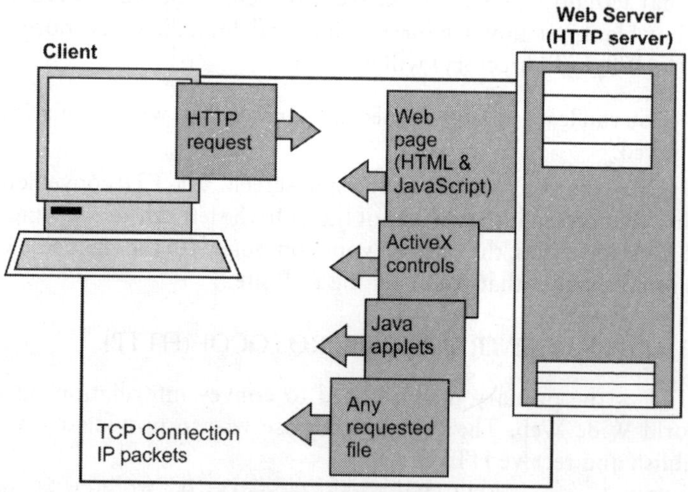

Fig. 4.9: Webserver

HTTP is a request/response protocol between clients and servers. An HTTP client, such as a web browser, typically initiates a request by establishing a TCP connection to a particular port on a remote host (port 80 by default). An HTTP server listening on that port waits for the client to send a request string, such as "GET / HTTP/1.1" (which would request the default page of that web server), followed by an e-mail-like MIME message which has a number of informational header strings that describe aspects of the request, followed by an optional body of arbitrary data. Some headers are optional, while others (such as Host) are required by the HTTP/1.1 protocol. Upon receiving the request string (and message, if any), the server sends back a response string, such as "200 OK", and a message of its own, the body of which is perhaps the requested file, an error message, or some other information.

4.4.1 HTTP Request Methods

Get By far the most common, for statically requesting a resource by specifying a URL.

Post: Similar to GET, except that a message body, typically containing key-value pairs from an HTML form submission, is included in the request.

PUT: Used for uploading files to a specified URI on a web-server.

Delete Rarely implemented, and self-explanatory.

Head: Identical to GET, except that the page content is not returned; just the headers are. Useful for retrieving meta-information.

Trace: Echoes back the received request, so that a client can see what intermediate servers are adding or changing in the request.

Options: Returns the HTTP methods that the server supports. This can be used to check the functionality for a web server.

Connect: Rarely implemented, for use with a proxy that can change to being an SSL tunnel.

5.1 INTERNET INFORMATION SERVER (IIS)

IIS is Microsoft's business-class Web server. Although Apache boasts more total worldwide installations than IIS, many corporations choose IIS as their Web server because it is supported by a commercial organization. Some have criticized IIS in the past, though, for its alleged security weaknesses.

IIS has improved over the years, supports many Web standards including SSL, and integrates well with other Microsoft Web technologies such as FrontPage. An administrator can manage IIS through a graphical console, the Internet Services Manager.

5.2 APACHE WEB SERVER

Apache is generally recognized as the world's most popular Web server (HTTP server). Originally designed for Unix servers, Apache has been ported to Windows and other network operating systems (NOS). The name "Apache" derives from the word "patchy" that the Apache developers used to describe early versions of their software.

Apache, a server for Unix computers, is the most popular and is considered one of the most robust implementations. It does not do any logfile analysis. There is no vendor so there is no formal technical support. There is no real way to administrate the computer through a Web interface. On the other hand, Apache's modular approach allows for a great deal of customization. It includes strong links for database applications. It allows you to save your logfiles in formats you define. But most importantly, it runs on any Unix computer, and Unix is an operating system designed for client/server and TCP/IP applications.

As of version 1.3b, Apache now also runs under Windows 95 and NT. It will also be ported to Rhapsody (Macintosh) when that operating system becomes available this year. Consequently, Apache represents a good, all-around HTTP server. One that you can/will be able to take to just about any computer. Apache, based on the original NCSA httpd application, got its name from when its developers where trying to break httpd down into its original parts. Thus, it was "a patchy server."

There are two alternatives for acquiring the Apache software. One, you can download a precompiled binary for your particular version of Unix or Windows. Or two, you can download Apache's source code and compile it yourself. This section outlines the second option for Unix since it offers you greater flexibility.

Apache provides a full range of Web server features, including CGI, SSL, and virtual domains. Apache also supports plug-in modules for extensibility. Apache is reliable, free, and relatively easy to configure.

The Apache HTTP server is free software distributed by the Apache Software Foundation. The Apache Software Foundation promotes various free and open source advanced Web technologies. Often referred to as simply Apache, a public-domain open source Web server developed by a loosely-knit group of programmers. The first version of Apache, based on the NCSA httpd Web server, was developed in 1995.

Core development of the Apache Web server is performed by a group of about 20 volunteer programmers, called the Apache Group. However, because the source code is freely available, anyone can adapt the server for specific needs, and there is a large public library of Apache addtions. In many respects, development of Apache is similar to development of the Linux operating system.

The original version of Apache was written for UNIX, but there are now versions that run under OS/2, Windows and other platforms. The name is a tribute to the Native American Apache Indian tribe, a tribe well known for its endurance and skill in warfare. A common misunderstanding is that it was called Apache because it was developed from existing NCSA code plus various patches, hence the name a patchy server, or Apache server.

6.1 CLIENT-SIDE SCRIPTING

Client-side scripting generally refers to the class of computer programs on the web that are executed client-side, by the user's web browser, instead of server-side (on the web server). This type of computer programming is an important part of the Dynamic HTML (DHTML) concept, enabling web pages to be scripted; that is, to have different and changing content depending on user input, environmental conditions (such as the time of day), or other variables.

Web authors write client-side scripts in languages such as JavaScript (Client-side JavaScript) or VBScript, which are based on several standards:

- HTML scripting
- HTTP
- Document Object Model (DOM)

Client-side scripts are often embedded within an HTML document, but they may also be contained in a separate file, which is referenced by the document (or documents) that use it. Upon request, the necessary files are sent to the user's computer by the web server (or servers) on which they reside. The user's web browser executes the script, then displays the document, including any visible output from the script. Client-side scripts may also contain instructions for the browser to follow if the user interacts with the document in a certain way, *e.g.*, clicks a certain button. These instructions can be followed without further communication with the server, though they may require such communication.

By viewing the file that contains the script, users may be able to see its source code. Many web authors learn how to write client-side scripts partly by examining the source code for other authors' scripts.

In contrast, server-side scripts, written in languages such as Perl and PHP, are executed by the web server when the user requests a document. They produce output in a format understandable by web browsers (usually HTML), which is then sent to the user's computer. The user cannot see the script's source code (unless the author publishes the code separately), and may not even be aware that a script was executed. The documents produced by server-side scripts may, of course, contain client-side scripts.

6.2 PERL

Perl is short for Practical Extraction Report Language. It is a programming language written by Larry Wall that combines syntax from several Unix utilities and languages. Introduced in 1987, Perl is designed to handle a variety of system administrator functions and provides comprehensive string handling functions. It is widely used to write Web server programs for such tasks as automatically updating user accounts and newsgroup postings, processing removal requests, synchronizing databases and generating reports. Perl has also been adapted to non-Unix platforms.

Perl is a general-purpose programming language originally developed for text manipulation and now used for a wide range of tasks including system administration, web development, network programming, GUI development, and more.

Perl is a simple language, easy to learn, yet powerful enough to accomplish the most difficult tasks. It is widely available, and is probably already installed on your Unix server. Perl is an interpreted language, meaning you don't need to compile your script—you simply write your script and run it (or have the web server run it). The script itself is just text code; the Perl interpreter does all the work. The advantage to this is you can copy your script with little or no changes to any machine with a Perl interpreter. The disadvantage is you won't discover any bugs in your script until you run it.

The language is intended to be practical (easy to use, efficient, complete) rather than beautiful (tiny, elegant, minimal). Its major features are that it's easy to use, supports both procedural and object-oriented (OO) programming, has powerful built-in support for text processing, and has one of the world's most impressive collections of third-party modules.

6.2.1 Language Features

The overall structure of Perl derives broadly from C. Perl is a procedural programming language, with variables, expressions, assignment statements, brace-delimited code blocks, control structures, and subroutines.

Perl also takes features from shell programming. Perl programs are interpreted. All variables are marked with leading sigils. Sigils unambiguously identify variable names, thus allowing Perl to have a rich syntax. Importantly, sigils allow variables to be interpolated directly into strings. Like the Unix shells, Perl has many built-in functions for common tasks, like sorting, and for accessing system facilities.

Perl takes associative arrays from awk and regular expressions from sed. These simplify and facilitate all manner of parsing, text handling, and data management tasks.

In Perl 5, features were added that support complex data structures and an object-oriented programming model. These include references, packages, and class-based method dispatch. Perl 5 also saw the introduction of lexically scoped variables, which make it easier to write robust code, and modules, which make it practical to write and distribute libraries of Perl code.

All versions of Perl do automatic data typing and memory management. The interpreter knows the type and storage requirements of every data object in the program; it allocates and frees storage for them as necessary. Legal type conversions are done automatically at run time; illegal type conversions are fatal errors. It is not possible, within the language, to leak memory, crash the interpreter, or corrupt its internal data representation.

6.2.2 Applications

Perl has many and varied applications. Perl has been used since the early days of the web to write CGI scripts, and is a component of the popular LAMP (Linux/Apache/MySQL/(Perl/PHP/Python)) platform for web development. Perl has been called "the glue that holds the web together". Large systems written in Perl include Slashdot and early implementations of PHP [1].

Perl finds many applications as a "glue language", tying together systems and interfaces that were not specifically designed to interoperate. Systems administrators use Perl as an all-purpose tool; short Perl programs can be entered and run on a single command line.

Perl is widely used in finance and bioinformatics, where it is valued for rapid application development, ability to handle large data sets, and the availability of many standard and 3rd party modules.

6.2.3 Perl's Implementation

Perl is implemented as a core interpreter, written in C, together with a large collection of modules, written in Perl and C. The source distribution is currently 12 MB when packaged in a tar file and compressed. The interpreter is 150,000 lines of C code and compiles to a 1 MB executable on typical machine architectures. Alternately, the interpreter can be compiled to a link library and embedded in other programs. There are nearly 500 modules in the distribution, comprising 200,000 lines of Perl and an additional 350,000 lines of C code. Much of the C code in the modules consists of character encoding tables.

The interpreter has an object-oriented architecture. All the elements of the Perl language—scalars, arrays, hashes, coderefs, file handles—are represented in the interpreter by C structs. Operations on these structs are defined by a large collection of macros, typedefs and functions; these constitute the Perl C API. The Perl API can be bewildering to the uninitiate; however, its entry points follow a consistent naming scheme, which provides guidance to those who use it.

The execution of a Perl program divides broadly into two phases. First, the interpreter parses the program text into a syntax tree. Then, it executes the program by walking the tree. The text is parsed only once, and the syntax tree is subject to optimization before it is executed, so the execution phase is relatively efficient.

Perl has a context-free grammar; however, it cannot be parsed by a straight Yacc/Lex parser/lexer combination. Instead, it implements its own lexer, which coordinates with a modified GNU bison parser to resolve ambiguities in the language. It is said that "only perl can parse Perl", meaning that only the Perl interpreter can parse the Perl language. The truth of this is attested by the imperfections of other programs that undertake to parse Perl, such as source code analyzers and auto-indenters.

Maintenance of the Perl interpreter has become increasingly difficult over the years. The code base has been in continuous development since 1994. The code has been optimized for performance at the expense of simplicity, clarity, and strong internal interfaces. New features have been added, yet virtually complete backwards compatibility with earlier versions is maintained. The size and complexity of the interpreter is a barrier to developers who wish to work on it.

Perl is distributed with some 90,000 functional tests. These run as part of the normal build process, and extensively exercise the interperter and its core modules. Perl developers rely on the functional tests to ensure that changes to the interpreter do not introduce bugs; conversely, Perl users who see the interpreter pass its functional tests on their system can have a high degree of confidence that it is working properly.

There is no written specification or standard for the Perl language, and no plans to create one. There has only ever been one implementation of the interpreter. That interpreter, together with its functional tests, stands as a de facto specification of the language.

Perl is widely favored for database applications. Its text handling facilities are good for generating SQL queries; arrays, hashes and automatic memory management make it easy to collect and process the returned data.

In early versions of Perl, database interfaces were created by relinking the interpreter with a client-side database library. This was somewhat clumsy; a particular problem was that the resulting perl executable was restricted to using just the one database interface that it was linked to. Also, relinking the interpreter was sufficiently difficult that it was only done for a few of the most important and widely used databases.

In Perl 5, database interfaces are implemented by modules. The DBI (Database Interface) module presents a single, database-independent interface to Perl applications, while the DBD:: (Database Driver) modules handle the details of accessing some 50 different databases. There are DBD:: drivers for most ANSI SQL databases.

You can edit your Perl scripts either on your local machine (using your favorite text editor - Notepad, Simpletext, etc.), or in the Unix shell. If you're using Unix, try pico - it's a very simple, easy to use text editor. Just type pico filename to create or edit a file. Type man pico for more information and instructions for pico. If you're not familiar with the Unix shell, see Appendix B for a Unix tutorial and command reference.

You can also use a text editor on your local machine, and upload the finished scripts to the Unix server. Be sure to use a plain text editor, such as Notepad (on PC) or BBEdit (on Mac), and turn off special characters such as smartquotes—CGI files must be ordinary text. Also, it is imperative that you upload your CGI as text, NOT binary! If you upload it as binary, it will come across with a lot of control characters at the end of the lines, and these will break your script. You can save yourself a lot of time and grief by just uploading everything as text (unless you're uploading pictures - for example, GIFs or JPEGs - or other binary data. HTML and Perl CGIs are not binary, they are plain text).

Once your script is uploaded to the Unix server, you'll want to be sure to move it to your public_html directory (or whatever directory you have set-up for web pages). Then, you will also need to change the permissions on the file so that it is "executable" (or runnable) by the system. In Unix, this command is:

```
chmod 755 filename
```

This sets the file permissions so that you can read, write, and execute the file, and all other users (including the webserver) can read and execute it. See Appendix B for a full description of chmod and its options.

Most FTP programs also allow you to change file permissions; if you use your FTP client to change perms, you'll want to be sure that the file is readable and executable by everyone, and writable only by the owner (you).

One final note: Perl scripts, Unix commands, and filenames are all case-sensitive. Please keep this in mind as you write your first Perl scripts, because in Unix, "perl" is not the same as "PERL".

Basics of a Perl Script

You probably are already familiar with HTML, and so you know that certain things are necessary in the structure of an HTML document, such as the <HEAD> and <BODY> tags, and that other tags like links and images have a certain allowed syntax. Perl is very similar; it has a clearly defined syntax, and if you follow those syntax rules, you can write Perl as easily as you do HTML.

If you're creating scripts on Unix, you'll need one statement as the first line of every script, telling the server that this is a Perl script, and where to find the Perl interpreter. In most scripts the statement will look like this

```
#!/usr/bin/perl
```

For now, there should generally not be anything else on the same line with this statement. (There are some flags you can use there, but we'll go into those later.) If you aren't sure where Perl lives on your system, try typing these commands:

```
which perl OR where is perl
```

If the system can find it, it will tell you the path name to Perl. That path is what you should put in the above statement.

After the above line, you'll write your Perl code. Most lines of Perl code must end in a semicolon (;), except the opening and closing lines of loops and conditional blocks. We'll cover those later.

Basics of a CGI Script

A CGI program is still a Perl script. But one important difference is that a CGI usually generates a web page (for example: a form-processing CGI, such as a guestbook, usually returns a "thank you for writing" page). If you are writing a CGI that's going to generate a HTML page, you must include this statement somewhere in the script, before you print out anything else:

```
print "Content-type:text/html\n\n";
```

This is a content header that tells the receiving web browser what sort of data it is about to receive—in this case, an HTML document. If you forget to include it, or if you print something else before printing this header, you'll get an "Internal Server Error" when you try to access the CGI. A good rule of thumb is to put the Content-type line at the top of your script (just below the #!/usr/bin/perl/ line).

6.3 TOOLS COMMAND LANGUAGE (TCL)

Tcl (originally from "Tool Command Language", but nonetheless conventionally rendered as "Tcl" rather than "TCL"; and pronounced like "tickle") is a scripting language created by John Ousterhout that is generally thought to be easy to learn, but powerful in competent hands. It is most commonly used for rapid prototyping, scripted applications, GUIs and testing.

Tcl's features include:

- Everything is a command, including language structures. They are in Polish notation.
- Everything can be dynamically redefined and overridden.
- All data types can be manipulated as strings, including code.
- Extremely simple syntactic rules.
- Event-driven interface to sockets and files. Time based and user defined events are also possible.
- Flexible scope, with variable visibility restricted to lexical (static) scope by default, but uplevel and upvar allowing process to interact with the enclosing functions' scopes.
- Readily extensible, via C, C++, Java, and Tcl.
- Interpreted language, code can be created and modified dynamically.
- Full Unicode support, first released 1999.
- Platform independent: Win32, UNIX, Mac, etc.

- Close integration with windowing (GUI) interface Tk.
- Easy to maintain code. Tcl scripts are often more compact and readable than functionally equivalent code in other languages.

Tcl proper does not provide an object-oriented framework, but it can be extended to provide new features as needed. Many C extensions have been written to give OO functionality, including the XOTcl and incr Tcl packages. Other OO extensions, including Snit, are written fully in Tcl.

Functional programming can easily be done in Tcl, as higher-order functions or functional abstractions are built into the language, though it is not widely used for this purpose. As an example, consider the ease with which two functions can be composed:

Syntax **`proc o {f g x} {$f [$g $x]}`**

Very simple and consistent syntax

Tcl has a very simple syntax which is applied in a consistent way. A Tcl script consists of several commands. A command is a list of words separated by whitespace.

```
word0 word1 word2 ... wordN
```

The first word is the name of a command, which is not built into the language, but which is in the library. The following words are arguments. So we have:

```
commandName argument1 argument2 ...
argumentN
```

Any argument may be replaced by another command in square brackets. The subcommand is evaluated first and the result is substituted as the argument. Alternatively, any argument placed in curly braces will not be evaluated, but rather will be handed directly back to the original command as an argument.

To summarize: there is one basic construct (the command) and a set of simple substitution rules, and only the block, the curly braces and the backslash have a special meaning besides the quotes. The single equality sign (=) for example is not used at all, and the double equality sign (==) is the test for equality, and even then only in the expr command.

All commands have the same structure, a keyword which is followed by several parameters. A command is terminated by a new line or a semicolon. Even comments are just commands which happen to do nothing. Tcl is not statically typed: each variable may contain integers, floats, strings, lists or any other value.

Extension Packages

The Tcl language has always supported extension packages, which provide additional functionality (such as a GUI, terminal-based application automation, database access, etc.).

The most popular Tcl extension is the Tk toolkit, which provides a graphical user interface library for a variety of operating systems. Each GUI consists of one or more frames. Each frame has a layout manager.

Another popular extension is Expect, which allows automated driving of terminal-based programs (such as passwd, ftp, telnet and command driven shells).

6.4 PYTHON PROGRAMMING LANGUAGE

Python is an interpreted, interactive programming language created by Guido van Rossum in 1990. Python is fully dynamically typed and uses automatic memory management; it is thus similar to Perl, Ruby, Scheme, Smalltalk, and Tcl. Python is developed as an open source project, managed by the non-profit Python Software Foundation. Python 2.4.2 was released on September 28, 2005.

Python was created in the early 1990s by Guido van Rossum at Stichting Mathematisch Centrum (CWI) in the Netherlands as a successor of the ABC programming language.

Python is a *multi-paradigm* language. This means that, rather than forcing coders to adopt one particular style of coding, it permits several. Object orientation, structured programming, functional programming, aspect-oriented programming, and more recently, design by contract are all supported. Python is dynamically type-checked and uses garbage collection for memory management. An important feature of Python is dynamic name resolution, which binds method and variable names during program execution.

While offering choice in coding methodology, Python's designers reject exuberant syntax, such as in Perl, in favor of a more sparse, less cluttered one. As with Perl, Python's developers expressly promote a particular "culture" or ideology based on what they want the language to be, favoring language forms they see as "beautiful", "explicit" and "simple". For the most part, Perl and Python users differ in their interpretation of these terms and how they are best implemented.

Another important goal of the Python developers is making Python fun to use. This is reflected in the origin of the name (after the television series Monty Python's Flying Circus); in the common practice of using Monty Python references in example code; and in an occasionally playful approach to tutorials and reference materials.

Although Python is sometimes classified as a "scripting programming language", it has been used to develop many large software projects such as the Zope application server and the Mnet and BitTorrent file sharing systems. It is also extensively used by Google. Python proponents prefer to call it a high level dynamic programming language, on the grounds that "scripting language" implies a language that is only used for simple shell scripts or that refers to a language like JavaScript: much simpler and, for most purposes, less capable than "real" programming languages such as Python.

Another important goal of the language is ease of extensibility. New built-in modules are easily written in C or C++. Python can also be used as an extension language for existing modules and applications that need a programmable interface.

Though the designer of Python is somewhat hostile to functional programming and the Lisp tradition, there are significant parallels between the philosophy of Python and that of minimalist Lisp-family languages such as Scheme. Many past Lisp programmers have found Python appealing for this reason.

Data Types and Structures

Since Python is a dynamically typed language, Python values, not variables, carry type. This has implications for many aspects of the way the language functions.

All values in Python are references to objects, and these references are passed to functions by value; a function cannot change the value a variable references in its calling function. Some people (including Guido van Rossum himself) have called this parameter-passing scheme "Call by object reference".

Among dynamically typed languages, Python is moderately type-checked. Implicit conversion is defined for numeric types, so one may validly multiply a complex number by a long integer (for instance) without explicit casting. However, there is no implicit conversion between (*e.g.*) numbers and strings; a string is an invalid argument to a mathematical function expecting a number.

Syntax

Python was designed to be highly readable. It has a simple visual layout, uses English keywords frequently where other languages use punctuation, and has notably fewer syntactic constructions than many structured languages such as C, Perl, or Pascal. For instance, Python has only two structured loop forms:

- For, which loops over elements of a list or iterator (like Perl for each)
- While, which loops as long as a boolean expression is true.

  It thus lacks C-style complex for, a do...while, though of course equivalents can be expressed. Likewise, it has only if...elif...else for branching—no switch or labeled goto (goto was implemented as a joke for April 1st 2004, in an add-on module).

Object-oriented Programming

Python's support for object oriented programming paradigm is vast. It supports polymorphism, not only within a class hierarchy but also by duck typing. Any object can be used for any type, and it will work so long as it has the proper methods and attributes. And everything in Python is an object, including classes, functions, numbers and modules. Python also has support for metaclasses, an advanced tool for enhancing classes' functionality. Naturally, inheritance, including multiple inheritance, is supported. It has limited support for private variables using name mangling. See the "Classes" section of the tutorial for details. Many Python users don't feel the need for private variables, though. The slogan "We're all consenting adults here" is used to describe this attitude. Some consider information hiding to be unpythonic, in that it suggests that the class in question contains unaesthetic or ill-planned internals.

7.1 SERVER-SIDE SCRIPTING

Server-side scripting is a web server technology in which a user's request is fulfilled by running a script directly on the web server to generate dynamic HTML pages. It is usually used to provide interactive web sites that interface to databases or other data stores. This is different from client-side scripting where scripts are run by the viewing web browser, usually in JavaScript. The primary advantage to server-side scripting is the ability to highly customize the response based on the user's requirements, access rights, or queries into data stores.

VBScript is used for both server-side and client-side scripting, but primarily server-side. This version of Visual Basic is supported primarily on Microsoft platforms.

In the "old" days of the web this was almost exclusively performed using a combination of C programs, Perl scripts and Shell scripts using the *Common Gateway Interface* (CGI). Those scripts were executed by the operating system, and the results simply served back by the web server. Nowadays, these and other online scripting languages such as ASP and PHP can often be executed directly by the web server itself or by extension modules (*e.g.* mod_perl or mod_php) to the web server. Either form of scripting (*i.e.* CGI or direct execution) can be used to build up complex multi-page sites, but direct execution usually results in lower overhead due to the lack of calls to external interpreters.

Dynamic websites are also sometimes powered by custom web application servers, for example the Python "Base HTTP Server" library, although some may not consider this to be server-side scripting.

7.1.1 List of Server-side Scripting Technologies

ASP: Microsoft designed soloution allowing various languages (though generally vbscript is used) inside a html like outer page, mainly used on windows but with limited support on other platforms.

ColdFusion : Cross platform tag based commercial server side scripting system.

JSP: A java based system for embedding code in html pages.

SSI : A fairly basic system which is part of the common Apache web server. Not a full programming environment by far but still handy for simple things like including a common menu.

Perl: Not specifically designed for web scripting but still often used either through CGI or a special apache module.

Python: Another language commonly used through CGI.

Tcl: Another language sometimes used through CGI.

PHP: Common opensource soloution based on including code in its own language into a html page.

Server-side JavaScript: A language mostly used on the client side but also occasionally on the server side.

7.2 PHP HYPERTEXT PRE PROCESSOR (PHP)

PHP is a popular open-source, reflective programming language used mainly for developing server-side applications and dynamic web content, and more recently, other software. Originally, PHP stood for "Personal Home Page". Today, the official meaning is the recursive acronym "PHP Hypertext Preprocessor".

PHP was originally designed as a small set of Perl scripts, followed by a rewritten set of CGI binaries written in C by the Danish-Canadian programmer Rasmus Lerdorf in 1994 to display his **rsum** and to collect certain data, such as how much traffic his page was receiving.

Famous examples of PHP applications include phpBB as well as MediaWiki, the software behind Wikipedia. The PHP model can be seen as an alternative to Microsoft's ASP.NET/C#/VB.NET system, Macromedia's ColdFusion system, Sun Microsystems' JSP/Java system, the Zope/Python system, and to the Mod perl/Perl system.

PHP's ease of use and similarity with the most common structured programming languages—most notably C and Perl (and from version 5, Java)-allows most experienced programmers to start developing complex applications with minimal learning. It also enables experienced developers to get involved with dynamic web content applications without having to learn a whole new set of functions and practices.

One of the more attractive parts of PHP is that it is more than just a scripting programming language. Due to its modular design PHP is also used to develop GUI applications (using PHP-GTK), and can be used from the command line just like Perl or Python.

PHP allows interaction with a large number of relational database management systems, such as MySQL, Oracle, IBM DB2, Microsoft SQL Server, PostgreSQL and SQLite while maintaining a simple and straightforward syntax. PHP runs on most major operating systems, including UNIX, Linux, Windows, and Mac OS X, and can interact with many major web servers. The official PHP website contains extensive documentation. The Linux, Apache, MySQL, PHP (LAMP) architecture has become popular in the Web industry as a way of deploying inexpensive, reliable, scalable, secure web applications. (The 'P' in LAMP can also stand for Perl or Python.)

Alternatively, the Windows, IIS, MySQL, PHP (WIMP) and Windows, Apache, MySQL, PHP (WAMP) architectures exist as alternatives for those wishing to use Windows as their server operating system.

7.2.1 Popularity

PHP is currently one of the most popular server-side scripting systems on the Web. It has been widely adopted since the release of Version 4.

One major part of PHP which has helped it become popular is that it is a very loose language; in particular, it is dynamically typed. That is, the rules are not as strict with variables—they do not have to be declared and they can hold any type of object. Arrays are heterogenous, meaning a single array can contain objects of more than one type.

According to Netcraft's April 2002 survey, PHP is now the most deployed server-side scripting language, running on around 9 million of the 37 million domains in their survey. This is confirmed by PHP's own figures, which show PHP usage (measured on a per-domain basis) growing at around 5 per cent per month. In May 2003, almost 13 million domains were using PHP, based on the same source.

Due to PHP's popularity in the web space, a new breed of programmer emerged who are only familiar with PHP. This encouraged the development of a command line interface for PHP, as well as GUI libraries such as GTK+ and text mode libraries like Ncurses and Newt. This was a major step for PHP, because it helped move it from being a language used only for CGI to a general-purpose programming language. On the desktop it has been favored by some new programmers as a rapid prototyping environment.

Code example

Here is a Hello World code example (See Basic syntax in the PHP manual):

```
<?php
echo 'Hello World!';
?>
```

7.2.2 Libraries

PHP includes a large number of free and open-source libraries with the core build. PHP is a fundamentally Internet-aware system with modules built in for accessing FTP servers, many database servers, embedded SQL libraries like embedded MySQL and SQLite, LDAP servers, and others. Many functions familiar to C programmers such as the printf family are available in the standard PHP build.

7.3 ACTIVE SERVER PAGES (ASP)

There are many ways you can script a Web page. This can be done on either client side or server side or both. Remember that clients computer is always slower than server's computer. One of them is Active Server Pages (ASP). Currently ASP works with Microsoft's Internet Information Server (IIS), but it is possible that other Web servers may soon support it. Although this section is mostly about ASP, you'll find you have to do some scripting on the user's computer (client) to get the job done. Therefore, you'll also find a great deal of information about client-side scripting in JavaScript and VBScript. Armed with these tools (and a good understanding of HTML), you'll be ready to conquer the Web.

Microsoft's Active Server Pages (ASP) technology has grown very popular as a Web server-side technique, combining HTML, embedded scripts, and external components to deliver dynamic content. One feature of ASP is that different languages can be used for its scripting portion, though the most widely used ASP scripting language is VBScript a subset of Microsoft's Visual Basic.

However, just because VBScript is the most popular language used by ASP developers doesn't mean it's the only one, or even the best one to use in a particular circumstance. For instance, Perl has long been synonymous with Web server development, beginning with the earliest uses of the language by CGI, and is still one of the most popular languages among Web developers. But there hasn't been much discussion of using Perl with ASP.

If your organization has been working with Perl, and you're interested in developing for the ASP environment, you don't have to give up your favorite language (or your existing Perl code) to make the transition to the new technology you can employ Perl for ASP scripting through the use of PerlScript.

7.3.1 What is ASP?

ASP is an abbreviation for Active Server Pages. It is free and already built into Win2000 FREE and also in NT4 or Win 95/98 if one installs the NT4 Option Pack.

Active Server Pages is a programming environment that provides the ability to combine HTML, scripting, and components to create powerful Internet applications that run on your server. If you are already creating Web sites that combine HTML, scripting, and some reusable components, you can use ASP to glue these items together. You can create an HTML interface for your application by adding script commands to your HTML pages and you can encapsulate your business logic into reusable components. These components can be called from script or other components.

Active Server Pages (ASP), is an open, compile-free application environment in which you can combine HTML pages, scripts, and ActiveX server components to create powerful Web-based business solutions. ASP 1.0 ships with IIS 63.40, while ASP 2.0 ships with IIS 4.0. There are other topics that cover ASP 2.0 related information such as ADO 63.5, IIS 4.0, ASP Applications, ASP Sessions, and Active Components.

Microsoft originally introduced ASP technology was with the company's own Web server, Internet Information Server (IIS). However, ASP has now been ported to other Web servers through the use of software such as ChiliSoft's platform-independent version of ASP. In addition, ASP was created originally to work in a Windows environment, but again thanks to ChiliSoft and other asp companies, ASP now runs in non-Windows environments such as UNIX and Linux.

ASP pages have an .asp extension, and are a mix of HTML and embedded script. When a client requests the page, the embedded script is accessed and processed. The results generated by the script are embedded into the Web page, which is then returned to the client browser.

7.3.2 How ASP Works?

When you incorporate ASP into your Web site, here's what happens:

- The user brings up a Web site where the default page has the extension .asp.
- The browser requests the ASP file from the Web server, instead of processing itself like in .html extension.
- The server-side script begins to run with ASP.
- ASP processes the requested file sequentially (top-down), executes any script commands contained in the file, and produces an HTML Web page. This processing is done on server side rather than client side.

The result, which is 100 per cent pure HTML code, is sent back to the browser. Because your script runs on the server, the Web server does all of the processing and standard HTML pages can be generated and sent to the browser. This means that your Web pages are limited only by what your Web server supports. Another benefit of having your script reside on the server is that the user cannot "view source" on the original script and code. Instead, the user sees only the generated HTML as well as non-HTML content, such as XML, on the pages that are being viewed.

The concept is pretty straightforward. Imagine that you have an office of 50 people. Each of them has their own phone line. You would like to show their phone usage (a very business-like way of saying "bill") over the Internet. You could do it by creating 50 different pages every night at midnight when the phone bill is compiled. That would work, but would be very labor-intensive. Or you could set up an ASP system to do the work for you.

You would create a text-based page and give it an .asp extension. That page would act as a template for the 50 different pages it will produce. Employee A logs in and offers a password to prove that he is who he says he is. The ASP page is then called upon. The template grabs the person's name, along with a few other basic items found in an employee database set up on the same server as the page. Then a second database, the one with all the phone records, is contacted and that person's phone usage is gathered.

All these pieces of information are then displayed to the viewer in the form of an HTML page. You can look at the source code, but you won't see the template. You'll see the HTML code produced by the template. All of the work is being done by the server rather than by coders. It's all very clever.

ASP really shines here in that it doesn't involve the browser to do the work. At the moment there are a lot of error messages being thrown around the Web because a person is attempting to look at a Web page designed specifically for Internet Explorer using Netscape Navigator. Silly surfer, CSS is for kids!

But ASP does all of its magic before the browser gets the page, so what browser the user has matters very little. The server does all the work rather than relying on the browser to read all the tags and see the CGI which contacts the database, which then replies with a code that is placed in the browser window—hopefully—in a scripting language the browser understands.

7.4 INTRODUCTION TO JSP

Java Server Pages or JSP for short is Sun's solution for developing dynamic web sites. JSP provide excellent server side scripting support for creating database driven web applications. JSP enable the developers to directly insert java code into jsp file, this makes the development process very simple and its maintenance also becomes very easy. JSP pages are efficient, it loads into the web servers memory on receiving the request very first time and the subsequent calls are served within a very short period of time.

In today's environment most web sites servers dynamic pages based on user request. Database is very convenient way to store the data of users and other things. JDBC provide excellent database connectivity in heterogeneous database environment. Using JSP and JDBC its very easy to develop database driven web application.

Java is known for its characteristic of "write once, run anywhere". JSP pages are platform independent. Your port your .jsp pages to any platform.

Java Server Pages technology (JSP) offers a simple way to create dynamic Web pages that are both platform-independent and server-independent, giving you more freedom through JavaTM technology's "Write Once, Run Anywhere TM" capability.

JSP technology separates content generation from presentation and takes advantage of reusable tags and objects, simplifying the maintenance of your Web applications. It's your choice — JSP technology provides the scripting ability you need to create simple interactive Web pages, or it scales to support complex Web sites that are fully integrated with enterprise class applications.

The biggest difference between the two is platform compatibility. ASP is available only on Windows, through Microsoft's IIS Web server (although one third-party vendor, ChiliSoft, offers an ASP implementation available for other platforms). JSP is available for any Web server and platform that support Java servlets, since the JSP engine itself is written entirely in Java.

Another difference is that JSP pages are compiled the first time they're requested — leading to faster page loading for subsequent requests — while ASP pages have to be translated each and every time. This helps JSP offset the fact that a lot of JVMs are still slower than native code, in which the ASP engine is written. With newer, faster JVMs, JSP may eventually be faster than ASP overall.

Finally, ASP uses either VBScript or JScript as its embedded language; JSP uses Java. Choosing the best solution for you will depend on your language preference.

7.4.1 Java Servlets

Servlets are Java technology's answer to CGI programming. They are programs that run on a Web server and build Web pages. Building Web pages on the fly is useful (and commonly done) for a number of reasons:

The Web page is based on data submitted by the user. For example, the results pages from search engines are generated this way, and programs that process orders for e-commerce sites do this as well.

The data changes frequently. For example, a weather-report or news headlines page might build the page dynamically, perhaps returning a previously built page if it is still up to date.

The Web page uses information from corporate databases or other such sources. For example, you would use this for making a Web page at an on-line store that lists current prices and number of items in stock.

Servlets are objects that conform to a specific interface that can be plugged into a Java-based server. Servlets are to the server-side what applets are to the client-side—object bytecodes that can be dynamically loaded off the net. They differ from applets in that they are faceless objects (without graphics or a GUI component). They serve as platform independent, dynamically loadable, pluggable helper bytecode objects on the server side that can be used to dynamically extend server-side functionality.

Similar to the way applets run on a browser and extend the browser's capabilities, servlets run on a Java-enabled Web server and extend the server's capabilities. Servlets are Java programs that use the Java Servlet Application Programming Interface (API) and the associated classes and methods. In addition to the Java Servlet API, servlets can use Java class packages that extend and add to the API.

Servlets extend server capabilities by creating a framework for providing request and response services over the Web. When a client sends a request to the server, the server can send the request information to a servlet and have the servlet construct the response that the server sends back to the client.

A servlet can be loaded automatically when the Web server is started, or it can be loaded the first time a client requests its services. After loading, a servlet continues to run, waiting for additional client requests.

Servlets are modules of Java code that run in a server application (hence the name "Servlets", similar to "Applets" on the client side) to answer client requests. Servlets are not tied to a specific client-server protocol but they are most commonly used with HTTP and the word "Servlet" is often used in the meaning of "HTTP Servlet".

Servlets make use of the Java standard extension classes in the packages javax.servlet (the basic Servlet framework) and javax.servlet.http (extensions of the Servlet framework for Servlets that answer HTTP requests). Since Servlets are written in the highly portable Java language and follow a standard framework, they provide a means to create sophisticated server extensions in a server and operating system independent way.

A servlet is, according to Sun Microsystems, "a standard approach to extending server functionality without the limitations of cgi-based or server-specific approaches"

- A servlet is a group of Java byte code that extends a javax.servlet. class, Generic Servlet
- A servlet overrides the default service() method
- A servlet works in conjunction with a web server
- A servlet is a reasonable alternative to cgi-bin functionality
- A servlet can act as a dynamic content provider

7.4.2 Advantage of Servlets Over "Traditional" CGI

Java servlets are more efficient, easier to use, more powerful, more portable, and cheaper than traditional CGI and than many alternative CGI-like technologies.

1. Efficient: With traditional CGI, a new process is started for each HTTP request. If the CGI program does a relatively fast operation, the overhead of starting the process can dominate the execution time. With servlets, the Java Virtual Machine stays up, and each request is handled by a lightweight Java thread, not a heavyweight operating system process. Similarly, in traditional CGI, if there are N simultaneous request to the same CGI program, then the code for the

CGI program is loaded into memory N times. With servlets, however, there are N threads but only a single copy of the servlet class. Servlets also have more alternatives than do regular CGI programs for optimizations such as caching previous computations, keeping database connections open, and the like.

2. Convenient: Hey, you already know Java. Why learn Perl too? Besides the convenience of being able to use a familiar language, servlets have an extensive infrastructure for automatically parsing and decoding HTML form data, reading and setting HTTP headers, handling cookies, tracking sessions, and many other such utilities.

3. Powerful: Java servlets let you easily do several things that are difficult or impossible with regular CGI. For one thing, servlets can talk directly to the Web server (regular CGI programs can't). This simplifies operations that need to look up images and other data stored in standard places. Servlets can also share data among each other, making useful things like database connection pools easy to implement. They can also maintain information from request to request, simplifying things like session tracking and caching of previous computations.

4. Portable: Servlets are written in Java and follow a well-standardized API. Consequently, servlets written for, say I-Planet Enterprise Server can run virtually unchanged on Apache, Microsoft IIS, or WebStar. Servlets are supported directly or via a plugin on almost every major Web server.

5. Inexpensive: There are a number of free or very inexpensive Web servers available that are good for "personal" use or low-volume Web sites. However, with the major exception of Apache, which is free, most commercial-quality Web servers are relatively expensive. Nevertheless, once you have a Web server, no matter the cost of that server, adding servlet support to it (if it doesn't come preconfigured to support servlets) is generally free or cheap.

JSP is a technology that lets you mix regular, static HTML with dynamically-generated HTML. Many Web pages that are built by CGI programs are mostly static, with the dynamic part limited to a few small locations. But most CGI variations, including servlets, make you generate the entire page via your program, even though most of it is always the same. JSP lets you create the two parts separately. Here's an example:

```
<!DOCTYPE HTML PUBLIC "-//W3C//DTD HTML
4.0                 Transitional//EN">
```

```
<HTML>
<HEAD><TITLE>Welcome to Our Store</TITLE></HEAD>
<BODY>
<H1>Welcome to Our Store</H1>
<SMALL>Welcome,
<!- User name is "New User" for first-time
visitors ->
<% out.println(Utils.getUserNameFromCookie(request)); %>
To access your account settings, click
<A HREF="Account-Settings.html">here.</A></SMALL>
<P>
Regular HTML for all the rest of the on-line
store's Web page.
</BODY></HTML>
```

The JavaServer Pages technology is based on Servlets. Servlets enable JavaServer Pages technology to provide Web application developers with a simple, consistent mechanism for extending and customizing Web server functionality. JavaServer Pages technology also integrates easily with Enterprise JavaBeansTM architecture-based technology to access transaction systems and enable re-usable, cross-platofrm capabilities using CORBA and XML.

Like CGI programs, servlets are easy to write and easy to run, and they play to the entire installed base of browsers. Servlets can do things applets can't — write to files, open sockets — and they can do them very quickly because they're invoked as threads in a demon process.

7.4.3 Advantages of JSP vs. Active Server Pages (ASP)

ASP is a similar technology from Microsoft. The advantages of JSP are two-fold. First, the dynamic part is written in Java, not Visual Basic or other MS-specific language, so it is more powerful and easier to use. Second, it is portable to other operating systems and non-Microsoft Web servers.

1. JSP vs. Pure Servlets: JSP doesn't give you anything that you couldn't in principle do with a servlet. But it is more convenient to write (and to modify!) regular HTML than to have a zillion println statements that generate the HTML. Plus, by separating the look from the content you can put different people on different tasks: your Web page design experts can build the HTML, leaving places for your servlet programmers to insert the dynamic content.

2. JSP vs. Server-side Includes (SSI): SSI is a widely-supported technology for including externally-defined pieces into a static Web page. JSP is better because it lets you use servlets instead of a separate program to generate that dynamic part. Besides, SSI is really only intended for simple inclusions, not for "real" programs that use form data, make database connections, and the like.

3. JSP vs. JavaScript: JavaScript can generate HTML dynamically on the client. This is a useful capability, but only handles situations where the dynamic information is based on the client's environment. With the exception of cookies, HTTP and form submission data is not available to JavaScript. And, since it runs on the client, JavaScript can't access server-side resources like databases, catalogs, pricing information, and the like.

4. JSP vs. Static HTML: Regular HTML, of course, cannot contain dynamic information. JSP is so easy and convenient that it is quite feasible to augment HTML pages that only benefit marginally by the insertion of small amounts of dynamic data. Previously, the cost of using dynamic data would preclude its use in all but the most valuable instances.

8.1 DATABASE CONNECTIVITY: AN OVERVIEW

The great thing about Active Server Pages is that they are parsed completely on the web server before any content is sent to the browser. For example, you can write an ASP page to display the current time, so that the page is unique for every different second in the day. ASP allows you great power, but its true capabilities shine through once you turn your ASP pages into data-driven ASP pages.

What are data-driven ASP pages? These are ASP pages which connect to databases and either retrieve information from the database and display it, or collect information from the user and store this information back into the database.

The next sensible question is, "Why use databases"? Databases give your web pages the ability to serve as a reporting program. Let's say that your company sells widgets, and the number of widgets currently in stock is in a database. You could write an ASP page which would hit this database and display the current number of widgets in stock. That way, the boss could see the inventory count from anywhere in the world!

That's just one example; there are many, many more real-world examples in which data-driven ASP pages can be used.

To connect to a database, ASP uses what is called ActiveX Data Objects, or ADO for short. ADO comes with ASP, and can be used easily from your ASP pages. ADO allows ASP to connect to any ODBC-compliant database. Nowadays, just about every database is ODBC-compliant. The database you are using is most likely ODBC-compliant; here is a short list of the more popular ODBC-compliant databases: MS-SQL Server, Access, Excel, Informix, Oracle, FoxPro.

8.2 OPEN DATABASE CONNECTIVITY

Open Database Connectivity (ODBC) is a standard software API for connecting to database management systems (DBMS). This API is independent of any one programming language, database system or operating system. ODBC is based on the Call Level Interface (CLI) specifications from SQL, X/Open (now part of the Open Group), and the ISO/IEC. ODBC was created by the SQL Access Group and first released in September, 1992.

ODBC is a native interface that is accessed through a language that can make calls into a native library. In case of the Windows platform, this library is a DLL. Microsoft Windows was the first to provide an ODBC product. Now versions exist for UNIX, OS/2, and Macintosh platforms as well.

In addition to the ODBC software, a separate module or driver is needed for each database to be accessed. The functions in the ODBC API are implemented by these DBMS-specific drivers. ODBC allows programs to use SQL requests that will access databases without having to know the proprietary interfaces to the databases. It handles the SQL request and converts it into a request the individual database system understands.

Most current DBMSs support ODBC which means that computer programs that rely on ODBC can connect to several different brands of DBMSs using the same basic code.

A JDBC-ODBC Bridge is a JDBC driver which employs the ODBC driver to connect to the database. This driver translates JDBC method calls into ODBC function calls. Such a bridge is usually used when there is no JDBC driver available for a particular database. Sun included one such bridge in the JVM, but this was intended as a stop-gap measure while JDBC Drivers were in short supply, was never meant for production use, and Sun generally recommends against its use. Independent data access vendors now deliver JDBC-ODBC Bridges which support current standards for both mechanisms, and far outperform the JVM built-in.

Is ODBC the Right way to Connect?

Some people believe that using ODBC is not as good as using a native database driver tuned to the specific database. Often this view is because ODBC is seen as being less efficient and less stable than a native driver. This is balanced by the flexibility of using ODBC making your application database independent and allowing a wider range of data sources.

Sources of ODBC Drivers

OpenLink Software ships ODBC Drivers for a number of target databases, including Bridges to other data access mechanisms (*e.g.*, ODBC, JDBC) which often provide more ODBC functionality than the targeted data source (*e.g.*, full cursor support).

Actual Technologies makes ODBC drivers for Macintosh OS X. Most popular databases are supported, including Microsoft SQL Server, Oracle and MySQL.

8.3 JAVA DATABASE CONNECTIVITY (JDBC)

JDBC is a programming interface that lets Java applications access a database via the SQL language. Since Java interpreters (Java Virtual Machines) are available for all major client platforms, this allows a platform-independent database application to be written. In 1996, JDBC was the first extension to the Java platform. JDBC is the Java counterpart of Microsoft's ODBC.

JDBC in effect is an API for the Java programming language that defines how a client may access a database. It provides methods for querying and updating data in a database. JDBC is oriented towards relational databases. The Java 2 Platform, Standard Edition includes the JDBC API together with an ODBC implementation of the API enabling connections to any relational database that supports ODBC.

JDBC allows multiple implementations to exist and be used by the same application. The API provides a mechanism for dynamically loading the correct Java packages and registering them with the JDBC Driver Manager. The Driver Manager is used as a connection factory for creating JDBC connections.

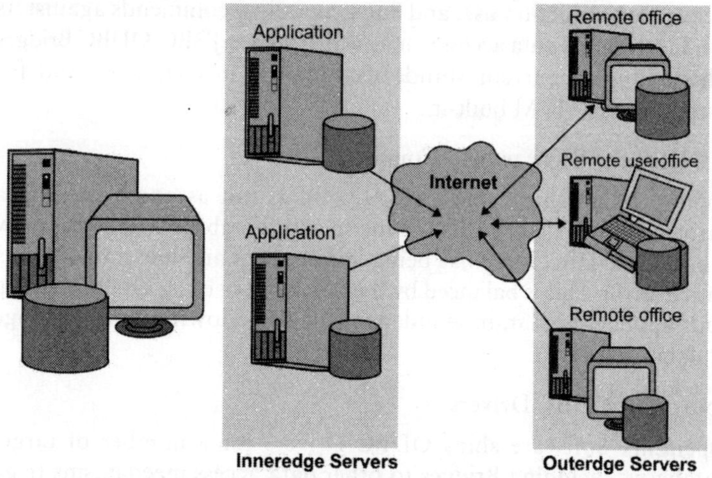

Fig. 4.10: Java Database Connecvity

JDBC is Sun Microsystems' standard SQL database access interface, providing uniform access to a wide range of relational databases. It consists of a set of classes and interfaces written in the Java programming language. The diagram below shows components of the SCO JDBC model. See below the diagram for an explanation of each component.

JDBC-Aware Java applet. A Java applet that uses the JDBC API can manipulate a database. This will typically be downloaded over the Internet.

JDBC-Aware Java application. A Java application that uses the JDBC API can manipulate a database. This will be installed on the client device.

JDBC driver manager. The JDBC API defines Java classes to represent database connections, SQL statements, result sets, database metadata, and so on. The JDBC API is implemented via a driver manager that can support multiple drivers connecting to different databases. This must be installed on the client device. The driver manager is provided by Sun Microsystems.

SCO JDBC driver. The driver is a set of Java classes that processes JDBC calls and returns results to the applet or application. The SCO JDBC driver is a JDBC-Net pure Java driver. This will typically be downloaded over the Internet. The driver is provided by SCO.

Server module. The SCO JDBC driver has a component based on the client device and a server module that are connected through a RPC mechanism built into the driver.

JDBCTM was designed to keep simple things simple. This means that the JDBC API makes everyday database tasks, like simple SELECT statements, very easy. This trail will walk you through examples of using JDBC to execute common SQL statements, letting you see for yourself how easy it is to use the basic JDBC API.

Leading database, middleware and tool vendors have been building support for JDBC technology into many new products. This ensures that customers can build portable Java applications while choosing from a wide range of competitive products for the solution best suited to their needs. The following list is a partial listing of companies that are shipping products with support for JDBC technology:

- Borland InterBase 4.0
- Caribou Lake Software Ingres, Oracle
- IBM IBM DB2
- IDS Software Oracle, Sybase, MS SQL Server, MS Access, Informix,
- Watcom, and others via ODBC
- Imaginary mSQL
- Information Builders over 70 databases
- Informix Informix
- Ingres, Informix, Oracle, Microsoft SQL Server, Sybase
- Oracle Oracle
- Sybase, Inc Sybase SQL Server, SQL Anywhere, Sybase IQ, Replication Server and more than 25 enterprise and Legacy database servers via Sybase

8.3.1 The JDBC API: Universal Data Access for the Enterprise

The "Write Once, Run Anywhere" Java 2 Platform is a safe, flexible, and complete cross-platform solution for developing robust Java applications for the Internet and corporate intranets. The open and extensible Java Platform APIs are a set of essential interfaces that enable developers to build their Java applications and applets. The Java 2 Platform provides uniform, industry-standard, seamless connectivity and interoperability with enterprise information assets.

The JDBCTM API is the industry standard for database-independent connectivity between the Java programming language and a wide range of databases. The JDBC API provides a call-level API for SQL-based database access. JDBC technology allows developers using the Java programming language to exploit "Write Once, Run Anywhere" capabilities for applications that require access to enterprise data.

The JDBC API makes it possible to do three things:

- Establish a connection with a database or access any tabular data source
- Send SQL statements
- Process the results

8.3.2 JDBC Architecture

The JDBC API contains two major sets of interfaces: the first is the JDBC API for application writers, and the second is the lower-level JDBC driver API for driver writers. JDBC technology drivers fit into one of four categories. Applications and applets can access databases via the JDBC API using pure Java JDBC technology-based drivers, as shown in the following figure:

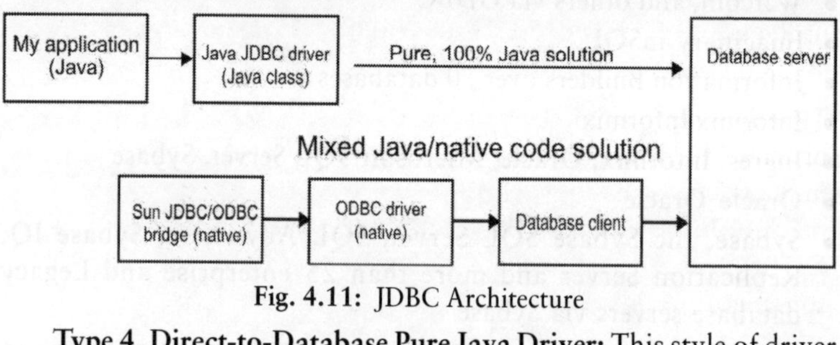

Fig. 4.11: JDBC Architecture

Type 4. Direct-to-Database Pure Java Driver: This style of driver converts JDBC calls into the network protocol used directly by DBMSs, allowing a direct call from the client machine to the DBMS server and providing a practical solution for intranet access.

Type 3. Pure Java Driver for Database Middleware: This style of driver translates JDBC calls into the middleware vendor's protocol, which is then translated to a DBMS protocol by a middleware server. The middleware provides connectivity to many different databases.

The graphic below illustrates JDBC connectivity using ODBC drivers and existing database client libraries:

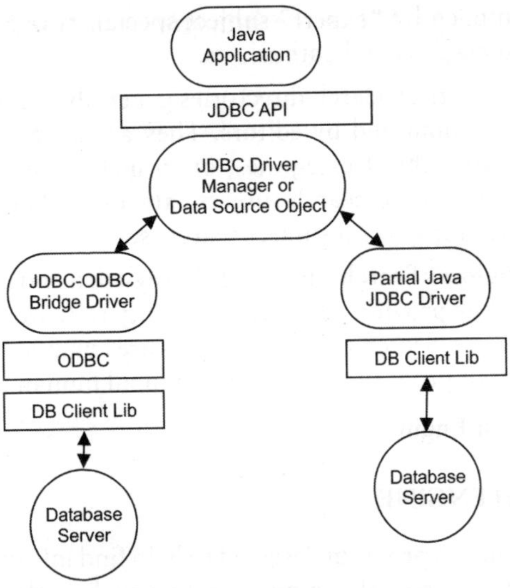

Fig. 4.12: JDBC Connectivity using ODBC Drivers

Type 2. A Native-API partly Java Technology-enabled driver: This type of driver converts JDBC calls into calls on the client API for Oracle, Sybase, Informix, DB2, or other DBMS. Note that, like the bridge driver, this style of driver requires that some binary code be loaded on each client machine.

Type 1. JDBC-ODBC Bridge plus ODBC Driver: This combination provides JDBC access via ODBC drivers. ODBC binary code—and in many cases, database client code—must be loaded on each client machine that uses a JDBC-ODBC Bridge. Sun provides a JDBC-ODBC Bridge driver, which is appropriate for experimental use and for situations in which no other driver is available.

9.0 INTERNET SEARCH TOOLS

There are many different types of search tools that help you search for information on the Internet. Each have different strengths and weaknesses and should be used accordingly. A description of each type of Internet search tool is provided to help you decide which may be the most appropriate for your search. Links are also provided to sites that provide search tips for some of the various search tools.

1. Subject Guides and Gateways: Subject Guides are Web pages containing collections of hypertext links on a particular subject area.

They are compiled by "expert" subject specialists (*e.g.* Librarians), agencies, associations and enthusiasts.

2. **Search Directories:** Search directories generally are hand-selected sites picked or submitted by editors. They are generally organized into hierarchical subject categories, often include descriptions and allow you to browse or search subject categories. These directories are most useful for general subject searches.

3. **Search Engines:** Search engines full keyword index of selected Web pages. They provide you with the ability to search by keyword but not to browse. Search engine databases are compiled by "spiders" (computer-robot programs) with very minimal human input.

4. **Meta Search Engines**

9.1 SEARCH ENGINE

A search engine is a program designed to help find information stored on a Internet or personal computer system such as the World Wide Web, or a personal computer. The search engine allows one to ask for content meeting specific criteria (typically those containing a given word or phrase) and retrieving a list of references that match those criteria. Search engines use regularly updated indexes to operate quickly and efficiently.

Without further qualificiation, search engine usually refers to a Web search engine, which searches for information on the public Web. Other kinds of search engine are enterprise search engines, which search on intranets and personal search engines, which search individual personal computers.

A search engine is a service that indexes, organizes, and often rates and reviews Web sites. It helps you find the needle that one Web site you've got to see in the Internet haystack. Different search engines work in different ways. Some rely on people to maintain a catalog of Web sites or pages. Some use software to identify key information on sites across the Internet. Some combine both types of service. So when you search their "holdings," you're bound to get different results.

Search engines are online utilities that quickly search thousands of Web documents for an entered word or phrase. Although there are some subscription-based search engines, most operate off of profits from advertisements. It should be noted that no single search engine has the contents to every Web page on the Internet. Instead, each search engine has the contents to only the Web pages that are manually entered into that search engine by the Web page operators. This is why you might get different results from different search engines.

Search engines are usually accessed through Web browser software. Each search engine provides different searching options and has its own look. Search engines also differ greatly in the number of resources they allow you to search some search engines have both searching and browsing capabilities.

Once you get connected to the Internet, you need to know how to find the Web sites and newsgroups that interest you. You may find, as many people do, that it's fun just to wanderor surf from site to site, going wherever the links take you.

Listed below are some search engines available on the Web. This is in no way intended to be a comprehensive list. Rather, it is to provide you with the names and URL's (addresses) of a few reputable search engines.

- **Infoseek http://www.infoseek.com**
- **Excite http://www.excite.com**
- **Lycos http://www.lycos.com**
- **Yahoo! http://www.yahoo.com**
- **Magellan http://www.mckinley.com**
- **OpenText Index http://index.opentext.net**
- **Accufind http://nln.com**
- **AltaVista http://altavista.digital.com**
- **HotBot http://www.hotbot.com**
- **Webcrawler http://www.webcrawler.com**

MSN Web Search: Using this search page, you can conduct a basic Web search or an advanced search, specifying one or more of the following: the type of information, the language, the origin, the domain, the date created, and the types of files contained in the Web page. AltaVista. At AltaVista, you can search for keywords or type your query in the form of a question. Other features include Babelfish, which translates sites in several languages; Photo Finder, which searches for images; Family Filter, which helps you limit the types of results produced by search.

- **Yahoo!**: This is an all purpose search engine that provides a wide range of indexed links. This was the premiere search engine a few years ago, but has digressed considerably over the last 12 months. Several links are old and no longer in existence, and it is almost impossible to get a link added to their database. This is however, a good location to start, because they have the most comprehensive system to cross-reference their lesser results with other more up to date search engine services.

This is actually a directory rather than a search engine, meaning humans compile and categorize the sites it searches. So you may get fewer results. But Yahoo! makes it easy to search for items by category and to continue your search using a search engine, if you want to.

People Search: Looking for an old war buddy? Old class-mate? Childhood friend? Movie star? This is the great place to start. You'll have to have at least a clue where the person lives, but even if you don't, you'll be amazed at what you find. To the degree you can enter in information about the person you're looking for, you will have a greater chance of finding them. This site is owned and operated by Yahoo!.

City Map: Have a party to go to, but you don't know how to get there? This search engine will use an incredible amount of city map information to find exactly how to get there. You can even find data on to and from routes. This site is owned and operated by Yahoo!.

Image Surfer: Owned an operated by Yahoo!, this site allows you to find images of nearly anything. If you're looking for clipart, they have your image!

Excite: Similar to Yahoo!, this is an all purpose search engine that is very up to date and very easy to add a URL to. The search results are very effective and come with a percentile rating of how close their results match your search words.

It is a hybrid search engine, Excite fulfills the traditional search engine definition (it is always visiting Web sites and cataloging them using a software program) but also has a directory—a subsection of its cataloged sites, which have been reviewed by people.

• **Alta Vista:** This was the premiere search engine of all time. Created by the Digital Corporation, Yahoo! uses their index databanks for some of their information, although it would appear that Yahoo! no longer uses their services. Alta Vista is a very large source of information. Results vary in accuracy, but you can usually find what you're looking for. Expect to sort through several links that don't apply, but this is the price of a huge database.

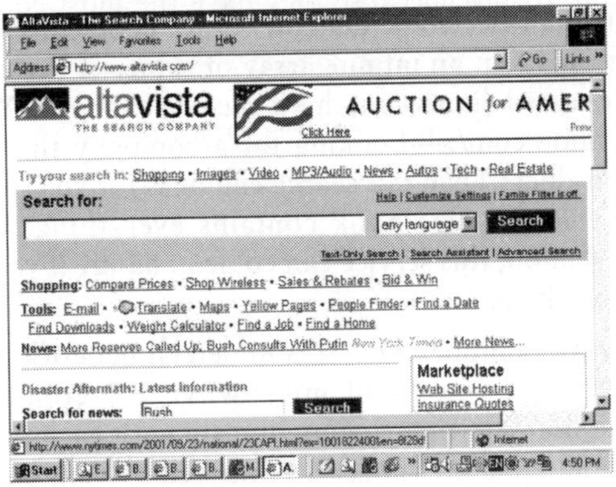

Fig. 4.13: Main Page of Altavista

• **InfoSeek:** This is an all purpose search engine that provides not only a global search engine for any subject you might be looking for, it also provides an online Dictionary and thesaurus for quick word look-ups. The results are rated and adding a URL is very simple.

Infoseek's search engine boasts an extensive catalog of sites. Infoseek has a directory, too, that categorizes sits by topic using software. Some sites have been reviewed and are recommended.

• **Webcrawler:** Webcrawler has a slightly different spin than other all purpose search engines, in that it offers links into Yellow Pages, Stock Quotes, Classifieds, Weather, People Finding, and many other interesting services. If you don't find what you're looking for, they have a cross-reference link to Excite for more searching. Webcrawler also provides a service that allows you to enter a profile of your favorite type of websites to surf to. Each time you visit, you can view only the pages that you like. Webcrawler will continue to find new webpages that match your interests.

• **HotBot:** HotBot is another all purpose search engine that provides similar services like Webcrawler. Unlike other engines that allow cross-referencing, this service is a one try searching experience with no cross-reference helper links. Lycos <http://www.lycos.com/>—is a very nice search engine that contains all the stimulus.com results you'll ever need! (laugh) There aren't any cross-reference features, but it does provide a large amount of pre-determined search categories to search from.

• **DejaNews:** This is a unique search engine that allows you to search newsgroup postings. If you're not familiar with newsgroups, they are very similar to an electronic bulletin board. People discuss all kinds of information about an infinite array of subjects. If you're looking to get into a heated discussion, this is the place for you. Yellow Pages <http://yp.yahoo.com/>—Looking for a company that produces a particular product? Then this search engine is for you. Just like your local Yellow Pages, this book contains everything. Owned and operated by Yahoo, this service can come in handy if you're looking for specific products.

• **CompuServe:** CompuServe is a commercial network based in US. It uses telephone lines and microwave for communication. CompuServe user can easily communicate with each other, around the world. Following are the services offered by CompuServe

- Electronic Mail box for message transfer
- Bulletin Board Services
- News report alongwith report on sports and weather
- Information on computer hardware and software.

• **Lycos:** Like Excite and Infoseek, Lycos is a search engine with a directory, organized by subject. You can also search the Lycos service "Top 5", sites considered by its reviewers to be among the top five per cent of sites on the Internet.

• **Snap:** Snap is a directory of Web sites that the Snap directory team has chosen, divided into topic areas, and summarized. Within each topic area, sites are divided into more specific subjects to help you locate the type of information you want. Snap's top picks are marked with a "Best" icon.

9.1.1 Challenges Faced by Search Engines

The web is growing much faster than any present technology search engine can possibly index. The following challanges are there:

- Many web pages are updated frequently, which forces the search engine to revisit them periodically.
- The queries one can make are currently limited to searching for key words, which may result in many false positives.
- Dynamically generated sites may be slow or difficult to index, or may result in excessive results from a single site.
- Many dynamically generated sites are not indexable by search engines; this phenomenon is known as the invisible web.
- Some search engines do not order the results by relevance, but rather according to how much money the sites have paid them.

- Some sites use tricks to manipulate the search engine to display them as the first result returned for some keywords. This can lead to some search results being polluted, with more relevant links being pushed down in the result list.

9.1.2 How Search Engines Work

Web search engines work by storing information about a large number of web pages, which they retrieve from the WWW itself. These pages are retrieved by a web crawler (sometimes also known as a spider)—an automated web browser which follows every link it sees. The contents of each page are then analyzed to determine how it should be indexed (for example, words are extracted from the titles, headings, or special fields called meta tags). Data about web pages is stored in an index database for use in later queries.

Some search engines, such as Google, store all or part of the source page (referred to as a cache) as well as information about the web pages, whereas some store every word of every page it finds, such as Altavista. This cached page always holds the actual search text since it is the one that was actually indexed, so it can be very useful when the content of the current page has been updated and the search terms are no longer in it. This problem might be considered to be a mild form of linkrot, and Google's handling of it increases usability by satisfying user expectations that the search terms will be on the returned web page. This satisfies the principle of least astonishment since the user normally expects the search terms to be on the returned pages. This relevance to the search makes these cached pages very useful, even beyond the fact that they may contain data that may no longer be available elsewhere.

When a user comes to the search engine and makes a query, typically by giving key words, the engine looks up the index and provides a listing of best-matching web pages according to its criteria, usually with a short summary containing the document's title and sometimes parts of the text. Most search engines support the use of the boolean terms AND, OR and NOT to further specify the search query. An advanced feature is proximity search, which allows you to define the distance between keywords.

The usefulness of a search engine depends on the relevance of the results it gives back. While there may be millions of Web pages that include a particular word or phrase, some pages may be more relevant, popular, or authoritative than others. Most search engines employ methods to rank the results to provide the "best" results first. How a search engine decides which pages are the best matches, and what order the results should be shown in, varies widely from one engine to another. The methods also change over time as Internet usage changes and new techniques evolve.

Most Web search engines are commercial ventures supported by advertising revenue and, as a result, some employ the controversial practice of allowing advertisers to pay money to have their listings ranked higher in search results.

The vast majority of search engines are run by private companies using proprietary algorithms and closed databases, the most popular currently being Google, MSN Search, and Yahoo!.

9.2 META SEARCH ENGINES

A meta search engine is a search engine that sends user requests to several other search engines and returns the results from each one. They allow users to enter their search criteria only one time and access several search engines simultaneously. Since is hard to catalog the entire web, the idea is that by searching multiple search engines you are able to search more of the web in less time and do it with only one click. The ease of use and high probability of finding the desired page(s) make metasearch engines popular with those who are willing to weed through the lists of irrelevant 'matches'.

Metasearch engines do not compile databases or catalog the web. They act as a middle-man, taking a user's request and passing it to several other search engines. No two metasearch engines are alike. Some only search the most popular search engines while others also search lesser known engines, newsgroups and other databases. They also differ in how the results are presented. Some will list results according to search engine. This allows the user to see which search engines are returning hits along the lines of what they are looking for. If the user wants to pursue deeper, they can then go to that particular search engine's page and repeat the search there. Others return results according to relevance, often concealing which search engine returned which results. This benefits the user by eliminating duplicate hits and clumping the most relevant ones at the top of the list. One metasearch engine, Dogpile, allows you to choose how results are presented.

Search engines frequently have different ways they expect requests submitted. For example, some search engines let you use the word

"AND" while others require "+" and others only require a space to combine words. The better metasearch engines try to synthesize requests appropriately when submitting them.

Results can vary between metasearch engines based on a large number of variables. Still, even the most basic metasearch engine will allow more of the web to be searched at once than any one stand-alone search engine.

9.2.1 Some Popular Search Engines

1. Dogpile: Popular metasearch site that sends a search to a customizable list of search engines, directories and specially search sites, then displays results from each search engine individually. Owned by InfoSpace, which also owns MetaCrawler.

Dogpile is a metasearch engine that "fetches" results from About, Ask Jeeves, FindWhat, LookSmart, Google, Teoma, Yahoo! and several other popular search engines.

Dogpile began operating in January of 1996 and became part of the InfoSpace Network (which also owns Metacrawler, Webcrawler and Excite) in 2000.

2. Clusty: Clusty is a metasearch engine developed by Vivisimo which offers clusters of results. It is currently in a beta that began 30 September 2004. Clusty adds new features and a new interface to the previous Vivisimo clustering web metasearch. Different tabs also offer metasearches for news and blogs. Special tabs can be added to search eBay, Wikipedia, or Slashdot. Customized tabs allow users to select sources for their own metasearch to create personalized tabs.

Clusty has free toolbars for Internet Explorer and Mozilla Firefox as well as a Mycroft Project search plugin for Mozilla and Firefox.

Clustering: A cluster is a group of similar topics that are related to the original query. The clusters are presented to the user through the Desktop metaphor of Folders. The aim of this search engine technique is to organize numerous search results into several meaningful categories (clusters).

The user benefits include:

(a) Get an overview of the available themes or topics. Via one or two clicks on a folder and subfolders, arrive at interesting search results that may be too far down in the ranking.

(b) View similar results together in folders rather than scattered throughout a list.

3. **Kartoo:** A next generation search engine which shows results in visual interface.

4. **MetaCrawler:** One of the oldest meta search services it is a metasearch engine from InfoSpace that was developed at the University of Washington and released on the Internet in 1995. Its technology was acquired by Go2Net, Inc. in 1997 and by the InfoSpace Network in 2000. Other Internet properties from InfoSpace in the area of online directory search include Dogpile.com, Webcrawler.com and webFetch.com.

5. **WebCrawler:** A search portal that searches other search engines for results, provides yellow and white page lookup and offers three levels of searching: general Web pages, news and photos. It also provides featured listings and site recommendations by category.

6. **Excite:** Formerly a crawler based search engine with its own index, it was transformed into a meta search engine when purchased by InfoSpace in 2001.

9.3 EVALUATING META SEARCH ENGINES

Meta search engines look pretty much the same up front, but their approach to presenting results varies widely. Here's a list of Search Engine Watch's pick of the best and most popular metas for searching the web. In evaluating a meta search engine, the quality of results is obviously important. Because meta search engines query multiple sources, they can often cull the best results the web has to offer. But if they don't work well, they can also have a unpleasant multiplier effect, actually amplifying poor or irrelevant search results.

While quality of results is important, there are other key factors Search Engine Watch takes into consideration when evaluating any search service. First, users prefer established services that have a degree of name recognition and reputation. Users want to see some evidence of "sustainability," that a service we review today will still be around tomorrow.

Part of the way users do this is to check the link popularity of a site. But we also look for comprehensive "about us" information, including data about the company behind the engine, a list of the engines searched, the approach to handling queries and presenting results, and so on.

Even if a site isn't well known or doesn't have a lot of link popularity, if it serves quality results and provides convincing "about us" information users will give the site careful consideration.

Above all, meta search is true search, not just providing search forms for multiple engines. Some sites that call themselves meta search engines actually just list a wide-variety of search engines and allow you to search at your choice without having to go directly to that search engine. See the All-In-One Search Pages for a list of these type of resources.

No search engine keeps track of all the content on the Internet. Even the major search engines such as MSN Web Search, AltaVista, Infoseek, Lycos, and Snap won't give you everything. (Some studies indicate that even the top search engines find less than half of what's really out there!) So try several search engines to see which produce the most useful results for the types of information you usually look for. Whenever a search fails to produce the results you wanted, try another search engine.

Here are two guidelines for picking a search engine:

- If you're looking for a broad, common topic, such as how to buy a personal computer, use a search engine that tends to give you fewer results of higher quality (usually the type that relies on people to catalog sites), such as *Yahoo*!
- If you're looking for a rare topic, such as 19th-century Hungarian authors, use a site that may be less discriminate but yields more results (usually the type compiled by a software program), such as *Excite*.

10.1 WEB DESIGN

Web development incorporates all areas of creating a Web site for the World Wide Web. This includes Web design (graphic design, XHTML, CSS, usability and semantics), programming, server administration, content management, marketing, testing and deployment. The term can also specifically be used to refer to the "back end", that is, programming and server administration. There are usually more than one member that works on a given Web Development team, each specialising in his or her own field.

A Web site is a collection of Web pages, which are documents coded in HTML that are linked to each other and very often to pages on other Web sites. A Website is hosted on a server by its owner or at an ISP. It may share space on the server with other Web sites, reside on a server dedicated to that Website only or be on multiple dedicated servers. To qualify as a Web site, the Web server must be available on the Internet 24 hours a day.

10.2 WEBSITE

A website, Web site or WWW site (often shortened to just site) is a collection of webpages, that is, HTML/XHTML documents accessible via HTTP on the Internet; all publicly accessible web sites in existence comprise the World Wide Web. The pages of a web site will be accessed from a common root URL, the homepage, and usually reside on the same physical server. The URLs of the pages organize them into a hierarchy, although the hyperlinks between them control how the reader perceives the overall structure and how the traffic flows between the different parts of the site.

Some (parts of) websites require a subscription, with a fee to be paid periodically (*e.g.*, every month), or just a free registration. Examples include many Internet pornography sites, parts of many news sites, gaming sites, message boards, web-based e-mail services and sites providing real-time stock market data.

A website will often be the work of a single individual, a business or organization, or dedicated to a particular topic or purpose. This is quite a blurry definition, given the hypertext nature of the web: the whole of Wikipedia forms a website, but whether the Meta-Wikipedia pages are part of the same website or a sister website is open to debate.

10.2.1 WebSite Content

Content is critical in the design of your site. You must decide what works best with your audience and goals; text or graphics. You need to weed out the extraneous material while being sure to leave the information necessary to sell your product.

Once you have the potential customer at your website, you want to be sure that you have the information available that will turn him into a paying customer.

The content is essentially the makeup of your Web site design. Content in this case is really the style in which you wish to present your presence. So the content of your Web site may very well be composed of a variety of items such as:

- Information
- Graphical Images
- Multimedia
- Demo products

So what does style have to do with anything? Well, the answer is everything and nothing. What I mean by that is, if you want to make a lasting impression, then consider how style goes hand in hand with the content and presentation. Several things will determine the style you choose, for instance:

- The bottomline is that you will want to make your Web site interesting, attractive and above all useful in content, if you want a good shot at return visits. If your Web site is boring, poorly designed or just plain lacks any useful content, you can pretty much be assured that you won't get many repeat visitors. Here's something to keep in mind, everybody is a critic and it is virtually impossible to please everybody. So before you settle on a style, do some research.
- In case you decide not to employ the advice given above or perhaps your just in an information overload mode. Here are some basic tips to remember.
- You should have a basic understanding of GUI design.
- Be consistent, only apply a few basic document design principles.
- Incorporate proper typographic design.

And don't forget to include.

- Hypertext or links
- Contact information
- Date of creation and revision
- Statement of Copyright

10.3 MARKUP LANGUAGE

A set of labels that are embedded within text to distinguish individual elements or groups of elements for display or identification purposes. The labels are typically known as "tags".

For rendering, mark-up languages indicate where font and other layout changes start and stop. For content identification, markup languages turn a text document into the equivalent of a database record in which individual data elements can be located for processing.

SGML is the grand-daddy mark-up language that served as the foundation for HTML and XML. HTML is used for rendering the document, and XML is used for identifying the content of the document

Web sites are written in, or dynamically converted to HTML (Hyper Text Markup Language) and are accessed using software called a web browser. Websites consist of static HTML pages or pages that are created dynamically using technologies such as Active

Server Pages (ASP), or Java Server Pages (JSP). A website also requires software known as an HTTP Server, such as Apache (http://www.apache.org), the most commonly used web server software used on the Internet, or Microsoft's Internet Information Server (IIS). Often websites will also consist of content stored in one or more databases.

Plugins are also available for browsers which allow them to show active content, such as Flash, Shockwave or applets written in Java. Dynamic HTML also provides for user interactivity and realtime element updating within webpages (*i.e.*, pages don't have to be loaded or reloaded to effect any changes), mainly using the DOM and JavaScript, support for which is built-in for most modern browsers.

10.4 WEB DESIGN

Web design is the design or designing of a Web page, Website or Web application. The term generally refers to the graphical side of Web development using images, CSS and XHTML. When the Internet was first invented, Web design consisted of a basic mark-up language that included some formatting options, and the unique ability to link pages together using hyperlinks. It was this feature that characterized the Web among other communication methods, and characterized Web design among other design methods. Because of this unique behaviour of the World Wide Web, and the unique behaviour it encouraged in users, Web design would prove to be unlike any other form of design before or since, with the possible exception of interactive CD-ROM design.

As the Web and Web design progressed, the mark-up language used to make it, known as HTML, became more complex and flexible. Things like tables, which could be used to display tabular information, were soon subverted for use as invisible layout devices. With the advent of Cascading Style Sheets (CSS), table based layout is increasingly regarded as outdated. Database integration technologies such as server-side scripting (see PHP, ASP.NET, ASP, JSP, and ColdFusion) and design standards like CSS further changed and enhanced the way the Web was made.

The introduction of Macromedia Flash into an already interactivity-ready scene has further changed the face of the Web, giving new power to designers and media creators, and offering new interactivity features to users.

Web applications dynamically generate a series of web-pages in a standard format supported by common browsers such as HTML/XHTML. Client-side scripting in a standard language such as JavaScript is commonly included to add dynamic elements to the user interface. Generally, each individual web page is delivered to the client as a static document, but the sequence of pages can provide an interactive experience, as user input is returned through web form elements embedded in the page markup. During the session, the web browser interprets and displays the pages, and acts as the universal client for any web application.

The web interface places some limits on client functionality. Application-specific methods such as drawing on the screen, and more general-purpose techniques such as drag and drop are not supported by standard browser technology. Web developers often use client-side scripting to add functionality, especially to create an interactive experience that does not require page reloading (which many users find disruptive). Recently, technologies have been developed to coordinate client-side scripting with server-side technologies such as PHP; AJAX, a technology that works with JavaScript, is an example.

A significant advantage of building web applications to support standard browser features is that they should perform as specified regardless of the operating system or OS version installed on a given client. Rather than creating clients for Windows, Mac OS X, Linux, and other operating systems, the application can be written once and deployed almost anywhere. However, inconsistent implementations of the HTML, CSS, DOM and other browser specifications can cause problems in web application development and support. Additionally, the ability of users to customize many of the display settings of their browser (such as selecting different font sizes, colors, and typefaces, or disabling scripting support) can interfere with consistent implementation of a web application.

Another (less common) approach is to use *Macromedia Flash* or *Java applets* to provide some or all of the user interface. Since most web browsers include support for these technologies (usually through plug-ins), Flash- or Java-based applications can be implemented with much of the same ease of deployment. Because they allow the programmer greater control over the interface, this overcomes many browser-configuration issues, but incompatibilities between Java or Flash implementations on the client can introduce different complications. Because of their architectural similarities to traditional client-server applications, with a somewhat "thick" client, there is some dispute over whether to call systems of this sort "web applications"; an alternative term is "rich internet application".

Though many variations are possible, a web application is commonly structured as a three-tiered application. In its most common form, a web browser is the first tier, an engine using some dynamic web content technology (*e.g.*, CGI, PHP, Java Servlets or Active Server Pages) is the middle tier, and a database is the third tier. The web browser sends requests to the middle tier, which services them by making queries and updates against the database and generating a user interface.

Web interfaces have increasingly been used for applications that have previously been thought of as traditional, single-user applications. For example, Microsoft HTML Help replaced Windows Help as the primary help system in Microsoft Windows. Like their networked brethren, such applications generate web pages as their user interface and send them (sometimes via an embedded HTTP server) to a local web browser component, which then renders the pages for the user and returns user input to the application. Web applications powered by embedded web servers have also become commonplace as the user interfaces for configuring network components such as servers, routers, and gateways.

An emerging strategy for application software companies is to provide web access to software previously distributed as local applications. Depending on the type of application, it may require the development of an entirely different browser-based interface, or merely adapting an existing application to use different presentation technology. These programs allow the user to pay a monthly or yearly fee for use of a software application without having to install it on a local hard drive. A company which follows this strategy is known as an application service provider (ASP), and ASPs are currently receiving much attention in the software industry.

10.5 STANDARD GENERALIZED MARK-UP LANGUAGE (SGML)

The Standard Generalized Mark-up Language (SGML) is a metalanguage in which one can define mark-up languages for documents. SGML is a descendant of IBM's Generalized Mark-up Language (GML), developed in the 1960s by Charles Goldfarb, Edward Mosher and Raymond Lorie (whose surname initials also happen to be GML).

SGML provides a variety of mark-up syntaxes that can be used for many applications. By changing the SGML Declaration one does not even need to use "angle brackets" although they are the norm, the so-called concrete reference syntax.

SGML was originally designed to enable the sharing of machine-readable documents in large projects in government and the aerospace industry, which have to remain readable for several decades—a very long time in information technology. It has also been used extensively in the printing and publishing industries, but its complexity has prevented its widespread application for small-scale general-purpose use.

SGML syntax example:

```
<QUOTE TYPE="example">
typically something like <ITALICS>this</
ITALICS>
</QUOTE>
```

SGML is an ISO standard: "ISO 8879:1986 Information processing-Text and office systems-Standard Generalized Mark-up Language (SGML)"

XML is derived from SGML. XML is a profile—a specific subset of SGML, designed to be simpler to parse and process than full SGML. Another markup language originally created as an application of SGML is DocBook, designed for authoring technical documentation. DocBook is now also available as an XML application. XML is an attempt to simplify SGML for general-purpose applications, such as the Semantic Web. XML has been used for a large number of applications, including notably XHTML, RSS, XML-RPC and SOAP.

HTML, although originally designed independently, was later reformulated (at version 2.0) to be an application of SGML, although there's some debate on whether it ever actually became one. There are also a number of languages that are related in part to SGML and XML, but, because they cannot be parsed or validated or otherwise processed using standard SGML and XML tools, cannot be considered to be applications of SGML or XML. One example is the Z Format, a language designed for typesetting and documentation.

10.6 HYPERTEXT MARKUP LANGUAGE (HTML)

HTML is the simple markup system used to create hypertext documents. HTML is not intended to be a comprehensive page-layout system. Instead, HTML aims to let you describe the structure of your document by indicating headings, emphasis, links to other documents and so forth. The more you work with HTML rather than against it, the happier you'll be.

You can include images and other multimedia objects in your documents, but it should be remembered that not all web users have graphical clients, and many web users voluntarily turn graphics _off_ to save downloading time! If you try to spite such users, you will only lose readers (and customers). You can in fact specify a great deal about the appearance of your document in the latest web browsers. There is no harm in taking advantage of these features, but as a rule of thumb, always make sure your document looks good in a text-based browser such as Lynx as well as in the graphical browser of your dreams. This is more than a simple matter of taste. Keep in mind that not all users can see!

10.6.1 HTML: The Infrastructure of Cybersystems

HTML, or hypertext mark-up language, is a programming language used to build Web sites. It contains standard codes, or tags, that determine how a Web page looks when your browser displays it. HTML tags also make possible the hyperlinks that connect information on the World Wide Web.

HTML is an SGML DTD.Document Type Definition (DTD)—this is the formal specification of a mark-up language, written using SGML. In practical terms, HTML is a collection of platform-independent styles (indicated by mark-up tags) that define the various components of a World Wide Web document. HTML was invented by Tim Berners-Lee while at CERN, the European Laboratory for Particle Physics in Geneva. Currently (1999) HTML Version 4.0-plus is in vogue.

HTML documents are plain-text (also known as ASCII) files that can be created using any text editor (*e.g.*, Emacs or vi on UNIX machines; SimpleText on a Macintosh; Notepad on a Windows machine). You can also use word-processing software if you remember to save your document as "text only with line breaks".

There are three ways to produce HTML documents:

- Writing them yourself, which is not a very difficult skill to acquire,
- Using an HTML editor, which assists in doing the above, and
- Converting documents in other formats to HTML.

10.6.2 Starting learning HTML

The following are the essential elements that are need to create a Web page:

HTML

This element tells your browser that the file contains HTML-coded information. The file extension .html or simply .htm also indicates that this an HTML document. (If you are restricted to 8.3 filenames (*e.g.* in DOS mode, use only .htm for your extension as in Buddha.htm.)

HEAD: The head element identifies the first part of your HTML-coded document that contains the title. The title is shown as part of your browser's window (see below).

TITLE: The title element contains your document title and identifies its content in a global context. The title is typically displayed in the title bar at the top of the browser window, but not inside the window itself. The title is also what is displayed on someone's hotlist or bookmark list, so choose something descriptive, unique, and relatively short. A title is also used to identify your page for search engines (such as HotBot <http://www.hotbot.com/> or Infoseek <http://www.infoseek.com/>).

For example, you might include a shortened title of a book along with the chapter contents: NCSA Mosaic Guide (Windows): Installation. This tells the software name, the platform, and the chapter contents, which is more useful than simply calling the document Installation. Generally you should keep your titles to 64 characters or fewer.

BODY: The second—and largest—part of your HTML document is the body, which contains the content of your document (displayed within the text area of your browser window). The tags explained below are used within the body of your HTML document.

10.6.3 Creating Web Pages with HTML

Tutorial 1: Tags and the basic structure of an HTML document

Open up Notepad from windows by clicking start > accessories > Notepad and then type

```
<HTML>
</HTML>
```

Each one of those things is called a tag. There is a starting tag and a closing tag. To make a closing tag just add a / to the starting tag but after <. Most, but not all tags have a closing tag.

Think of tags as talking to the browser, or better yet giving it instructions. What you have just told the browser is 'this **(<HTML>)** is the start of a HTML document' and 'this **(</HTML>)** is the end of a HTML document'. Now we need to put some information

Every HTML document needs a pair of HEAD tags.

```
<HTML>
<HEAD>
</HEAD>
</HTML>
```

The only thing we have to concern ourselves with in the HEAD tags (for now) are the TITLE tags.

```
<HTML>
<HEAD>
<TITLE></TITLE>
</HEAD>
</HTML>
```

The Head tag incorporates the matter which shall appear in the title bar of the window which is going to appear when the program is hosted to WWW and browse through any browser. And the bulk of the page is going to be within the BODY tags.

```
<HTML>
<HEAD> .....
<TITLE></TITLE> | Head
</HEAD> .....|
<BODY> .....
</BODY> .....| Body
</HTML>
```

Don't forget give your document a title.

```
<HTML>
<HEAD>
<TITLE> My Very Own Web Page! </TITLE>
</HEAD>
<BODY>
</BODY>
</HTML>
```

Now save it, not as a text document, but as a html document. In your note pad window click File then Save As. Save it as page1.html in a new folder somewhere (Win3.x users save it as page1.htm).

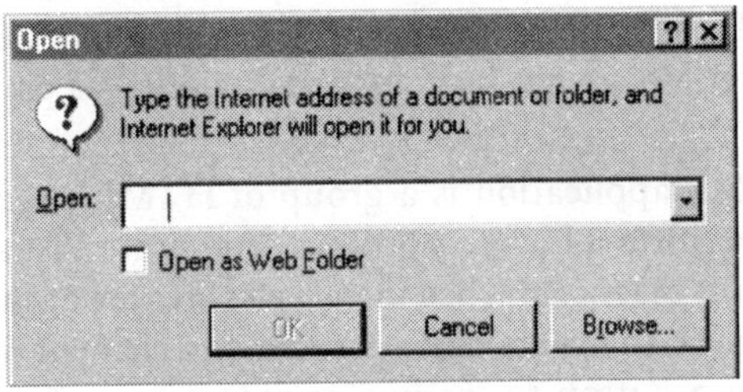

Fig. 4.14: "Open" Dialog Box

You will be presented with a dialog box. If you are working on a computer in a lab or if you do not want to save your work to the hard drive, change to the (A:) drive by using the drop down window at the top and make sure you have a formatted disk in the drive. If you are working on your own computer, migrate to the spot on your drive you will want to save the files you create for.

Make a new folder by clicking this. Name it whatever you want. Then double click on it to open it up. Where it says File name: type in page1.html where it says Save as type: make sure it says All Files(*.*). Press enter.

Congratulations! You are the proud parent of a fully functional Web Page! You could upload it to a server and the whole world can see your creation! If you are using Internet Explorer, the file you made should look something like this...

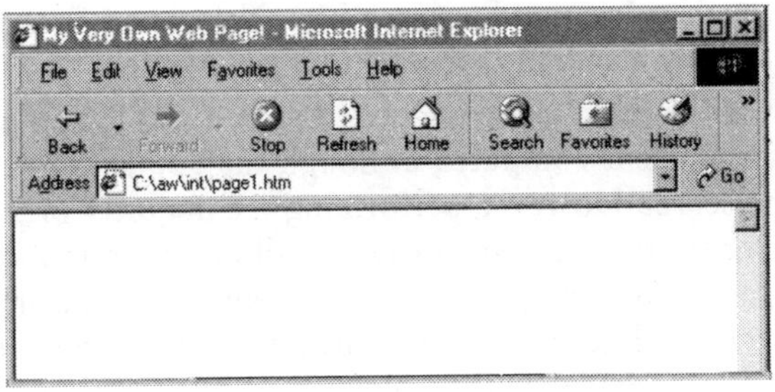

Fig. 4.15

You can double click on it now and see the results of your handiwork. This should open a new window in your browser.

10.7 DYNAMIC HTML (DHTML)

A set of innovative additions to HTML that allows page authors and developers to dynamically change the style and attributes of elements on an HTML page, as well as insert, delete, or modify elements and their text after a page has been loaded. Included as part of Internet Explorer 4.x.

A DHTML application is a group of HTML pages that work together to perform a business process. You write Visual Basic code to handle events that occur when these pages are viewed in the browser. You can respond to events that occur on any element on the page _ from clicking a button to loading an image to passing the mouse over a certain part of the page.

Dynamic HTML or DHTML designates a technique of creating interactive web sites by using a combination of the static markup language HTML, a client-side scripting language (such as JavaScript) and the style definition language Cascading Style Sheets. It may be used to create applications in a web browser: for example to ease navigation, to create interactive forms or to create interactive exercises to use in e-learning. Competing techniques include Macromedia Flash for animation, Java applets and Scalable Vector Graphics (SVG), though SVG is not yet well-supported by the major web browsers.

Some disadvantages of DHTML are that it is difficult to develop and debug due to varying degrees of support among web browsers of the aforementioned technologies and that the variety of screen sizes means the end look can only be fine-tuned on a limited number of browser and screen-size combinations. Development for recent browsers, such as Internet Explorer 5.0+, Netscape 6.0+, and Opera 7.0+, is aided by a shared Document Object Model.

Most of the processing associated with a DHTML application occurs on the client computer, although the application can make calls to the server. However, performing the majority of processing on the client (in this case, the browser) allows your applications to respond quickly to user actions without making time-consuming trips to the(server. In addition, corporations can lower the cost of training and support associated with their applications by presenting commonly used information and applications through the browser.

The end user views a DHTML application with Internet Explorer 4.0 or later. DHTML applications are designed to work most optimally on Intranets. An Intranet is a Web site or series of Web sites that belong to an organization and can be accessed only by the organization's members. Many corporations use an Intranet, rather than the Internet, to offer their employees easy access to corporate information, such as customer order information, sales data, or performance figures, while preventing outside access to that data.

In addition to writing Visual Basic code to hook up to a Web page, you can use a special designer in Visual Basic to create Web pages that act as the user interface of your application. You can also pull an existing Web page into Visual Basic and modify it. This allows you to tailor the process to your preferences _ if you are familiar with HTML and want to create your own pages you can do so using the DHTML Page designer. If not, you can have an experienced Web developer create the pages and simply add them to your application.

10.7.1 Web Pages vs. Forms

DHTML applications are structured differently than forms-based Visual Basic applications. In a DHTML application, the user interface consists of a series of HTML pages rather than forms. An HTML page is like a form in that it contains all the visual elements that make up your application's user interface. You can place the some of the same items into a page as you do a form, including text, buttons, checkboxes, and radio buttons.

An HTML page is stored in an .htm file that is analogous to a .frm file, in that it is used to render and display the form to the end user. You can create these pages yourself, or a Web designer or a developer can create the pages and you can then link the finished pages into your project.

The following table sums up the differences between forms-based applications and Web-based applications:

- Forms-based application
- Web-based application
- User interface Visual Basic forms HTML pages
- UI elements (for example, text boxes, buttons, images) controls elements
- File format .frm files .htm or .html files, or generated from Visual Basic code

- Creator developer Web designer or developer
- Run time Visual Basic run-time DLL, msvbvm60.dll Web browser or Web browser control, with msvbvm60.dll

10.7.2 Structure of DHTML Applications

DHTML applications are made up of the following pieces:

- One or more HTML pages.
- Visual Basic code that handles the events generated from the HTML pages.
- A run-time component that hosts the page in the Web browser or Web browser control.

A project DLL that contains your Visual Basic code and is accessed by the run-time component, generated automatically when you debug or compile.

There is a one-to-one relationship between the designers and the HTML pages in your project. For each page in your application, there is a page designer.

10.7.3 Advantages of DHTML Applications

Building a DHTML application in Visual Basic provides several advantages over other methods of Internet development. DHTML applications give you:

1. Dynamic HTML: When you create a DHTML application, you have full access to the richness of Dynamic HTML, integrated with the power of Visual Basic code and controls. See "Dynamic HTML in Visual Basic" for more information about the Dynamic HTML features you can access.

2. Lessened server load: DHTML applications conserve server resources because each request or user action does not have to be routed through the Web server.

3. Fewer refreshes, faster responses: When an end user's actions initiate changes to a typical Web page, the browser must refresh the page from the server. In a DHTML application, the browser can process user data, make changes to the page's layout and appearance, and process code all without refreshing the page.

4. Dynamic interaction: Visual Basic code on a Web page can directly manipulate any element on the page and create and manage new elements on the fly, allowing for truly dynamic user interfaces.

5. **Improved state management:** Typically, HTML pages are stateless, that is, no information about an HTTP request is maintained after the response is received from the server. Visual Basic DHTML applications allow you to store state between requests, without using the server. Therefore, multiform or multipage applications are possible without requiring server interaction, complex URL-based state, or cookies.

6. **Offline capability:** For a DHTML application, users can browse to and use a DHTML application on their corporate Intranet. Later, when disconnected, the same users can still make use of their Web-based application through the browser's cached storage.

7. **Code security:** When you embed scripts within an HTML page, anyone can access your page, read the script, and make changes to it. Using Visual Basic to develop your DHTML application, your code is compiled, is not part of the HTML page itself, and cannot be tampered with as easily.

10.7.4 HTML Elements as Objects

An HTML page consists of a series of objects that are called elements. Elements on a page can include simple items such as headings, paragraphs of text, and images, or more complicated items such as form fields, or tables and their rows, columns, and cells. On the surface, the page's elements correspond very closely to the items in a Visual Basic form. However, in past versions of HTML you could not access properties and events on all of these objects: only certain elements were available to be programmed.

With Dynamic HTML, you can now access and manipulate every element on a Web page. This allows you to treat an HTML page much like a form you can change the attributes for any element, write code or script that performs actions on it, and access its methods and events.

10.7.5 Dynamic Styles

Originally, HTML was intended to provide a way to separate the structure and content of a Web page from its presentation. That is, the HTML code that defined a page's content and structure was not intended to also define the physical appearance of elements on the page. In earlier version of HTML, this separation was not possible.

The only way to allow users to control the formatting of page elements was to introduce tags such as bold <B> and italic <I> and a variety of attributes that controlled such things as font style and size.

With Dynamic HTML, the original intent of HTML is realized in that you can use cascading style sheets (CSS) to format your page elements. The code controlling the appearance of page elements can be stored in a style sheet. This style sheet is kept separate from the HTML code controlling the page's content and structure.

A style sheet is a collection of properties that control the appearance of elements on a Web page. Style sheets can apply a style to a group of elements, or to a single element. In addition, you can apply multiple styles to each element on the page. There are three main types of styles you can use in your DHTML applications:

- Inline styles are styles applied to a single element and applied within the element's HTML tag. This is the least flexible way to use styles.
- Global style sheets are defined at the beginning of an HTML file and set and apply styles for all elements within the page. This is a flexible way to set all styles for a single page.
- Linked style sheets are stored in a separate file and can apply to multiple pages. This is the most powerful and flexible way to set styles for an application that uses more than one page.

The advantage of using style sheets is that it is very easy to make changes to a page's appearance without writing code. That is, you can change a single line of code in the style sheet— such as increasing the font size of an element — and have that change be reflected in multiple locations throughout your page or in multiple pages. In previous version of HTML, you would have had to make the same change in each location where you wanted to alter the page's appearance. Because of this, it is preferable to avoid using inline styles as much as possible, and store your style information in either a global or a linked style sheet.

Note to define global style sheets in your DHTML applications, you need to edit your application's HTML pages in an external editor, as Visual Basic does not provide a way to define styles sheets within the designer. However, if you use the designer to open a page that references a style sheet, the styles for that document appear in the toolbar.

10.8 EXTENSIBLE MARKUP LANGUAGE (XML)

XML is a text-based format that lets developers describe, deliver and exchange structured data between a range of applications to clients for local display and manipulation. XML also facilitates the transfer of structured data between servers themselves. Vast stores of legacy information exist today, distributed across disparate, incompatible databases. XML allows the identification, exchange and processing of this data in a manner that is mutually understood, using custom formats for particular applications if needed.

XML provides a significant advance in how data is described and exchanged by Web-based applications using a simple, flexible standards-based format. HyperText Markup Language (HTML) enables universal methods for viewing data; XML provides universal methods for working directly with data.

XML provides a format for describing data. This facilitates more precise declarations of content and more meaningful search results across multiple platforms. In addition, XML enables the separation of presentation from data. For example, in HTML you use tags to tell the browser to display data as bold or italic; in XML you use tags only to describe data, such as city name, temperature, and barometric pressure. In XML, you use style sheets such as Extensible Stylesheet Language (XSL) and cascading style sheets (CSS) to present the data in a browser. XML separates the data from the presentation and the process, enabling you to display and process the data as you wish by applying different style sheets and applications.

10.8.1 XML and Existing Protocols

Internet Protocol (IP), the HyperText Mark-up Language (HTML) and the Hypertext Transport Protocol (HTTP) have revolutionized the manner in which information is distributed, displayed and searched for. Organizations rapidly embraced browsers and search engines with the creation of corporate Intranets, then extended these to customers, suppliers, and business partners with extranets.

XML which complements HTML, promises to increase the benefits that can be derived from the wealth of information found today on IP networks around the world. This is because XML provides a uniform method for describing and exchanging structured data. The ability to describe structured data in an open text-based format and deliver this data using standard HTTP protocol is significant for two reasons. XML will facilitate more precise declarations of content and more meaningful search results across multiple platforms. And once the data is located it will enable a new generation of viewing and manipulating the data.

Although XML differs from HTML in some fundamental ways, learning to author XML documents does not require a great deal of effort, once you are familiar with HTML. Because HTML and XML both share a common heritage (SGML), their syntax is very similar. As an HTML author, you will already be familiar with many aspects of XML's syntax before you even begin to read any XML code.

10.8.2 XML: An Subset of SGML

XML derived from the Standard Generalized Mark-up Language (SGML). You can use both XML and SGML to create self-describing documents. Both languages use textual mark-up (tags) to describe data so that other applications or tools (like an SGML or XML parser) can correctly read the information and then do interesting things with it. XML is a simplified version of SGML, more suitable for use on the Web.

Microsoft Corporation is working with the W3C and other organizations to create a powerful standard definition of XML to significantly advance how people can take advantage of Web-based data. HTML enables a universal method for displaying data; XML provides a universal method for describing data. There is a rapidly growing need to go beyond simply viewing data, as is the case with Web browsers today. People need to be able to work with and manipulate the data. And this is what XML allows.

XML is optimized for delivery over the Web. It is defined by the World Wide Web Consortium (W3C). This standardization ensures that structured data will be uniform and independent of applications or vendors.

XML is at the core of many features of Visual Studio.NET and the .NET Framework. The topic list below shows some of the places you can find information about XML as it relates to various feature areas, including data, Web Services, Web Forms, and reference materials.

10.8.3 Structure of XML

Fortunately, if all you want to do is read and parse an XML file from someone else, and you can make the assumption that the document is valid XML, you don't have to know the entire standard. In fact, there are just a few key terms you need to understand to get started in the XML world. You can always pick up more later, but for now, just concentrate on these four ideas:

- A document is a chunk of XML. You can think of an XML document as the equivalent of an HTML page. It's a bit of XML that's served up by some Web server, or other data feed, as a single chunk.
- An XML declaration is a processing directive that identifies the document as being XML, and includes the XML version.
- Elements are the building blocks of the document. If you're familiar with HTML, HTML tags are the rough equivalent of XML elements. One of the key distinctions is that in XML, new elements may be defined by any document.
- Attributes describe some feature of an element.

You probably know that XML is basically structured ASCII with embedded tags that allow data to be self-describing and extensible. XML is relatively easy to read whether you're a computer, or human, XML documents are reasonably clear to the lay person and because it's so simple, every modern platform is capable of working with it, making it a very Web-friendly technology.

Element Mark-up

XML, like HTML, is made up of elements. Within HTML, elements are often referred to as "tags." However, a distinction between tags and elements must be made clear, for treating them as one and the same could cause you to run into some problems when creating sound XML documents. A tag is a singular entity that opens or closes an element. For instance, the <P> HTML tag opens a paragraph element. Likewise, the </P> HTML tag closes a paragraph element. These two tags, plus the content between them, represent the HTML element. A tag is only part of an element, not the element itself.

Like HTML, XML's purpose is to describe the content of a document. Unlike HTML, XML does not describe how that content should be displayed. Instead, it describes what that content is. Using XML, the Web author can semantically mark up the contents of a document, describing that content in terms of its relevance as data.

For example, the following HTML element

```
<P>Devdas</P>
```

describes the contents within the tags as a paragraph. This is fine if all we are concerned with is displaying the words " Devdas" within a Web page. But what if we want to access those words as data? Using XML, we can mark-up the words "The Phantom of the Opera" in a way that better reflects their significance as data:

```
<film>Devdas</film>
```

Note: Although XML does not describe a display structure, you can display your XML code using a style sheet language, such as XSL.

XML does not limit you to a set library of tags. When marking up documents in XML, you can choose the tag name that best describes the contents of the element. For instance, in the above case, <film> may not be accurate enough. You may need to differentiate between the silent classic, starring Ashok Kumar, and the 2002 remake, starring Sahrukh Khan. This can be achieved a number of ways. You could create a more specific name for the element:

```
<silent-film>Devdas</silent-film>
```

That would adequately serve your purposes in this particular case. But what if you wanted to access all films, regardless of whether they have sound, and still differentiate between versions of the film? In such a case, referring to the contents as a "silent-film" would not be as useful as referring to it as simply a film. However, if we were to change it back we would be left with the same ambiguities as before.

One solution is to use an attribute to describe the type. XML allows you to place attributes on elements. The naming of such attributes, just as with tag names, is the domain of the author. Using an attribute, you could describe "Devdas" as a film and then specify whether that film has sound:

```
<film sound="no">Devdas</film>
```

You are now faced with another dilemma: There is more than one version of the movie with sound. The following element

```
<film sound="yes">Devdas</film>
```

remains ambiguous. Because XML allows for multiple attributes, however, you can resolve such ambiguities by adding a second attribute:

```
<film sound="yes" year="1979">Devdas</film>
```

The above markup might be a bit more complex than <P>, but we are now able to distinguish when the words "Devdas" refer to one film and when they refer to another. In the future, this could enable the cataloging of such films in a variety of ways.

10.8.4 Uses of XML

XML is an extremely flexible way to pass around data. The following are all examples where XML can be used:

- An ordinary document.
- A structured record, such as an appointment record or purchase order.
- Internet/Intranet Web applications that move data.
- An object with data, such as the persistent format of an object or ActiveX control.
- A data record, such as the result set of a query.
- Meta-content about a Web site, such as Channel Definition Format (CDF).
- Graphical presentation, such as an application's user interface.
- Links between information and people on the Web.
- Discovery documents used to locate available Web Services.

The following lists several advantages XML has over other formats when storing information:

- XML formats are text-based, making them more readable, easier to document, and sometimes easier to debug.
- XML documents can use much of the infrastructure already built for HTML, including the HTTP protocol and some browsers. HTTP allows XML to be transferred across firewalls.
- XML parsing is well defined and widely implemented, making it possible to retrieve information from XML documents in a variety of environments.
- Applications can rely on XML parsers to do some structural validation, as well as data type checking (when schemas are used).
- XML is built on a Unicode foundation, making it easier to create internationalized documents.
- XML is not appropriate for every situation, however. XML documents tend to be more verbose than the binary formats they replace. They take up more network bandwidth and storage space, or require more processor time for compression. XML parsing can be slower than parsing highly optimized binary formats and can require more memory. However, careful application design can avoid some of these problems.

Data can be uniquely tagged with XML, allowing a customer to specify books by, rather than about, Winston Churchill, for example. By contrast, searches using present methods would probably yield both types of books mixed together. Without XML, it is necessary for the searching application to understand the schema of each database, which describes how it is built. This is virtually impossible because every database describes its data differently. With XML, however, books could be easily categorized in a standard way by author, title, ISBN number, or other criteria. Agents could then search these identified bookstore sites in a consistent way for books about Winston Churchill.

11.1 INTERNET SECURITY

Internet is not a very a safe place and that there is a reason for you to protect your password. Unfortunately, there are MANY people out there who would LOVE to break into your account and "use your account as a base for operations".The main defense against people who want to break into your account— a.k.a. "*hackers*" or "*crackers*" —is your password. Keep your password secure and you should never have anything to worry about. Give your password to others, or write your password down and put it near your computer, and... well, you get the picture.

HTTP sets up the connection between the client and the server, and in many cases that connection must be secure. For example, assume your company sets up an Internet Web server to provide sensitive documents to its mobile work force. These documents should not be accessible to anyone else (since the site is connected to the Internet, anyone could attempt to access them).

As the use of e-mail and electronic commerce becomes more widely adopted, the amount of confidential information being exchanged over the Internet is growing rapidly. As a result, there is a need to make e-mail messages more secure and private. In addition, with the growing popularity of ActiveX controls, scripts and Java applets, there is an increased chance that the HTML content you receive in an e-mail message could access or modify files on your computer without your knowledge or consent.

Outlook Express includes tools to help protect you from fraud, help increase your electronic privacy, and help reduce the risk of unauthorized access to your computer. These tools enable you to send and receive secure e-mail more securely and to control potentially harmful e-mail messages through security zones.

A list of the major challenges of Internet Security and simple explanations of how they affect a network are explained below.

1. **Protection from hackers and vandals on the Internet:** Hacker attacks can cause irreparable damage to sensitive and confidential information. Critical data resides on a company's private network and businesses must actively prevent the theft, destruction and corruption of this indispensable information. For some companies, the confidential information residing on the business network belongs to corporate clients, raising issues of legal liability as well as financial loss.
2. **Protection from DoS (Denial of Service Attacks':** As businesses rely more and more on networked resources, maintaining their availability is becoming more and more critical. A common type of attack referred to as a Denial of Service attack, is designed to deny access to networked resources by crashing servers or routers or by overwhelming the network with enough traffic to degrade service. These DoS attacks hinder normal network activity, making productivity nearly impossible.
3. **The ability to filter inappropriate materials on the Internet, such as pornography or racially intolerant sites:** The valuable information so readily available online also includes material inappropriate for the workplace, classroom, and home. Companies can face losses in productivity due to the vast array of non-work related material available at the click of a mouse. Objectionable content can create an uncomfortable work environment, creating a vulnerability to harassment lawsuits. Schools, libraries and home offices risk children accessing pornography or racially intolerant sites.
4. **Protection against viruses:** Virus attacks are the greatest threat to Internet Security today! The outbreak of viruses has increased in the past several years and statistics show the number will continue to rise.
5. **Protection from active content, such as ActiveX and Java:** ActiveX is a programming language that is used to embed small programs in web pages. It is generally considered an insecure protocol to allow into a network since it is possible for malicious programmers to write controls that can delete files, compromise security, or cause other damage. Java is also used to embed

small programs, also called applets, in web pages. It is generally considered safer than ActiveX, but some administrators may want to filter out Java since there have been instances of bugs in these safety mechanisms.

Crackers, (sometimes known incorrectly as hackers) sometimes write a program called a computer virus. Whatever the reason, these viruses infect your computer, possibly damage your computer, and begin to infect other computers. Viruses can slow down system performance, cause strange system behavior or simply reproduce. In order to detect a virus, you need antivirus software. Unlike firewalls, most antivirus software is not free. This is mainly because it is a subscription based service. Once you have the antivirus software running on your machine, updates are released for the software allowing it to better detect new viruses. When a virus is detected, the company who makes the antivirus software creates rules on how to detect it based on behavior of the virus, and files that contain the virus. Then, once you download the update, your antivirus software is prepared and able to detect that virus if it is on your computer. Because your antivirus can only detect viruses that were known up to the date it was last updated, it must constantly be updated. Because of this ongoing support required by the company to maintain the usefulness of the software, the software usually requires a subscription fee.

Browser Choice

Internet Explorer is one of the most used web browsers in the world, but this has made it the prime target for phishing and many other possible attacks. With all these problems, many are switching to different browsers, such as those offered by Netscape, Opera and Mozilla (Mozilla Firefox). Some consider using IE is the prime mistake most users make about internet security.

Spyware

Spyware is software that runs on a computer without the explicit permission of its user. It often gathers private information from a users computer and sends this data over the internet back to the software manufacturer.

Adware

Adware is software that runs on a computer without the owner's consent, much like spyware. However, instead of taking information, it installs pop-up advertisements and similar programs onto the system.

This slows the computer down and shows the computer's user unwanted advertisements.

11.2 FIREWALL

A *firewal* is a piece of hardware and/or software which functions in a networked environment to prevent some communications forbidden by the security policy, analogous to the function of firewalls in building construction. A firewall is also called a Border Protection Device (BPD), especially in NATO contexts, or packet filter in BSD contexts.

A firewall has the basic task of controlling traffic between different zones of trust. Typical zones of trust include the Internet (a zone with no trust) and an internal network (a zone with high trust). The ultimate goal is to provide controlled connectivity between zones of differing trust levels through the enforcement of a security policy and connectivity model based on the least privilege principle.

Proper configuration of firewalls demands skill from the administrator. It requires considerable understanding of network protocols and of computer security. Small mistakes can render a firewall worthless as a security tool.

A firewall is any one of several ways of protecting one network from another untrusted network. The actual mechanism whereby this is accomplished varies widely, but in principle, the firewall can be thought of as a pair of mechanisms: one which exists to block traffic, and the other which exists to permit traffic. Some firewalls place a greater emphasis on blocking traffic, while others emphasize permitting traffic.

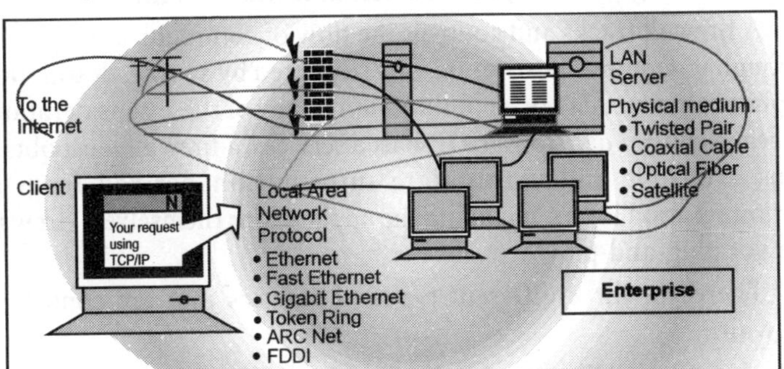

Fig. 4.16: Firewall

The Internet, like any other society, is plagued with the kind of jerks who enjoy the electronic equivalent of writing on other people's walls with spraypaint, tearing their mailboxes off, or just sitting in the street blowing their car horns. Some people try to get real work done over the Internet, and others have purpose is to keep the jerks out of your network while still letting you get your job done.

Many traditional style corporations and data centers have computing security policies and practices that must be adhered to. In a case where a company's policies dictate how data must be protected, a firewall is very important, since it is the embodiment of the corporate policy. Frequently, the hardest part of hooking to the Internet, if you're a large company, is not justifying the expense or effort, but convincing management that it's safe to do so. A firewall provides not only real security—it often plays an important role as a security blanket for management.

Lastly, a firewall can act as your corporate "ambassador" to the Internet. Many corporations use their firewall systems as a place to store public information about corporate products and services, files to download, bug-fixes, and so forth. Several of these systems have become important parts of the Internet service structure (*e.g.* UUnet.uu.net, gatekeeper.dec.com) and have reflected well on their corporate sponsors.

A firewall is a component or set of components that controls access between enterprise and public networks like the Internet. Firewalls are often used in conjunction with virtual private networks (VPNs) to allow secure remote access to a corporate network. Today, more and more firewalls are being implemented within enterprise networks for greater control of internal traffic as well.

A firewall tracks and controls the flow of communication passing through it. Communication may be restricted by source, destination, application, time, day or a combination of any of these considerations. To reach control decisions for IP-based services, a firewall must obtain, store, retrieve and manipulate information derived from all communication layers, including the applications themselves. Firewalls are not plug-and-play.

There are three different types of services that are considered firewalls:

- **Packet filters:** These were the first firewalls and are just routers with statically created access control lists (ACLs) for specific layer 3/4 addresses. Communication is limited based on source or destination IP addresses.

Application gateways or proxies—These are platforms running firewall software to implement security at the application level.

Stateful Inspection Systems: These systems look at all communication layers, extract the relevant data, and build connection tables that allow higher performance. Stateful inspection supports a large number of protocols/applications and provides extensibility to new applications.

A firewall is only as good (or bad) as the overall corporate security policy and the configuration of the firewall itself. Firewalls are also slow, as all packets must be examined; today there is no such thing as a wirespeed firewall (except to the WAN).

A firewall can do the following:

- Be a focal point for security decisions (choke point)
- Enforce security policy (traffic cop)
- Log activity (Big Brother watching who you communicate with)
- Limit your exposure (keep intrusions from spreading)

11.2.1 Types of Firewalls

There are three basic types of firewalls depending on

- Whether the communication is being done between a single node and the network, or between two or more networks.
- Whether the communication is intercepted at the network layer, or at the application layer.
- Whether the communication state is being tracked at the firewall or not with regard to the scope of filtered communication there exist:
- Personal firewalls, a software application which normally filters traffic entering or leaving a single computer through the Internet
- Network firewalls, normally running on a dedicated network device or computer positioned on the boundary of two or more networks or DMZs (demilitarized zones). Such a firewall filters all traffic entering or leaving the connected networks.

The latter definition corresponds to the conventional, traditional meaning of "firewall" in networking. In reference to the layers where the traffic can be intercepted, three main categories of firewalls exist:

- Network-layer firewalls
- Application-layer firewalls
- Application firewalls

These network-layer and application-layer types of firewall may overlap, even though the personal firewall does not serve a network; indeed, single systems have implemented both together.

There's also the notion of application firewalls which are sometimes used during wide area network (WAN) networking on the world-wide web and govern the system software. An extended description would place them lower than application layer firewalls, indeed at the Operating System layer, and could alternately be called operating system firewalls.

Lastly, depending on whether the firewalls track packet states, two additional categories of firewalls exist:

- Stateful firewalls
- Stateless firewalls

11.2.2 Network layer firewalls

Network layer firewalls operate at a (relatively low) level of the TCP/IP protocol stack as IP-packet filters, not allowing packets to pass through the firewall unless they match the rules. The firewall administrator may define the rules; or default built-in rules may apply (as in some inflexible firewall systems).

A more permissive set-up could allow any packet to pass the filter as long as it does not match one or more "negative rules", or "deny rules". Today network firewalls are built into most computer operating system and network appliances.

Modern firewalls can filter traffic based on many packet attributes like source IP, source port, destination IP or port, destination service like WWW or FTP. They can filter based on protocols, TTL values, netblock of originator, domain name of the source, and many other attributes.

11.2.3 Application-layer Firewalls

Application-layer firewalls work on the application level of the TCP/IP stack (*i.e.*, all browser traffic, or all telnet or ftp traffic), and may intercept all packets traveling to or from an application. They block other packets (usually dropping them without acknowledgement to the sender). In principle, application firewalls can prevent all unwanted outside traffic from reaching protected machines.

By inspecting all packets for improper content, firewalls can even prevent the spread of the likes of viruses. In practice, however, this becomes so complex and so difficult to attempt (given the variety of applications and the diversity of content each may allow in its packet traffic) that comprehensive firewall design does not generally attempt this approach.

The XML firewall exemplifies a more recent kind of application-layer firewall

11.3 PROXIES

A proxy device (running either on dedicated hardware or as software on a general-purpose machine) may act as a firewall by responding to input packets (connection requests, for example) in the manner of an application, whilst blocking other packets.

Proxies make tampering with an internal system from the external network more difficult, and misuse of one internal system would not necessarily cause a security breach exploitable from outside the firewall (as long as the application proxy remains intact and properly configured). Conversely, intruders may hijack a publicly-reachable system and use it as a proxy for their own purposes; the proxy then masquerades as that system to other internal machines. While use of internal address spaces enhances security, crackers may still employ methods such as IP spoofing to attempt to pass packets to a target network.

Network Address Translation

Firewalls often have network address translation (NAT) functionality, and the hosts protected behind a firewall commonly use so-called "private address space", as defined in RFC 1918. Administrators often set-up such scenarios in an effort (of debatable effectiveness) to disguise the internal address or network.

11.3.1 Proxy Server

A proxy server is a computer network service that allows clients to make indirect network connections to other network services. A client connects to the proxy server, then requests a connection, file, or other resource available on a different server. The proxy provides the resource either by connecting to the specified server or by serving it from a cache. In some cases, the proxy may alter the client's request or the server's response for various purposes.

Web Proxies

A common proxy application is a caching Web proxy. This provides a nearby cache of Web pages and files available on remote Web servers, allowing local network clients to access them more quickly or reliably.

When it receives a request for a Web resource (specified by a URL), a caching proxy looks for the resulting URL in its local cache. If found, it returns the document immediately. Otherwise it fetches it from the remote server, returns it to the requester and saves a copy in the cache. The cache usually uses an expiry algorithm to remove documents from the cache, according to their age, size and access history. Two simple cache algorithms are Least Recently Used (LRU) and Least Frequently Used (LFU). LRU removes the leastm recently used documents, and LFU removes the least frequently used documents.

Web proxies can also filter the content of Web pages served. Some censorware applications—which attempt to block offensive Web content —are implemented as Web proxies. Other Web proxies reformat web pages for a specific purpose or audience; for example, Skweezer reformats Web pages for cell phones and PDAs. Network operators can also deploy proxies to intercept computer viruses and other hostile content served from remote Web pages.

5

UNIT

Database Management Systems

1.1 DATABASE (DB)

DB is a collection of information organized in such a way that a computer program can quickly select desired pieces of data. You can think of a database as an electronic filing system.

Traditional databases are organized by fields records and files. A field is a single piece of information; a record is one complete set of fields; and a file is a collection of records. For example, a telephone book is analogous to a file. It contains a list of records, each of which consists of three fields: name address and telephone number.

An alternative concept in database design is known as Hypertext. In a Hypertext database, any object, whether it be a piece of text, a picture, or a film, can be linked to any other object. Hypertext databases are particularly useful for organizing large amounts of disparate information, but they are not designed for numerical analysis.

To access information from a database, you need a database management system (DBMS). This is a collection of programs that enables you to enter, organize and select data in a database.

1.2 DATABASE MANAGEMENT SYSTEM (DBMS)

A database management system (DBMS) is a computer program (or more typically, a suite of them) designed to manage a database, a large set of structured data, and run operations on the data requested by numerous users. Typical examples of DBMS use include accounting, human resources and customer support systems. Originally found only in large organizations with the computer hardware needed to support large data sets, DBMSs have more recently emerged as a fairly standard part of any company back office.

DBMSs are found at the heart of most database applications. Sometimes DBMSs are built around a private multitasking kernel with built-in networking support although nowadays these functions are left to the operating system.

A database management system (DBMS) is a system, usually automated and computerized, for the management of any collection of compatible, and ideally normalized, data. Examples include Oracle database, DB2, Microsoft SQL Server, MySQL and PostgreSQL. Nowadays a small number of DBMSs are used by the great majority of database applications to manage practically all the world's databases.

A database application is computer software written to manage the data of a particular application or problem set. This software usually employs a DBMS for the generalised data management issues. Examples include Microsoft Access, Oracle Human Resources, Sage Payroll and most of the countless other database applications written either in-house or by the thousands of software houses around the world.

A database or dataset is a particular collection of related and structured data. Examples include social security numbers, voter rolls, lists of houses for sale, and accounting records. There are many millions of databases. Each database application might be implemented many times, each implementation managing its own particular database or databases.

Databases have been in use since the earliest days of electronic computing, but the vast majority of these were custom programs written to access custom databases. Unlike modern systems which can be applied to widely different databases and needs, these systems were tightly linked to the database in order to gain speed at the expense of flexibility.

1.2.1 Purpose of Database Management Systems

Functions of a DBMS:

- To store data
- To organize data
- To control access to data
- To protect data

1.2.2 Uses of a DBMS

- To provide decision support
- Managers and analysts retrieve information generated by the DBMS for inquiry, analysis, and decision-making.
- To provide transaction processing
- Users input, update, and process data with transactions that generate information needed by managers and other users or by other departments.

1.2.3 Requirements of a Well-Designed DBMS

Data Integrity

1. Elimination (or reduce) Redundancy:
   - Separate parts of the database must correlate to reduce data redundancy and provide efficient data updates. Duplication of data should at best be eliminated. If not the DBMS should be aware of it, and control it.
2. Maintain Consistency:
   - If there is duplication of data, *e.g.* a person's telephone number may be stored in two tables (it shouldn't), then a change will automatically be reflected in both tables.
3. Data Sharing:
   - A single copy of the data can be used by many users.
4. Standards can be enforced:
   - Both in presentation and storage
   - Useful to aid data interchange.
5. Provide Security of the Data:
   - Restrict access of data to authorised users
   - Permissions
   - Passwords
   - Log transactions
   - Provide backup/recovery facilities
6. Maintain Integrity:
   - Ensure the data in the DB is correct *e.g.*
   - range checks
   - Deletion check
7. Data Independence:
   - Different applications need different views of the same data

- Storage structure can change without having to alter existing applications (programs) *e.g.* new fields added.

8. Reduced program maintenance.
9. Interface with a Data Dictionary facility—METADATA:
   - Provides information about the database
   - Schemas and users
10. Provide a Query Language to access the data
    - SQL is the standard
    - Ingres has SQL, QUEL and QBF
    - Embedded in at least one language (*e.g.* C, C++, COBOL ...)
11. Avoid possible deadlock situations in on-line DB's
12. Downloading facility to a PC
13. Allow a program on PC to access the data
14. Sensitive data must be protected.
15. Concurrency control
16. Ability to lock records
17. Transaction processing
18. An individual transaction either is completed or not completed.
19. Recovery functions
20. Efficient data recovery after system or power failures.

1.2.4 Characteristics of DBMS

Typically, a data base management system (based upon controlled activities) has the following characteristics:

- It is a computerized record-keeping system.
- It contains facilities that allow the user to.
  - Add and delete files to the system.
  - Insert, retrieve, update, and delete data in existing files.
- It is collection of databases.
- A DBMS may contain many databases that can be used for separate purposes or combined to provide useful information.
- Operating environment independence capability to run on many computers with certain types of operating systems.
- User-oriented provision for an English-like language which enables the user to consider logical entities in place of physical entities, such as hardware and system software.
- Data base independence from application programs-use of a data base language and "call" commands from the application program.

- Security features provision for controlling data base access and control over the data base elements.
- DBMS overcomes limitations

Eliminates Separation and Isolation of Data

In a DBMS, all data is stored in the database. The DBMS allows for complex relationships between data files, providing efficient data integration. Programmers specify how the data is to be combined; the DBMS performs the functions to provide the information.

DBMS Reduces Data Redundancy: Minimal duplication of data provides more storage space in the system and allows for efficient updates when changes in records occur, maintaining data integrity. For example, if a customer's address is stored in only one file and a change is made to that file, all affected applications are updated automatically.

Eliminates Dependence between Programs and Data: In a DBMS record formats are defined in the database and accessed by the DBMS. Application programs only define which data items are needed, not data formats. Therefore, format changes are made to the database by the DBMS, and, since most fields are stored in only one file, the changes are made only to the affected file. Also, because structure formats and access methods are independent of the database, there is little impact on application programs if structure/access changes are made.

Allows for representation of data from the user's view: Relationships between data objects in a user's environment are stored in the DBMS. Data elements from any number of files can be combined to create useful forms, reports, and other applications.

Increases Data Flexibility: Because of the data-independent structure of DBMS, applications can be created without intensive programming. Thus, ad hoc requests for information and applications can be filled with less time and labor from programmers than with file processing systems.

1.2.5 Components of a DBMS

The basic components of a DBMS can be divided into three subsystems:

1. **Design Tools:** Provide features for creating the database and various applications, forms and reports.
2. **Run-time Facilities:** Process the applications created by the design tools.

3. **DBMS Engine:** Translates between the design tools and run-time facilities and the data.

Although they are all excellent advantages to DBMS users, the initial investment of system personnel time as well as software and equipment cost can be high; however, increasing experience with data base management systems is reducing these costs. Another disadvantage of DBMS is that incorrect transaction data tend to precipitate additional problems and errors throughout the system. Therefore, various tests should be incorporated to catch errors entering the data base before they are stored. Overall, these disadvantages of DBMS are capable of being overcome to a great degree if the management information system is properly designed and managed.

DBMS Components

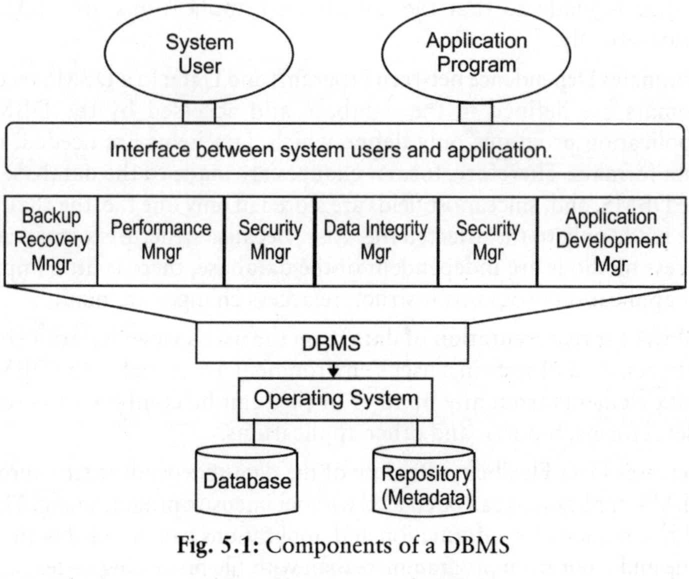

Fig. 5.1: Components of a DBMS

1.2.6 Advantages and Disadvantages of DBMS

Because the new management information system design will include a data base management system, the advantages of this arrangement should be carefully considered. First, DBMS is not only effective for generating and maintaining a wide variety of routine management and operating reports, but also adaptable to meeting the new and emerging requirements of management to answer a myriad of "What if?" questions. The latter capability means data base management systems will be important aids to manager seeking to explore and understand new relationships among various data elements.

- Second, data elements can be structured in a manner more suitable to their application, allowing their retrieval with a minimum of effort.
- Third, DBMS keeps redundancy of data elements to a minimum, since one data file serves many users; as a result of this "single record" concept, a transaction is entered once for all users. The DBMS also allows two or more files to be updated with the entry of a single transaction.
- Fourth, application programs are independent of the changes in the data base, so that their maintenance is kept to a bare minimum.
- Fifth, DBMS provides data protection not only for accessing one data base record at a time, but also for preventing data base access by unauthorized personnel.

The main advantage of using a DBMS is that the formalism of the model of data underlying the DBMS is imposed upon the data set to yield a logical and structured organisation of the data. Given a fuzzy, real-world data set, when a model's formalism is imposed on that data set the result is easier to manage, define and manipulate.

A DBMS can be an extremely complex set of software programs that controls the organization, storage and retrieval of data (fields, records and files) in a database. It also controls the security and integrity of the database. The DBMS accepts requests for data from the application program and instructs the operating system to transfer the appropriate data.

When a DBMS is used, information systems can be changed much more easily as the organization's information requirements change. New categories of data can be added to the database without disruption to the existing system.

Data security prevents unauthorised users from viewing or updating the database. Using passwords, users are allowed access to the entire database or subsets of it called subschemas (pronounced "sub-skeema"). For example, an employee database can contain all the data about an individual employee, but one group of users may be authorized to view only payroll data, while others are allowed access to only work history and medical data.

The DBMS can maintain the integrity of the database by not allowing more than one user to update the same record at the same time. The DBMS can keep duplicate records out of the database; for example, no two customers with the same customer numbers (key fields) can be entered into the database.

Database query languages and report writers allow users to interactively interrogate the database and analyse its data.

If the DBMS provides a way to interactively enter and update the database, as well as interrogate it, this capability allows for managing personal databases. However, it may not leave an audit trail of actions or provide the kinds of controls necessary in a multi-user organisation. These controls are only available when a set of application programs are customised for each data entry and updating function.

A business information system is made up of subjects (customers, employees, vendors, etc.) and activities (orders, payments, purchases, etc.). Database design is the process of deciding how to organize these data into record types and how the record types will relate to each other. The DBMS should mirror the organization's data structure and process transactions efficiently.

Organizations may use one kind of DBMS for daily transaction processing and then move the detail onto another computer that uses another DBMS better suited for random inquiries and analysis. Overall systems design decisions are performed by data administrators and systems analysts. Detailed database design is performed by database administrators.

The three most common organizations are the hierarchical, network and relational models. A database management system may provide one, two or all three methods. Inverted lists and other methods are also used. The most suitable structure depends on the application and on the transaction rate and the number of inquiries that will be made.

The dominant model in use today is the ad hoc one embedded in SQL, a corruption of the relational model by violating several of its fundamental principles. Many DBMSs also support the Open Database Connectivity API that supports a standard way for programmers to access the DBMS.

Database servers are specially designed computers that hold the actual databases and run only the DBMS and related software. Database servers are usually multiprocessor computers, with RAID disk arrays used for stable storage. Connected to one or more servers via a high-speed channel, hardware database accelerators are also used in large volume transaction processing environments.

1.3 HIERARCHICAL MODEL

The hierarchical data model organizes data in a tree structure. There is a hierarchy of parent and child data segments. This structure implies that a record can have repeating information, generally in the child data segments. Data in a series of records, which have a set of field values attached to it. It collects all the instances of a specific record together as a record type.

These record types are the equivalent of tables in the relational model, and with the individual records being the equivalent of rows. To create links between these record types, the hierarchical model uses Parent child relationships. These are a 1:N mapping between record types. This is done by using trees, like set theory used in the relational model, "borrowed" from maths. For example, an organization might store information about an employee, such as name, employee number, department, salary. The organization might also store information about an employee's children, such as name and date of birth. The employee and children data forms a hierarchy, where the employee data represents the parent segment and the children data represents the child segment. If an employee has three children, then there would be three child segments associated with one employee segment. In a hierarchical database the parent-child relationship is one to many. This restricts a child segment to having only one parent segment. Hierarchical DBMSs were popular from the late 1960s, with the introduction of IBM's Information Management System (IMS) DBMS, through the 1970s.

1.4 NETWORK MODEL

A network data structure can be regarded as an extended form of the hierarchic data structure—the principal distinction between the two being that in a hierarchic structure, a child record has exactly one parent whereas in a network structure, a child record can have any number of parents (possibly even zero).

A network database consists of two data sets, a set of records and a set of links, where the record types are made up of fields in the usual way.

Networks are complicated data structures. Operators on network databases are complex, functioning on individual records, and not sets of records. Increased complexity does not mean increased functionality and the network model is no more powerful than the relational model.

However, a network-based DBMS can provide good performance because its lack of abstraction means it is closer the storage structured used though this is at the expense of good user programming. The network model also incorporates certain integrity rules.

1.4.1 Network or Linked Databases

Network or linked databases organize dissimilar records using linked lists to implement relationships between the records. These are constructed based on pointers and links between data records (many-to-many relationship).

In many ways, the Network Database model was designed to solve some of the more serious problems with the Hierarchical Database Model. Specifically, the Network model solves the problem of data redundancy by representing relationships in terms of sets rather than hierarchy. The model had its origins in the Conference on Data Systems Languages (CODASYL) which had created the Data Base Task Group to explore and design a method to replace the hierarchical model.

In fact, the hierarchical model is a subset of the network model. However, instead of using a single-parent tree hierarchy, the network model uses set theory to provide a tree-like hierarchy with the exception that child tables were allowed to have more than one parent. This allowed the network model to support many-to-many relationships.

Visually, a Network Database looks like a Hierarchical Database in that you can see it as a type of tree. However, in the case of a Network Database, the look is more like several trees which share branches. Thus, children can have multiple parents and parents can have multiple children.

Nevertheless, though it was a dramatic improvement, the network model was far from perfect. Most profoundly, the model was difficult to implement and maintain. Most implementations of the network model were used by computer programmers rather than real users. What was needed was a simple model which could be used by real end users to solve real problems.

Like hierarchical databases, network databases are principally used on mainframe computers. Since more connections can be made between different types of data, network databases are considered more flexible. However, two limitations must be considered when using this kind of database. Similar to hierarchical databases, network databases must be defined in advance. There is also a limit to the number of connections that can be made between records.

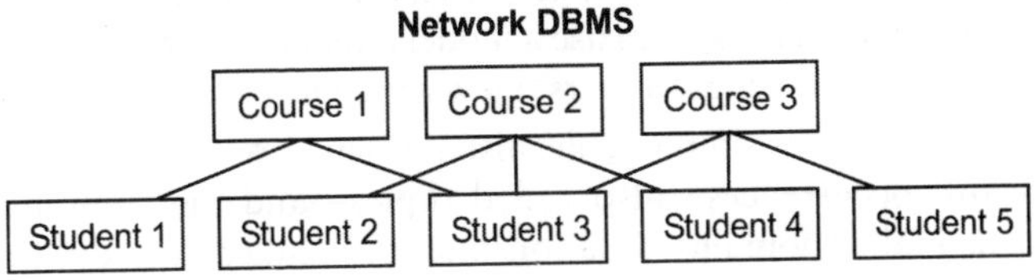

Fig. 5.2: Network

The popularity of the network data model coincided with the popularity of the hierarchical data model. Some data were more naturally modeled with more than one parent per child. So, the network model permitted the modeling of many-to-many relationships in data. In 1971, the Conference on Data Systems Languages (CODASYL) formally defined the network model. The basic data modeling construct in the network model is the set construct. A set consists of an owner record type, a set name, and a member record type. A member record type can have that role in more than one set, hence the multiparent concept is supported. An owner record type can also be a member or owner in another set. The data model is a simple network, and link and intersection record types (called junction records by IDMS) may exist, as well as sets between them. Thus, the complete network of relationships is represented by several pairwise sets; in each set some (one) record type is owner (at the tail of the network arrow) and one or more record types are members (at the head of the relationship arrow). Usually, a set defines a 1:M relationship, although 1:1 is permitted. The CODASYL network model is based on mathematical set theory.

1.5 RELATIONAL DBMS

Short for relational database management system and pronounced as separate letters, a type of database management system (DBMS) that stores data in the form of related tables. Relational databases are powerful because they require few assumptions about how data is related or how it will be extracted from the database. As a result, the same database can be viewed in many different ways.

An important feature of relational systems is that a single database can be spread across several tables. This differs from flat-file databases, in which each database is self-contained in a single table. Almost all full-scale database systems are RDBMS's. Small database systems, however, use other designs that provide less flexibility in posing queries. Relational databases use a model intended to greatly simplify the end-users' and programmers' view of a database.

Files are seen as simple **table**, also known as **relations**. The rows are record occurrences—also called tuples—and the columns are **fields**, also called **domains**. Relationships are handled quite differently from the way they are in hierarchical and network models. Instead of physically denoting relationships by using pointers, location, or indexes, true relational databases do not store relationships. Instead, relationships are inferred when needed. How? Relationships are determined from intentionally redundant fields, usually keys, common to the different tables. This concepts requires further explanation.

The DML for a relational DBMS doesn't navigate paths and pointers, which is what happens with hierarchical and network databases. Instead, to write reports and answer inquiries, the DML lets the programmer or end-user perform simple table operations to create temporary tables. These operations include

- Selecting specific records from a table and creating a new, but temporary, table that contains only those occurrences. Criteria can be set to determine which records to select from the initial table.
- Projecting out specific fields from a table, creating a temporary table that has fewer fields.
- Joining two or more table across a common field (this is the same as navigating relationship paths in a hierarchical or network database). Again, a temporary table is created.
- In all cases, the above relational operations—which, unfortunately, have different names in the various relational DBMSs on the market—create temporary, working tables that will go away when you exit the program. If changes are made to the data in those tables, those changes will be updated into the permanently stored tables before the working tables are discarded (assuming the end-user has update authority).

Also, note that the relational commands can be combined. For instance, let's say we need a table of orders for certain customers, perhaps to produce a report. First, we can use SELECT on the CUSTOMER table to create a working table of those customers needed. Second, we can PROJECT out only those fields needed in both the CUSTOMER and ORDER working table to reduce their size. Finally, we JOIN the two working tables to give us the final working table needed for the report.

Relational databases are definitely the trend. They present a simpler view-point to both programmers and end-users. Query language and report writers that are easy to learn and use have been built around relational databases. Most relational databases are currently converging to supersets of a de facto standard language called SQL that is utilized to create, update and use tables. SQL includes the SELECT, PROJECT and JOIN operations discussed earlier.

Examples of relational DBMSs include IBM's DB2 and SQL/DL, ADRs DATACOM, Relational Technology's Ingress, Oracle Corporation's Oracle, and Information Builder's FOCUS (which is also hierarchical). Additionally, most microcomputer DBMSs are relational. Examples include Ashton-Tate's dBASE-IV (which recently embraced true SQL relational standards) and Microrim's R:BASE for DOS. Many relational dBMSs are offered in both mainframe and micro versions. FOCUS is one such DBMS.

That's about all we want to say about the three common data models database textbooks and courses will offer entire chapters and units on each of the three models. We do, however, I want to demonstrate a simplified translation of a logical data model into each of the three physical data structures we have introduced. But first we need to describe the roles of the database design participants.

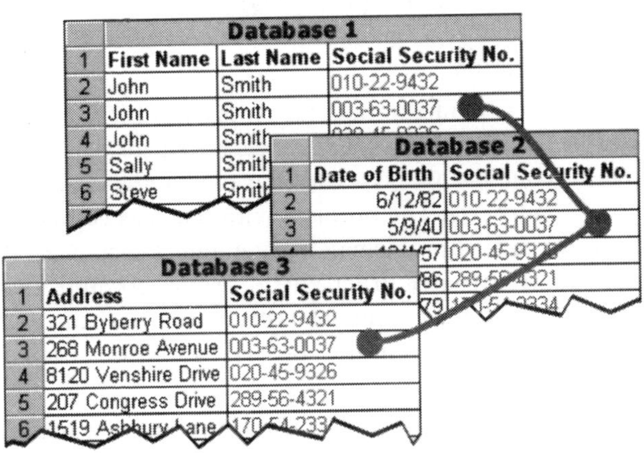

Fig. 5.3: Relational DBMS

A relational DBMs depicts record types as simple tables, much like those seen in spreadsheets that exhibit relational-like qualities. Tables are related to one another via intentionally redundant fields, usually keys. Relational DBMSs provide operators that build temporary working tables from the tables illustrated that are physically stored.

1.5.1 The Relational Model

The relational model was the first example of a formal data model. With it, user data is represented and manipulated intuitively. The model also uses techniques that help developers spot and correct possible design problems as data is prepared for implementation to the DBMS. Relational models involve.

- Determining data structure.
- Data is stored in a structure of relations (tables) defined by a data definition language (DDL). The elements of data structure used in relational models are relations, attributes, tuples, and domains.

The relational database model has become the de-facto standard for the design of databases both large and small. While the concepts involved are not terribly complex, it can be difficult at first to get a handle on the concept. I'll try and give you a little introduction to the following topics on this page:

- Primary and Foreign Keys
- Queries
- Structured Query Language (SQL)
- Referential Integrity
- Normalization

The simplest model for a database is a flat file. You have only a single table which includes fields for each element you need to store. Nearly everyone has worked with flat file databases, at least in the form of spreadsheets. The problem with flat files is that they waste storage space and are problematic to maintain. Let consider the classic example of a customer order entry system. Assume that you're managing the data for a company with a number of customers, each of which will be placing multiple orders. In addition, each order can have one or more items.

Before moving on, lets describe the data that we wish to record for each component of the application:

- Customers
  - Customer Number
  - Company Name
  - Address
  - City, State, ZIP Code
  - Phone Number
- Orders
  - Order Number
  - Order Date
  - PO Number
- Order Line Items
  - Item Number
  - Description
  - Quantity
  - Price

It doesn't take a database design expert to see what the problem is in using a flat file to represent this data. Each time an order is placed, you'll need to repeat the customer information, including the Customer Number, Company Name, etc. What's worse is that for each item, you not only need to repeat the order information such as the Order Number and Order Date, but you also need to continue repeating the customer information as well. Lets say there's one customer who has placed two orders, each with four line items. To maintain this tiny amount of information, you need to enter the Customer Number and Company Name eight times. If the company should send you a change of address, the number of records you need to update is equal to the sum of product of orders and order line items. Obviously this will quickly become unacceptable in terms of both the effort required to maintain the data and the likelihood that at some point there will be data entry errors and the customer address will be inconsistent between records.

The solution to this problem is to use a relational model for the data. Don't let the terminology get you down - the concept isn't that hard to understand. It simply means that in this example each order entered is related to a customer record, and each line item is related to an order record. A relational database management system (RDBMS) is then a piece of software that manages groups of records which are related to one another. Let's take our flat file and break it up into three tables: Customers, Orders, and Order Details. The fields are just as they are shown above, with a few additions. To the Orders table, we will add a Customer Number field, and to the Order Details table we will add an Order Number field. Here's the list again with the required additional fields and modified field names.

- Customers
  - CustID
  - CustName

- CustAddress
- CustCity
- CustState
- CustZIP
- CustPhone

- Orders
  - OrdID
  - OrdCustID
  - OrdDate
  - OrdPONumber
- OrderDetails
  - ODID
  - ODOrdID
  - ODDescription
  - ODQty
  - ODPrice

I'll digress here a little to describe how I came up with the field names. This is totally a personal preference, but I've found it works well for me. Every time I create a table, I decide on a field name prefix to be used. As you can see, I've used "Cust" for the Customers table, "Ord" for the Orders table, and "OD" for the OrderDetails table. I do this for a couple of reasons. First, by using a field name prefix, I can identify the table associated with a field by simply reading the field name. Second, I avoid name collisions in queries because every field in a database, no matter how large and complex the structure, will always be unique—I'll never end up with two fields in a query sharing a single name. I find this particularly helpful in distinguishing between primary and foreign keys—a subject I'll take up a bit later. One other thing you may have noticed is a table naming convention I use, which is to name tables using the plural form of the data they contain.

OK, back to the tables. What we've done besides the name change is to add fields to the Orders and Order Details tables. Each have key fields used to provide a link to the associated Customers and Orders records, respectively. These additional fields are called foreign keys.

Codd's paper was picked up by two people at Berkeley, Eugene Wong and Michael Stonebraker. They started a project known as INGRES using funding that had already been allocated for a geographical database project, using student programmers to produce code. Beginning in 1973, INGRES delivered its first test products which were generally ready for widespread use in 1979. During this time a number of people had moved "through" the group—perhaps as many as 30 people worked on the project, about five at a time. INGRES was similar to System R in a number of ways, including the use of a "language" for data access, known as QUEL - QUEL was in fact relational, having been based on Codd's own Alpha language, but has since been corrupted to follow SQL, thus violating much the same concepts of the relational model as SQL itself.

IBM itself did only one test implementation of the relational model, PRTV, and a production one, Business System 12, both now discontinued. Honeywell did MRDS for Multics, and now there are two new implementations: Alphora Dataphor and Rel. All other DBMS implementations usually called relational are actually SQL DBMSs.

1.5.2 Properties of Relational Tables

The following are the prime properties of the relational tables:

- Values are atomic
- Each row is unique
- Column values are of the same kind
- The sequence of columns is insignificant
- The sequence of rows is insignificant
- Each column has a unique name

Certain fields may be designated as keys, which means that searches for specific values of that field will use indexing to speed them up. Where fields in two different tables take values from the same set, a join operation can be performed to select related records in the two tables by matching values in those fields. Often, but not always, the fields will have the same name in both tables. For example, an "orders" table might contain (customer-ID, product-code) pairs and a "products" table might contain (product-code, price) pairs so to calculate a given customer's bill you would sum the prices of all products ordered by that customer by joining on the product-code fields of the two tables. This can be extended to joining multiple tables on multiple fields. Because these relationships are only specified at retrieval time, relational databases are classed as dynamic database management system. The RELATIONAL database model is based on the Relational Algebra.

1.6 OBJECT-ORIENTED DBMS

Multidimensional DBMS did have one lasting impact on the market: they led directly to the development of object database systems. Based on the same general structure and concepts as the multidimensional systems, these new systems allowed the user to store objects directly in the database. That is, the programming constructs being used in the object-oriented (OO) programming world could be used directly in the database, instead of first being converted to some other format.

This could happen because of the multidimensional system's concepts of ownership. In an OO program a particular object will typically contain others; for example, the object representing Bob may contain a reference to a separate object referring to Bob's home address. Adding support for various OO languages and polymorphism recreated the multidimensional systems as object databases, which continue to serve a niche today.

1.7 OBJECT-ORIENTED MODEL

Object DBMSs add database functionality to object programming languages. They bring much more than persistent storage of programming language objects. Object DBMSs extend the semantics of the C++, Smalltalk and Java object programming languages to provide full-featured database programming capability, while retaining native language compatibility. A major benefit of this approach is the unification of the application and database development into a seamless data model and language environment. As a result, applications require less code, use more natural data modeling, and code bases are easier to maintain. Object developers can write complete database applications with a modest amount of additional effort.

According to Rao (1994), "The object-oriented database (OODB) paradigm is the combination of object-oriented programming language (OOPL) systems and persistent systems. The power of the OODB comes from the seamless treatment of both persistent data, as found in databases, and transient data, as found in executing programs". In contrast to a relational DBMS where a complex data structure must be flattened out to fit into tables or joined together from those tables to form the in-memory structure, object DBMSs have no performance overhead to store or retrieve a web or hierarchy of interrelated objects. This one-to-one mapping of object programming language objects to database objects has two benefits over other storage approaches: it provides higher performance management of objects, and it enables better management of the complex interrelationships between objects. This makes object DBMSs better suited to support applications such as financial portfolio risk analysis systems, telecommunications service applications, world wide web document structures, design and manufacturing systems, and hospital patient record systems, which have complex relationships between data.

2.1 STRUCTURED QUERY LANGUAGE (SQL)

SQL is a mature, powerful, and versatile relational query language. The history of SQL extends back to IBM research begun in 1970. The next few sections discuss the history of SQL, its predecessors, and the various SQL standards that have developed over the years.

2.1.1 A Brief History of SQL

The relational model, from which SQL draws much of its conceptual core, was formally defined in 1970 by Dr. E.F. Codd, a researcher for IBM, in a paper entitled A Relational Model of Data for Large Shared Data Banks. This article generated a great deal of interest in both the feasibility and practical commercial application of such a system.

In 1974, IBM began the System/R project and with the work of Donald Chamberlin and others, developed SEQUEL, or Structured English Query Language. System/R was implemented on an IBM prototype called SEQUEL-XRM in 1974-75. It was then completely rewritten in 1976-1977 to include multi-table and multiuser features. When the system was revised it was briefly called "SEQUEL/2", and then re-named "SQL" for legal reasons.

In 1978, methodical testing commenced at customer test sites. Demonstrating both the usefulness and practicality of the system, this testing proved to be a success for IBM. As a result, IBM began to develop commercial products that implemented SQL based on their System R prototype, including SQL/DS (introduced in 1981), and DB2 (in 1983).

Several other software vendors accepted the rise of the relational model and announced SQL-based products. These included Oracle (who actually beat IBM to market by two years by releasing their first commercial RDBMS, in 1979), Sybase, and Ingres (based on the University of California's Berkeley Ingres project).

Note: PostgreSQL's name is, as you might have guessed, a play on the name Ingres. Both PostgreSQL and Ingres trace their roots back to the UC Berkeley's Ingres RDBMS system.

2.1.2 SQL and Its Predecessors

SQL is based largely on relational algebra and tuple relational calculus. Relational algebra, introduced by E.F. Codd in 1972, provided the basic concepts behind computing SQL syntax. It is a procedural way to construct data-driven queries, and it addresses the how logic of a structured query. The tuple relational calculus (TRC), on the other hand, affects the underlying appearance of SQL. Relational calculus uses declarative expressions, addressing the what logic of a structured query.

There are additional features that set SQL apart from those that merely implement features that are part of relational algebra or calculus. These features include:

- Support for data insertion, modification and deletion
- Users are allowed to insert, delete, and modify stored data records.

Arithmetic Operators

Arithmetic operations such as addition, subtraction, multiplication, and division (*e.g.*, (value1 *5) + value2) are allowed, as well as comparison operators (*e.g.*, value3 >= value4).

Display of Data

Users may display query-generated relationships (such as a table's contents).

Assignment

Users may rename a relation that is computed by a query instead of forcing the use of the default relationship name, which may be derived from a column or function name, depending on the query.

Aggregate Functions

User may group related rows together and evaluate averages, sums, counts, maximums and minimums.

2.2 SQL STANDARDS

The American National Standards Institute (ANSI) standardized SQL in 1986 (X3.135) and the International Standards Organization (ISO) standardized it in 1987. The United States government's Federal Information Processing Standard (FIPS) adopted the ANSI/ISO standard. In 1989, a revised standard known commonly as SQL89 or SQL1, was published.

Due partially to conflicting interests from commercial vendors, much of the SQL89 standard was intentionally left incomplete, and many features were labeled implementor-defined. In order to strengthen the standard, the ANSI committee revised its previous work with the SQL92 standard ratified in 1992 (also called SQL2). This standard addressed several weaknesses in SQL89 and set forth conceptual SQL features which at that time exceeded the capabilities of any existing RDBMS implementation. In fact, the SQL92 standard was approximately six times the length of its predecessor. As a result of this disparity, the authors defined three levels of SQL92 compliance: Entry-level conformance (only the barest improvements to SQL89), Intermediate-level conformance (a generally achievable set of major advancements), and Full conformance (total compliance with the SQL92 features).

More recently, in 1999, the ANSI/ISO released the SQL99 standard (also called SQL3). This standard addresses some of the more advanced and previously ignored areas of modern SQL systems, such as object-relational database concepts, call level interfaces, and integrity management. SQL99 replaces the SQL92 levels of compliance with its own degrees of conformance: Core SQL99 and Enhanced SQL99.

PostgreSQL presently conforms to most of the Entry-level SQL92 standard, as well as many of the Intermediate- and Full-level features. Additionally, many of the features new in SQL99 are quite similar to the object-relational concepts pioneered by PostgreSQL (arrays, functions, and inheritance).

2.3 ORACLE BASED DATABASES

A database is a structured collection of data. Data refers to the characteristics of people, things, and events. Oracle stores each data item in its own field. For example, a person's first name, date of birth, and their postal code are each stored in separate fields. The name of a field usually reflects its contents. A postal code field might be named POSTAL-CODE or PSTL_CD. Each DBMS has its own rules for naming the data fields.

A field has little meaning unless it is seen within the context of other fields. The postal code T6G 2H1, for example, expresses nothing by itself. To what person or business does it belong? The postal code field is informative only after it is associated with other data. In Oracle, the fields relating to a particular person, thing, or event are bundled together to form a single, complete unit of data, called a record (it can also be referred to as a row or an occurrence). Each record is made up of a number of fields. No two fields in a record can have the same field name.

During an Oracle database design project, the analysis of your business needs identifies all the fields or attributes of interest. If your business needs change over time, you define any additional fields or change the definition of existing fields.

- Extra processing overhead
- As data is centralised there is a reliance on a central site
- More complex to develop.
- If developed incorrectly it can have disastrous consequences

2.4 POSTGRES

Postgres is a database research project under Prof. Michael Stonebraker at U.C. Berkeley. To facilitate research efforts, a software test-bed was created; this is the "Postgres" DBMS software. The Postgres DBMS is extended relational or object oriented, depending on the buzzword du jour.

The major purpose of this software is to provide a platform and a basis for the testing of implementations of new ideas in database research.

Postgres is

- **Relational:** One of the major goals of Postgres is to show that an essentially relational DBMS can be extended to handle complex objects, rules, and be highly extensible.
- **Highly extensible:** Postgres allows user-defined operators, user-defined objects, and user-defined functions.
- While Postgres is relational, object-oriented ideas have been implemented in Postgres (inheritance, etc).

2.4.1 PostgreSQL

While PostgreSQL is commonly considered an RDBMS, or a "database," it may not be commonly understood what is meant specifically by the word database. A database within PostgreSQL is an object-relational implementation of what is formally called a schema in SQL99.

Put simply, a database is a stored set of data that is logically interrelated. Typically, this data can be accessed in a multiuser environment. This is the case with PostgreSQL, though there are well-defined rights and restrictions enforced with that access.

It may not be commonly understood that PostgreSQL can have several databases concurrently available, each with their own owner, and each with their own unique tables, views, indices, sequences, and functions.

In order to create a table, function, or any other database object, you must connect to a specific database via a PostgreSQL client. Once connected, you can create an object, which is then owned by the connected database, and therefore is inaccessible from any other database (though a client may have several connections open to different databases).

By keeping fundamental data objects segregated into their own databases in this fashion, you run a smaller risk of running into a naming collision by choosing a table name already chosen for another purpose (*e.g.*, if two users each wanted to have a table called products for two separate applications). This is because neither database has any knowledge of the other database's components, and will not attempt to make any kind of logical relationship between them. Furthermore, as the same rule applies to object-relational data objects, users may even create functions and language definitions within their database that are inaccessible to other users connected to other databases running within PostgreSQL.

By default, PostgreSQL installs only one functional database, which is called template1 to represent the template nature of the database. Any database created after template1 is essentially a clone, inheriting any of its database objects, including table structure, functions, languages, etc. It is not uncommon to create a default database for new PostgreSQL users with the same name as their PostgreSQL username, as PostgreSQL will attempt to connect to a database with the same name as the connecting user if a database name is not specified.

2.4.2 Understanding Tables

Tables are quite possibly the most important aspect of SQL to understand inside and out, as all of your data will reside within them. In order to be able to correctly plan and design your SQL data structures, and any programmatic routines toward accessing and applying that data, a thorough understanding of tables is an absolute pre-requisite.

A table is composed of columns and rows, and their intersections are fields. If you have ever used spreadsheet software before (such as Excel), these two terms are visually represented in the same manner, and the fields within a table are equivalent to the cells within a spreadsheet. From a general perspective, columns within a table describe the name and type of data that will be found (and can be entered) by row for that column's fields. Rows within a table represent records composed of fields that are described from left to right by their corresponding column's name and type. Each field in a row is implicitly correlated with each other field in that row. In this sense, columns can be thought of as descriptors for the discrete, sequential elements of a row, and each row can be thought of as a stored record matching that description.

Table 5.1 illustrates a simple table called books, used by our imaginary Bookstore, Book Town. We will frequently refer to this table in later examples. Each of its stored records describes a book by a numeric identifier, title, author identifier, and subject identifier. These characteristics, from left to right, are described by its columns (id, title, author_id, and subject_id).

Table 5.1: An example SQL table

Id	*Title*	*Author_id*	*Subject_id*
7808	The Shining	4156	9
156	The Tell-Tale Heart	15	9
4513	Dune	1866	15
4267	2001: A Space Odyssey	2001	15
1608	The Cat in the Hat	1809	2
1590	Bartholomew and the Oobleck	1809	2

As you can see, this describes a table with four columns, in a fixed, left-to-right order, currently populated by six rows (also known as tuples, or records). It is essential to understand that in a relational database, while a table has a fixed column order, rows themselves are inherently unordered. You will see later, as the SQL's query structure is explained in Chapter 4, that there are ways within SQL to order selected rows. However, the rows in the database itself are not automatically ordered in any consistently predictable way. When order is meaningful for a SQL query, you must carefully consider and explicitly order records.

Every table must have at least one column, but tables may at times contain no rows, because each vertical column corresponds to a relatively fixed attribute of the data represented in that table (such as the title column in the previous example's books table). Without a column, a row's contents would be ambiguous; without a row, a table is merely lacking recorded data. As of PostgreSQL 7.1, there is a maximum of 1600 columns to a table, and an unlimited number of rows (*i.e.*, you are limited only by hardware limitations, such as disk space).

In Table 5.1, the column names fairly clearly indicate the significance of each column. The decision of how to name columns is fairly arbitrary, though, and care must be taken in planning table names and conventions to avoid ambiguity.

Though it may not be immediately obvious, each of the columns of a table have an associated data type. While a column's data type helps to further describe the sort of information it contains, it constrains the kind of data that may be inserted into the column. For example, the author_id column is of type integer; this signifies that any insertion attempts not consisting of pure a integer (*e.g.*, 110a) will fail.

PostgreSQL is an Object-Relational Database Management System (ORDBMS) that has been developed in various forms since 1977. It began as a project named Ingres at the University of California at Berkeley. Ingres itself was later developed commercially by Relational Technologies/Ingres Corporation.

In 1986, another team led by Michael Stonebraker from Berkeley continued the development of the Ingres code to create an object-relational database system called Postgres. In 1996, due to a new open source effort and the enhanced functionality of the software, Postgres was renamed to PostgreSQL, after a brief stint as Postgres95. The PostgreSQL project is still under very active development worldwide from a team of open source developers and contributors.

PostgreSQL is widely considered to be the most advanced open source database system in the world. It provides many features that are traditionally seen only in enterprise-caliber commercial products.

Open Source Free Version

PostgreSQL is an open source project. Open source by definition means that you can obtain the source code, use the program, and modify it freely without the confines of proprietary software. In the database world, open source means that you have honest access to benchmarking numbers and performance statistics, which companies such as Oracle prohibit. Open source also means that you are free to modify PostgreSQL to fit your particular needs.

However, there is a misconception that since open source software is free from distribution restrictions, it is always free of cost to your company. This is not necessarily the case. It is true that you can, without external cost, download and install open source software, but there will always be costs associated with the time and energy your company puts into support and research of the application. As such, if you do not have those resources to spend, there are several commercial vendors and consultants who deal specifically with PostgreSQL.

2.4.3 PostgreSQL Feature Set

As stated previously in this chapter, PostgreSQL is widely considered the most advanced open source database in the world. PostgreSQL provides a wealth of features that are usually only found in commercial databases such as DB2 or Oracle. The following is a brief listing of some of these core features, as of PostgreSQL 7.1.x.

2.4.4 Postgress: Object-Relational DBMS

PostgreSQL approaches data with an object-relational model, and is capable of handling complex routines and rules. Examples of its advanced functionality are declarative SQL queries, multi-version concurrency control, multi-user support, transactions, query optimization, inheritance, and arrays.

- **Highly extensible:** PostgreSQL supports user-defined operators, functions, access methods, and data types.
- **Comprehensive SQL support:** PostgreSQL supports the core SQL99 specification and includes advanced features such as SQL92 joins.
- **Referential integrity:** PostgreSQL supports referential integrity, which is used to insure the validity of a database's data.
- **Flexible API:** The flexibility of the PostgreSQL API has allowed vendors to provide development support easily for the PostgreSQL

RDBMS. These interfaces include Object Pascal, Python, Perl, PHP, ODBC, Java/JDBC, Ruby, TCL, C/C++, and Pike.

Procedural Languages

PostgreSQL has support for internal procedural languages, including a native language called PL/pgSQL. This language is comparable to the Oracle procedural language, PL/SQL. Another advantage to PostgreSQL is its ability to use Perl, Python, or TCL as an embedded procedural language.

- **MVCC:** MVCC, or Multi-Version Concurrency Control, is the technology that PostgreSQL uses to avoid unnecessary locking. If you have ever used another SQL capable DBMS, such as MySQL or Access, you have probably noticed that there are times when a reader has to wait for access to information in the database. The waiting is caused by people who are writing to the database. In short, the reader is blocked by writers who are updating records.

  By using MVCC, PostgreSQL avoids this problem entirely. MVCC is considered better than row-level locking because a reader is never blocked by a writer. Instead, PostgreSQL keeps track of all transactions performed by the database users. PostgreSQL is then able to manage the records without causing people to wait for records to become available.
- **Client/server:** PostgreSQL uses a process-per-user client/server architecture. This is similar to the Apache 1.3.x method of handling processes. There is a master process that forks to provide additional connections for each client attempting to connect to PostgreSQL.
- **Write Ahead Logging (WAL):** The PostgreSQL feature known as Write Ahead Logging increases the reliability of the database by logging changes before they are written to the database. This ensures that, in the unlikely occurrence of a database crash, there will be a record of transactions from which to restore. This can be greatly beneficial in the event of a crash, as any changes that were not written to the database can be recovered by using the data that was previously logged. Once the system is restored, a user can then continue to work from the point that they were at before the crash occurred.

PostgreSQL is a sophisticated object-relational database management system (ORDBMS). An ORDBMS is an extension of the more traditional relational database management systems

(RDBMS). An RDBMS enables users to store related pieces of data in two-dimensional data structures called tables. This data may consist of many defined types, such as integers, floating-point numbers, character strings, and timestamps. Data inserted in the table can be categorized using a grid-like system of vertical columns, and horizontal rows. The relational model was built on a strong premise of conceptual simplicity, which is arguably both its most prominent strength and weakness.

The object-relational aspect of PostgreSQL adds numerous enhancements to the straight relational data model. These include support for arrays (multiple values in a single column), inheritance (child-parent relationships between tables), and functions (programmatic methods invoked by SQL statements). For the advanced developer, PostgreSQL even supports extensibility of its data types and procedural languages.

Due to this object-relational concept, tables are sometimes called classes, while rows and columns can be referred to as object-instances and object-attributes, respectively. We will use this terminology interchangeably in this book. Other SQL data structures, such as indices and views, can be referred to as database objects.

2.5 MICROSOFT'S SQL SERVER 2000

Microsoft SQL Server 2000 (originally given the code name Shiloh) is the widest used SQL server after Oracle's SQL. Microsoft's development team included enhanced security mechanisms, simplified administrative tools and support for advanced web integration technologies. Many database professionals believe that this latest release marks SQL Server's entry into the big leagues, ready to take on Oracle in the high-reliability DBMS world.

It includes greatly enhanced analytical features. Full-fledged data mining support is provided through the use of classification, regression and clustering algorithms. Microsoft's implementation of online analytical processing (OLAP) technology allows data warehousers to store and manipulate large amounts of data with newfound ease. Additionally, database administrators concerned with importing data from diverse sources will be interested in SQL Server 2000's enhanced Data Transformation Services.

Most significantly, Microsoft adopted the security industry standard Kerberos authentication algorithm to facilitate creation and enforcement of secure trust relationships. The National Security Agency recently announced their certification of SQL Server 2000 as a C2-compliant computing platform. Reliability improvements such as log shipping, online backups and failover clustering are discussed in the article Maximizing Uptime.

SQL Server 2000's debut marks the release of the first Back Office server product supporting Microsoft's Net strategy. Developers will discover inherent support for Extensible Markup Language (XML) programming as one of a series of improvements designed to enhance the SQL Server development environment.

2.5.1 Microsoft SQL Server 7.0

The database is the heart and soul of any company's business infrastructure. Small to medium-size businesses planning to make this critical investment must consider their specific business needs, support requirements, and budgets. The solution that offers the best value with the least administrative hassle is Microsoft SQL Server 7.0, and we award it our Editors' Choice.

In the small-business market, the differentiating factors are ease of database administration, Web connectivity, the speed and features of the database server engine, branch-office and mobile support, and the ability to warehouse data efficiently. SQL Server 7.0 shines in all of these areas except Web connectivity. Its administration tools include many wizards and self-tuning settings that make it the only database we reviewed that might not require a specially trained administrator.

For branch offices or distributed sales teams, SQL Server's per-seat client license includes the right to run its desktop version, and the database has excellent replication features that let you sync up with home base and vice-versa. An OLAP server in the box let you store and analyze all the data you have. All this comes at a fraction of the cost of competing databases.

Packed with more features than competing products and sporting a very attractive price, Microsoft SQL Server 7.0 is a modern, full-featured SQL database that's right on target for the small or midsize organization. Its complete set of tools, high-end engine features, and robust analysis capabilities provide most of what IBM and Oracle offer only in their Enterprise Edition databases (and a few key things that neither includes) for less than either one. In addition, we found SQL Server 7.0 amazingly easy to use, yet still powerful enough to crank through hundreds of complex transactions a second on our test hardware without choking. (SQL Server also comes in an Enterprise

Edition version, whose main benefit is to allow automatic failover to a secondary, standby machine.)

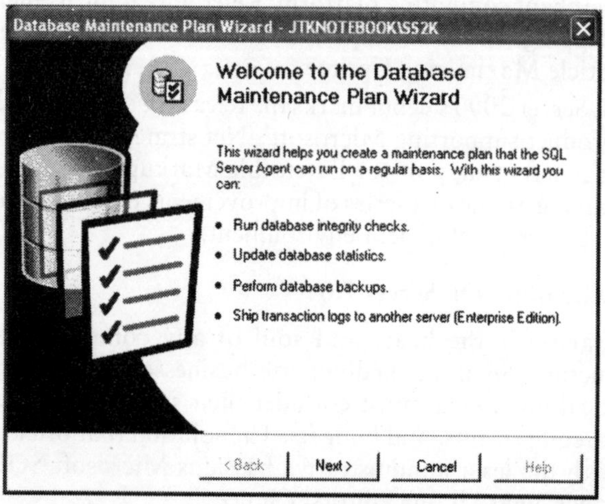

Fig. 5.4: Microsoft SQL

Yet SQL Server still lags behind the competition in two key areas: programmability and multimedia data support. For those who write their own business applications and put part of the code in the database itself (as opposed to putting all the code into applications that run on your Windows client systems), and for those who are moving to HTML or Java-based database client applications, SQL Server provides less built-in functionality and will be harder to use than DB2, Informix, Oracle, or Sybase.

SQL Server also runs only on Windows operating systems, so if you're running Unix-based servers, look elsewhere. But if you're running generic packaged Windows applications to access your database contents, or if you have custom applications that use only the ODBC standard to get information from the database, you'll be right in SQL Server's sweet spot.

Version 7.0 is more than just an upgrade of SQL Server 6.5. It has a new heart and brain in its latest incarnation. Think of 7.0 as a totally new database that just happens to have somewhat similar administration tools and to use the same programming language (Transact-SQL).

Microsoft has made a whole host of changes to bring its enterprise database up to technical snuff. SQL Server's engine—the heart of the product— now has a different data file format than its predecessors had.

One of the key reasons for changing the database file format is a switch to using 8K disk pages (the basic unit of database storage) instead of 2K pages. This permits storing more information in any database row and also makes the database faster when reading in many rows at a time, since the server has to retrieve fewer data pages from disk.

Also, SQL Server can now isolate data and index information at the row level (a feature called row-level locking). Previously, SQL Server could protect one user's changes from the activity of other users only by blocking other users' access to a group of perhaps 40 or 50 rows (depending on the size of a row) around the one being changed. SQL Server 7.0 locks down only the actual rows being changed and, as a result, can support more users at once. Some packaged business applications, like those from PeopleSoft and SAP AG, really need row-level locking to be at all usable.

Beyond all its engine changes, SQL Server includes an unmatched set of database tools. New with this release is a server product called SQL Server OLAP Services (OLAP stands for online analytical processing). It's a specialty-purpose database server that precalculates summary information for easy analysis and comparison.

For example, we used OLAP Services to view sales figures by store, by town, by region, and by country, and then compared last quarter's figures with those of this quarter. Since the totals had all been figured out ahead of time, we could flip instantly between high-level totals. With other OLAP servers currently priced at tens of thousands of dollars, OLAP Services is an incredible bargain.

OLAP Services is powerful, flexible, and easy to use, but it's not all roses. First, you have to buy separate, compatible software to use it, because OLAP Services uses a query language different from the one SQL Server uses. (Excel 2000 is one such client, but others are available from companies like Cognos Corp. and Seagate Software.)

Second, SQL Server doesn't make any use of OLAP Services (or of precalculation techniques in general) to speed up its own queries. The two are completely separate products; we had to load data from SQL Server into OLAP Services before we could do any OLAP analysis. By contrast, both IBM and Oracle have introduced some

OLAP features directly into their latest databases, in the form of query summary tables.

2.6 CDS/ISIS

UNESCO develops, maintains and disseminates two interrelated software packages for database management (CDS/ISIS) and data mining/statistical analysis (IDAMS). Both software are available on the "UNESCO Information Processing Tools" CD-ROM, distributed on a non-commercial basis, only to agreed distributors and Institutions upon request. CDS/ISIS and IDAMS are also individually distributed either directly from UNESCO Headquarters, or through a network of distributors at national, regional and specialized level.

CDS/ISIS is a generalized Information Storage and Retrieval system. The Windows version may run on a single computer or in a local area network. The JavaISIS client/server components allow remote database management over the Internet and are available for Windows, Linux and Macintosh. Furthermore, GenISIS allows the user to produce HTML Web forms for CDS/ISIS database searching. The ISIS_DLL provides an API for developing CDS/ISIS based applications.

IDIS is a tool for direct data exchange between CDS/ISIS and IDAMS. Knowledge and training in the use of information proces sing tools is as important as the tools themselves. At present, UNESCO offers various forms of traditional training in the use of CDS/ISIS and IDAMS. The production of a computerized tutorial available both on stand-alone PC configurations and in virtual courses through Internet is foreseen in the next biennium.

3.1 COMMON INTERFACE STANDARD FOR BIBLIOGRAPHIC DATABASES Z39.50

The Z39.50 standard was originally proposed in 1984 to provide a standard way of interrogating bibliographic databases. Since then, it has gone through 3 versions—in 1988 (v1),1992(v2) and 1995(v3). Version 2 in 1992 also incorporated and became compatible with an ISO standard (10162/3) called Search and Retrieve. Version 3 in 1995 extended the features of the protocol - it is this version that most suppliers are now implementing. It is maintained by the Z39.50 Maintenance Agency—administered by the Library of Congress.

Z39.50 — properly "Information Retrieval (Z39.50); Application Service Definition and Protocol Specification, ANSI/NISO Z39.50-

1995"— is a protocol which specifies data structures and interchange rules that allow a client machine (called an "origin" in the standard) to search databases on a server machine (called a "target" in the standard) and retrieve records that are identified as a result of such a search.

The rather forbidding name "Z39.50" comes from the fact that the National Information Standards Organization (NISO), the American National Standards Institute (ANSI)—accredited standards development organization serving libraries, publishing and information services, was once the Z39 committee of ANSI. NISO standards are numbered sequentially and Z39.50 is the fiftieth standard developed by NISO. The current version of Z39.50 was adopted in 1995, thus superseding earlier versions adopted in 1992 and 1988. It is sometimes referred to as Z39.50 Version 3.

The protocol is stateful and connection-oriented. The protocol defines interactions between two machines only. While some groups are now developing "broadcast search" applications that permit a client to search multiple servers in parallel, these are applications that are built on top of Z39.50 and use multiple concurrent Z39.50 connections to multiple machines. Z39.50 does not specify an applications program interface (API) to the services of the protocol on either the client or the server. It deals only with the interactions between the client and server machines. In addition, Z39.50 does not address any of the issues involved in user interfaces that the client may present or any of the issues involved in database management at the server.

The basic architectural model that Z39.50 uses is as follows: A server houses one or more databases containing records. Associated with each database are a set of access points (indices) that can be used for searching. This is a much more abstract view of a database than one finds with SQL, for example. Relatively arbitrary server-specific decisions about how to segment logical data into relations and how to name the columns in the relations are hidden; one deals only with logical entities based on the kind of information that is stored in the database, not the details of specific database implementations.

One of the basic Z39.50 functions allows the client to transmit a search to the server (a SEARCH request). A search produces a set of records, called a "result set", that are maintained on the server; the result of a search is a report of the number of records comprising the result set. The standard is silent as to whether the result set is materialized or maintained as a set of record pointers, and as to how the result set may interact with database updates that may be taking place at the server. Result sets can be combined or further restricted by subsequent searches. Note that this is substantially different from SQL servers, which do not employ result sets.

Records from the result set can be subsequently retrieved by the client using PRESENT requests. The PRESENT request offers elaborate options for controlling the contents and format of the records that are returned. The PRESENT request indicates specifically which records from the result set are to be retrieved. There are facilities for managing buffer space in the presence of very large records and also for transferring very large numbers of records from server to client without the need for repeated PRESENT requests and responses (and hence many round-trip interactions between client and server).

Z39.50 also contains functions for search management. For example, a server can provide progress reports for an active search, or can ask the client for authorization to continue a resource- intensive search; a client can abort an active search. The report for search completion can also return supplementary information such as how many records matched individual component terms in a search.

Z39.50 contains facilities for managing result sets, for sorting result sets, for browsing the values of access points associated with a database, for opening and closing connections, and also a general mechanism called "extended services", which is essentially an asynchronous remote procedure call mechanism that the client can use to invoke services on the server, optionally making reference to the contents of a result set as a parameter. Extended services (which will be discussed in more detail later) were originally intended as a means of saving result sets across sessions, queuing them for print or electronic mail processing at the server, or for registering and managing queries that would be executed periodically on the server.

The protocol also defines the following:

- A query language for specifying searches, which in turn builds upon registered definitions for attribute sets that specify the names of access points;
- Various record syntaxes that can be used for transferring records from the server to the client, including both some application domain specific syntaxes like MARC for bibliographic data and a massively

complex, very general purpose syntax called Generalized Record Syntax One (GRS-1);

- A language for describing how to construct records that are to be transferred from a result set back to the client; and
- A facility called EXPLAIN which allows clients to obtain a wide range of information from a server about what databases are available, what access points are supported in each database, and the like. EXPLAIN is modeled as a special purpose database which is searched using standard Z39.50 queries; the standard specifies the detailed structure of the records that can be retrieved from this database. EXPLAIN is intended to permit the development of clients that to at least some extent are dynamically self-configuring as they encounter various servers.

Z39.50 makes extensive use of registries for various types of objects, such as attribute sets used in queries and record syntaxes used in present requests. These are referred to via object identifiers which are used as parameters in the various protocol requests and responses that move between client and server. Some initial object identifiers are assigned by the standard; assignment of object identifiers on an ongoing basis is handled by the Z39.50 maintenance agency.

3.2 COMMON INTERFACING STANDARD Z39.50

Z39.50 is an international standard for communication between computer systems primarily, library and information related systems. Z39.50 is becoming increasingly important to the future development and deployment of inter-linked library systems. This technical briefing aims to explain the substance and significance of the standard to library and information system managers.

The typical (simplified) search process involved in a Z39.50 session is as follows:

- OPAC user selects Target library (Z-server) from an OPAC menu.
- OPAC user enters search terms.
- OPAC software sends search terms and Target library details to a "Z-client" a piece of software usually running as part of the library system.
- Z-client translates the search terms into "Z-speak" and contacts the Target library's Z-server software.

- There is a preliminary negotiation between the Z-client and Z-server to establish the rules for the "Z-Association" between the two systems.
- Z-server translates the "Z-speak" into a search request for the Target library's database and receives a response about numbers of matches etc.
- Z-client receives records.
- Records are presented to the OPAC interface for the user.

3.2.1 Basic Benefits

As is often the case, the basic technology is simple, but the ramifications are complex and far-reaching. Three key points have turned a process designed for simplifying a searcher's life into a powerful force for changing all aspects of library activity.

Modern Z-clients can send requests to several libraries simultaneously either the same request or different ones. This feature allows tremendous time saving when searching for rare items or for large numbers of records.

The basic record format used for interchange is MARC. The Z-client is presented with a MARC record for display and possible further processing. All libraries "trade" in bibliographic records one way or another. Z39.50 opens up that trade by making standardising the basic search and retrieve functions.

Extended services for ordering documents, updating databases and storing searches can be defined and controlled via Z39.50. By using Z39.50 as a basis, many other library processes, particularly ILL, can become "open".

3.3 SEARCH FEATURES

The latest version of Z39.50 (V.3 1995), allows extremely powerful search statements to be defined including:

- Complex Boolean statements involving all the standard operators AND, OR, NOT
- Comparison operators for dates *e.g.* Greater than, equal to *etc.*
- Proximity searching
- Truncation
- Completeness *i.e.* part of field, complete field *etc.*

3.4 ADDITIONAL FEATURES

Z39.50 enables

- Authentication allowing the Z-server to control who accesses their databases.
- Accounting/resource control to allow access to be charged for.
- "Explain" facility to allow information about the remote database services available *etc.* to be transmitted to the Z-client.
- Index browsing as typically available in OPAC systems.
- Defining record formats *e.g.* MARC format.

3.4.1 Extended Services

Version 3 also defines how to use the standard to implement what are called "Extended Services". These are not defined in the standard but may use Z39.50 as a control method. The types of tasks within the extended services area have been defined:

- Save a result set for later use:
- Save a query for later use
- Define a periodic search schedule
- Order an item
- Update a database
- Create an export specification

The implications for libraries as systems begin to implement these and the basic services are profound.

3.5 IMPLICATIONS FOR LIBRARIES

The implications for the library and information services are profound. Over the next few years, services will become "Z39.50 enabled" in much the same way that library OPACs have become "Web enabled". The process will be somewhat slower than the Web revolution but more far-reaching and structural. Somc possible effects on library operations are explored here:

3.6 OPAC

Many OPACs have been Z39.50 for a few years now. This is the basic benefit of Z39.50 operation for a typical user. Access any and all of the world's major library catalogues or just the local sources with a single search.

3.7 CATALOGUING

Searching for and downloading bibliographic records using a Z39.50 tool is simple and very efficient since multiple sources can be searched simultaneously and records easily compared. Currently libraries are often "locked in" via service agreement and proprietary software to a bibliographic utility. A Z39.50 world will allow users to establish relationships with a variety of sources without penalties of complexity generated by different software. Resource control how much have I spent? is built in to Z39.50 to make management easy. Note that the ease with which "free" records can be downloaded from libraries creates copyright issues that must be addressed. Again the mechanism for notifying a source library that a record has been used for cataloguing purposes is built into Z39.50 implementation is the real issue.

3.8 UNION CATALOGUES

Union catalogues— combined catalogues of several libraries—have been a valuable tool for decades within groups of otherwise separate libraries wanting to cooperate for inter-lending, cooperative purchase, and general reader service. They are, however, difficult and expensive to manage. Even using automated systems the coordination and maintenance required can be daunting from both an organisational and technical viewpoint. Using Z39.50 enabled catalogues and OPACs, a "virtual" union catalogue can be assembled without any changes to the individual organisation's methods and procedures. A user may sit at an OPAC screen and search several catalogues simultaneously as if they were one. Useful material and its location can be displayed with no additional work by library management apart from set-up of Z-clients. Ad-hoc groupings of libraries can be assembled to suit the needs of the users without any technical or administrative fuss. Colleges or companies suddenly merged or taken over, co-operative degrees, stock sharing schemes any scenario where it might be useful to have a consolidated view of library and information resources is simple to set up and administer.

3.9 INTER-LIBRARY LOAN (ILL)

The immediate benefit of a virtual union catalogue means that ILL is made easier the user can immediately identify the location of required items. The extended services of Z39.50 allows systems to arrange for the delivery, including account verification and billing, of an item to the enquirer. ILL services are currently either mostly manual or rely on disparate incompatible services from the large suppliers like OCLC and the British Library. In a Z39.50 enabled ILL future, libraries will be able to search and order items in one operation and deal directly with whichever library serves their needs all via their own library OPAC search tools.

3.10 CD-ROM ACCESS

Despite the steady migration of CD-ROM information providers to Web-based services, CD networks will be a feature of library services for some time. Apart from the idiosyncratic means required to force CD-ROMs to a networkable state, there remains the practical problem, when using them, of having to understand each different software interface and having to search each database separately. Using Z39.50, it would be possible to search each database using a single familiar interface and, additionally, several databases at the same time. Z 39.50 also solves the problems of being able to use different clients *e.g.* Macs, UNIX Workstations— even dumb terminals could be used. The SilverPlatter ERL technology actually provides similar facilities—but it is a proprietary standard and of limited application outside the CD-ROM area.

3.11 SELECTIVE DISSEMINATION OF INFORMATION

Version 3 of Z39.50 allows the user to specify search statements to be saved and run at intervals. Thus the user may for instance, identify useful libraries and information resources and set up SDI profiles using a single interface. Searches can be automatically run when required and the results downloaded from the database to a specified destination *e.g.* fax or e-mail. Z39.50 makes the much-vaunted Push Technology seem Stone Age in comparison.

3.12 COMMERCIAL INFORMATION DATABASES

Library catalogues are only a fraction of the searchable information available. There are hundreds of commercially available information service providers like Dialog, Lexis Nexis, FT Profile etc. These services allow very complex search statements and Z39.50 (version 3) contains equivalent search statements including proximity searching, term highlighting, image retrieval, individual chapter retrieval, specification of variant forms for downloading *e.g.* Word, Word Perfect etc. Accounting and authorisation controls are also built in. Again, by using Z39.50 protocols, the complexity of searching disparate databases can be reduced.

3.13 WEB SEARCHING AND FILTERING

Searching the Web is frustrating for some of the very reasons that Z39.50 was developed *i.e.* many different search engines and user interfaces. By adding an optional Z39.50 interface to Search Engines, much of the frustration and time wasting could be avoided. The much-discussed topic of filtering unwanted areas of Web content could be attacked through an extended service. Each library could set their own filter parameters on the Z39.50 client used to access the major Search Engines.

Given the rapid development and constant vying for competitive edge in the Web Search Engine arena Z39.50 server technology has probably a low priority. As the benefits of Z39.50 begin to be appreciated by professionals, however, it will only need one Search Engine to add the feature and there will be an avalanche of followers. See also a paper from Index Data for a useful discussion of Z39.50 on the Web.

3.14 DATABASE UPDATES

Z39.50 is not just a search and retrieve tool as mentioned, extended services such as ILL are now being linked into the standard by leading companies. Another task type that may be used as an extended service is updating a database. Thus a Z-client may for example retrieve a record from a database, edit it and then send it back to update the database GeoCAT product from Geac illustrates this function nicely since it may be used as a standard tool with either the ADVANCE or PLUS Library Management systems.

6
UNIT

Library Automation

1.0 Library

A library, by any definition, exists to provide economy in sharing of information resources, traditionally books and journals. But the scope of resources is ever expanding, especially in special libraries. It is:

- A place in which literary and artistic materials, such as books, periodicals, newspapers, pamphlets, prints, records, and tapes, are kept for reading, reference, or lending.
- A collection of such materials, especially when systematically arranged.
- A room in a private home for such a collection.
- An institution or foundation maintaining such a collection.
- A commercial establishment that lends books for a fee.
- A series or set of books issued by a publisher.
- A collection of recorded data or tapes arranged for ease of use.
- A set of things similar to a library in appearance, function, or organization: a library of computer programs.

1.1 THE MODERN LIBRARY

Modern libraries, in addition to providing patrons with access to books and other materials, often publish lists of accessions and may maintain a readers' advisory service. Interlibrary loan services, lecture series, public book reviews, and the maintenance of special juvenile collections are other important recent developments.

1.2 LIBRARY AUTOMATION

A library automation system is a computer system designed to automate tasks in a library. Such systems primarily host a catalog of the library's collection, replacing the old card catalog once used. Under the card catalog system, several cards may be prepared for each book, to be used in a sorted series of cards by Title, Author and Subject. Each card would also bear a call number, denoting the item's location within the library's collection (for example, according to the Dewey decimal system or other library classification scheme).

Automation of the catalog saves the labor involved in resorting the card catalog, keeping it up-to-date with respect to the collection, etc. Other tasks automated include checking out and checking in books, generating reports and borrowing, indexing journal articles and tracking interlibrary loans.

Library automation succeeds through the work performed by hardware, software, and people. Without the right human skills, no matter how advanced the technological tools, library automation will yield disappointing results. Alternatively, library automation systems may be extended to myriad applications, including a tool to support very sophisticated knowledge management projects, with the right people implementing and establishing goals and procedures. This can be accomplished even with severe limitations in the software or hardware resources.

Some remarkable results have been accomplished with very primitive software, all because of the skills of knowledgeable, confident, and practical professionals. Confidence comes from having the right mix of skill, know-how, and expertise. A confident professional is open and sharing and seeks cooperation and coordination. The latter are largely personal skills that depend on personality traits.

The essence of any type of knowledge management is sharing, peer-to-peer or through a formal information resource management process of coordinating knowledge. Librarians are often frustrated with the failure of potential users to seek them and their resources out. On the other hand, they are loath to promote and evangelize their services and resources. They tend to focus on tasks relating to physical management of collections instead of content management, which would more directly benefit the users. This is a generalization, but the scant availability of indexers and thesaurus developers illustrates the lack of interest in this area, especially among new librarians.

Given the challenge of working on a library automation initiative, there are two parallel opportunities for information professionals. One role is on the data processing and systems implementation side, working hand-in-hand with the IT group (people that control the hardware and networks). The other is on the content management side. Often, due to limited resources one person has to serve both roles.

1.3 MANAGING THE AUTOMATION PROJECT FORYOUR ORGANIZATION

The first imperative is to establish, why do it at all? It is going to be hard, require change, and require coordination among professionals, including other departments. One simple answer is that if you want to have a role in the information management, or knowledge management, arena of your organization you must perform all your work functions in accordance with business norms. Limiting your vision of how and what automation processes you can absorb or afford marginalizes the library severely and frequently results in total elimination. I, therefore have noted that often the people with extremely narrow views of what they want to automate, spend, and achieve are the ones whose jobs or libraries end up being eliminated. Limiting what you plan to automate versus totally integrating all library processes misses the principal reason to automate. Doing all your work to the full extent that automation supports is the accepted business norm in all organizations.

Can you accept the fact that if you want to follow normal business practices you must automate and it must be a complete solution to gain maximum efficiency? If so, you have to accept the difficulty, the change and the coordinating activities imposed by the automation process. The more accepting and open you are about the process, the more realistic and positive you can be in asserting your role; you will have conviction and the belief in the value of the outcome needed to embrace the challenge. If you can't go in with this view, go back to the drawing board and re-evaluate why automation needs to be done at all.

Questions you might ask are basic but important:

- What are the information needs of the organization?
- Who will access the database?
- What will the content be?

- What is our computing environment or can it be changed?
- Who will maintain the data?
- Who will determine the process and procedural controls and standards?

Even though you may not find a vendor who has all automation tools you need, you should still strive to develop a plan to coordinate all components as seamlessly as possible during this investigative period.

Regardless of the scope of the project, these questions need not only answers, but also planning and human resources. If you initially only plan for a limited use of automation, for example, serials check-in, with the idea of later creating a database of monographs for on-line retrieval, understand that you will have to repeat the planning phase. The second project will be complicated dramatically by the existence of the first system.

1.4 THE HUMAN FACE OF THE PROJECT

Formulating a vision of automation that assumes scalability is important. Planning for integration of processes and fit with the organization is equally important. The person who establishes the plan needs to have a vision that extends beyond the tasks of identifying and selecting a system. All organizations need someone on the library staff who grasps the full scope of the organization's mission, operational resources, and constraints. This cannot be learned by staying within the library. This person must spend 50 per cent of his or her time outside the department, building peer relationships with IT, management or any other department working with information resources.

The primary activity of the visionary is to discover what goals and resources the library currently, and potentially shares with the other groups. With the vision established, you can focus on the unique skills, expertise and resources the library has to offer. This is a good basis for building cooperation among the groups.

A person who works with other groups in planning for automation must look for needs that can be filled by his or her department. Take this planning phase as an opportunity to share expertise and educate other groups about the expertise you can offer. Information content management skills are lacking throughout all organizations. This is a library's opportunity to assert its place in information content management. A person with special competency, personal and professional, should be assigned to this role.

Once the project for implementing an integrated automation solution is defined, accepted and moving forward, the visionary/coordinator often continues as the project manager. Additionally, with the right technical expertise, that person may also be given the responsibility of handling networking, hardware, and operating system issues.

Where does the expertise originate for the roles described? It comes from having a solid grounding in several areas that aren't necessarily part of a single degree program. The following types of courses are crucial for a database administrator: systems analysis, fundamentals of computing, database management systems (addresses the types, range, and tools that are available), organizational structures, cataloging and indexing theory, thesaurus development. Hands on experience doing reference/research work, plus indexing experience will give the person realistic understanding of the challenges of building a database of quality, and the tools needed to do it efficiently.

Where does one find courses or continuing education to acquire this expertise? Not many schools offer them all; many of the best have closed in past years. The teaching staff is not there to sustain a program that addresses this multi-disciplinary field. Visit schools of information and library science, computer science, and business and describe what you are seeking. We can only hope that if the requests come frequently enough, especially in this age of knowledge management, higher education will begin to respond.

1.5 BUILDING THE DATABASE—THE DBA (DATA BASE ADMINISTRATOR)

The human skills needed to perform this job well are hard to come by. The DBA controls and manages the content, playing a vital role in establishing the data entry standards and ensuring that they are enforced. It is an exacting professional discipline to labor over database content and those who are good at cataloging/indexing are usually tapped for this job. However, there is another important professional role that may be more difficult because it also requires a level of coordination with the end-user community. There must be vigilance about what content is important and for how long. In most organizations this is constantly in flux, more with some disciplines than others. When organizational imperatives are expanded, new content must be sought and included. When other initiatives are limited or canceled, material may be extracted. The DBA must be aware of content that is getting out-of-date and terminology in the structured indexes that is outmoded. New terminology must be added constantly to the controlled vocabulary lists, but thought must also be given to compressing the terms when too many synonyms or unused language creep in. The language of the system (or "taxonomies", as Internet search systems call them) plays a crucial role in the success and usefulness of a database. Arcane or user unfriendly index entries will result in frustrating the user, often to the point that the database will never be used again.

Level of use of the database is the only true measure of the importance of its existence. The bottom line is, if it not being used, it is not valued. If it isn't valued, the people who build and defend it won't last very long. Staying in tune with your audience is your primary function, whether you are the planner/implementer or the content manager/DBA. In the end, the technology is really a minor part of the cost and effort. The best people produce the biggest benefits and successes.

1.6 PLANNING AND EVALUATING LIBRARY AUTOMATION SYSTEMS

Planning for an automated system, no matter how big or small, should be part of an overall long-range plan for the library. It is a mistake to let the automation project drive the library's priorities and become an end in itself. Automation should always be used as a means to achieve overall better patron service. Careful planning for technology will assure that your automation project is "sustainable", *i.e.* enhances the organization's ability to meet its service mission without disrupting the organizational stability of the institution.

Phase 1: Data Collection and Needs Assessment

Goal: To assess the library's needs for automation.

This data will be used for a variety of purposes throughout the planning process. Including, when you are preparing a preliminary cost estimate, providing information for an RFP, calculating the storage requirements for your new system.

If you do not have some of this information there are ways to reasonably estimate the data. Important to remember that large amounts of time need not be devoted to determining a perfectly exact number. Careful estimates can suffice for planning purposes.

Tasks: Study, analyze, and document the current system including review of policy and procedure manuals, training manuals, workflow diagrams.

Begin to communicate with staff and users about library automation (committee for involving staff in planning, staff survey, user survey or focus groups).

Collect and compile narrative and statistical information necessary to describe Library system and its operations

Assess the library's service mandates.

Assess the library's operating environment including facilities, budget, community relations, inter-library cooperative arrangements.

Phase 2: Examination of Automation Options

Goal: To explore what services could best benefit from automation.

Next step is to determine which library functions should be automated and in what order of priority. To make this determination you need to know how the service is currently organized and being performed and whether it is being done efficiently.

Need to Study and Analyze your Current Procedures

Question to ask is: which processes are repetitive and occupy large amounts of staff time? Which require retrieval of information from large difficult to maintain files? Which services place the biggest load on staff? Which are most popular with patrons?

For example, if the library circulates thousands of books a month, then circulation will be one of the first tasks to automate. If the library does not circulate a lot of books but manages thousands of periodicals then a strong serials control system will be most important. OPAC is important and because it depends on a quality database, the cataloging function will move up in order of priority.

Automation is complicated and disruptive. If you can evaluate the most important modules to bring up in a phased approach, you will have the ability of concentrating monetary and human resources according to a planned, coordinated implementation schedule.

For each functional area determine:

- Current workflow patterns and procedures
- How much space and equipment are used

- Volume of activity
- Cost to perform
- Current problems or needs in the area.

Tasks:

- Set priorities for what services to automate in a phased approach.
- Begin defining the steps for a phased and coordinated plan to automate.
- Analyze hardware, software, telecommunications needs
- Identify start-up and ongoing costs for automation plan
- Develop a budget for automation

Phase 3: Bibliographic Database Development

Goal: To establish a complete database of all library's holdings that can be ready to load when the new library system is installed.

Tasks:

- Identify, describe, and document existing shelflist files and begin to standardize the data they contain.
- Institute quality control measures to assure consistency of entries within the shelflist file.
- Inventory the collection.
- Verify existing bibliographic information in the shelflist records.
- Locate and add any necessary missing bibliographic information to the records.
- Implement uniform standardized cataloging practices and consistent cuttering practices.
- Prepare specifications for treatment of bibliographic data and item data (including authority control needs).
- Prepare and distribute a Request for Proposals for vendor supplied data conversion project (including authority control)
- Evaluate proposals and select a vendor.
- Organize staff and define procedures for working with vendor to resolve bibliographic record conflicts.

Phase 4: System Specifications and Requirements

Goal: To prepare an RFP with specifications for an automated system that will meet the service objectives and requirements for the County Library System.

Purpose of an RFP: To enable you to compare relative differences from one system to another. It is very difficult to do this only through vendor demonstrations, talking to sales representatives, reading literature, or comparing broad price quotations.

It is also very difficult to get comparable cost data from various vendors. A good RFP will force a vendor to detail costs in such a fashion as to make comparison possible, *i.e.* for specific configuration. Make sure you are comparing apples to apples.

This is where you take your analysis and operationalize it into a description of your dream system. It is necessary to take the data you have gathered and break it into functional specifications and technical specifications. Functional specifications describe the capabilities that you want in a system. Technical specifications deal with standards that must be adhered to guarantee minimum system performance.

Tasks:

- Write narrative to describe the Library System including statistical description of users, their use of services, community growth, projected five year growth of demand for library services.
- Write functional specifications for a system.
- Write technical specifications for a system.
- Develop criteria by which proposals will be analyzed and evaluated.
- Distribute RFP to appropriate vendors of automated systems.

Phase 5: Analyze Proposals and Select Vendor

Goal: To analyze vendor responses to RFP and to select a vendor to implement a system.

Tasks:

- Evaluate vendor proposals.
- Draft interview questions for vendor supplied contacts.
- Call Vendor references and interview them.
- Select top two or three proposals.
- Arrange for vendor onsite demonstrations/presentations.
- Have selection committee make site visits to systems that have actually implemented systems from top vendors and ask them to talk about their experience working with the vendor.
- Draft a list of questions to ask each vendor during the presentation.
- Analyze results of presentations/demonstrations.

For Presentation

- Request that high priority modules be demonstrated first.
- Request the display of a tagged full MARC record.
- Ask if the version that is being demonstrated for each module is the same as the version in current release, *i.e.* what you would be getting if you bought the system today.
- Watch for cumbersome or awkward operations within or between functions.
- In addition to the canned demo, ask that specific searches, operations, be performed that are analogous to real situations in your library. Ask each vendor to perform the same operations.

Phase 6: Begin Contract Negotiation with Top Vendor

Goal: To purchase a library automation system.

Purpose is to interpret and clarify the differences between a vendor's response and the library's specifications.

To formalize pricing and payment schedules, deal with non-performance issues and remedies as well as warranties, vendor bankruptcy, software infringement, and maintenance expectations.

Safeguards conformance to any legal requirements necessitated by the library's parent organization or governing board.

Tasks:

- Involve legal counsel, together with library personnel, in drafting or evaluating a contract.
- Compare list of necessary contract elements to actual contract.
- Reference RFP in contract.
- Set up a Contract Negotiation Agenda.
- Bring negotiations to a successful and favorable conclusion.

Phase 7: Implement the System.

Goal: to install the selected system.

Benchmark test is typically conducted after a vendor has been selected but before the system is actually installed in the library and prior to any payments by the library to the vendor. A Benchmark test may be necessary if the system being proposed exceeds the capabilities of any of the vendor's systems that are currently installed. Benchmark tests require careful planning and preparation and should be devised with legal counsel. A detailed "script" of transactions to be performed at each terminal should attempt to reflect the library's "real" operating environment.

Benchmark tests are costly and time consuming and should not be automatically required of every vendor responding to the RFP.

A performance bond may be required by the library as an alternative to, or in conjunction with a benchmark test. Performance bonds require that the vendor place a specified amount of money with a third party until the system has been installed and accepted by the library. The performance bond needs to be incorporated and fully described in the RFP.

Typical Performance Tests

System Reliability Test: Confirms that the system is fully operational for the percentage of time that has been specified in the RFP.

Functional Performance Acceptance Tests: Simply verify that the system capability can perform according to the functional requirements set out in the RFP and in the system's documentation.

Full-load Response Time Test: Determines whether the system can perform within required response-time tolerances under a full transaction load. The maximum response times required by the library will have to be carefully detailed in the RFP. The vendor and the library together usually develop the testing of the system during the contract negotiation and these tests result expectations are incorporated into the contract. Usually extend from 60 to 90 days after the library's bibliographic and patron databases have been loaded. These type of tests should never be performed with test data supplied by the vendor.

Tasks:

- Customize the vendor's system to the library's policies.
- Site Preparation.
- Installation of hardware and software with initial performance test.
- Acquire the necessary forms, supplies, and other peripheral equipment.
- Load and index the bibliographic database
- Train staff, realign workflow and space
- Activate system
- Evaluation operation and do appropriate testing.
- Accept System

The tasks involved in a complete implementation plan are:

1. Staff preparation and project management
2. Site preparation
3. Orientation and testing
4. Database creation
5. Database design and profiling
6. Database loading
7. Functional integration
8. System customization
9. Staff instruction
10. Documentation
11. Transition to production
12. Bibliography

Each task must be considered and incorporated in the general library automation plan. The level of effort required for each step in the plan will depend on the library's current environment and situation. Preliminary planning may begin before the system is installed or even selected.

Artificial Intlligence (AI)

Artificial Intelligence (AI) encompasses the following general areas of research:

1. Automatic programming,
2. Computer vision,
3. Expert systems,
4. Intelligent computer-assisted instruction,
5. Natural language processing,
6. Planning and decision support,
7. Robotics, and
8. Speech recognition.

Intelligent library systems utilize artificial intelligence technologies to provide knowledge-based services to library patrons and staff.

Wide Area Networking Considerations for Library Automation Systems

Wide Area Networks (WANs) allow related libraries to share automation systems and other local resources in addition to providing

Internet access. WAN connectivity options for Missouri public libraries as provided by MOREnet are outlined in the MOREnet document Wide Area Networking Options for REAL.

Libraries should review this document before considering a Wide Area Network to connect to remote libraries. Cost, distance and available technical support are primary considerations in considering WAN options. The technical aspects of specific automation systems must also be considered for optimum connectivity.

IPX Versus IP

Although Local Area Networks (LANs) can use a variety of network protocols, all data that traverses the Internet must use standard TCP/IP protocols. Because modern library Automation Systems typically communicate through TCP/IP communications over the Internet are simple to incorporate. Some automation systems may require communications that rely on the IPX protocol, (a proprietary networking protocol used by Novell Networks). If libraries need to establish connections between two or more sites that require transmission of the IPX protocol there are disadvantages to the libraries involved.

All data that travels the Internet (synonymous with the MOREnet backbone for this discussion) must do so via the TCP/IP protocol suite. For IPX to cross the Internet it must be "encapsulated" or "tunneled" inside of TCP/IP packets. This process requires configuration of the Novell server or the router at each site and will cause additional overhead on the file server and the Wide Area connection.

Another option would be a point-to-point connection as shown in Figure of Wide Area Networking Options for REAL. A point-to-point connection is a dedicated link that does not involve the Internet between two sites. Two libraries that need to exchange IPX data could simply route or bridge the IPX protocol across this connection. Unfortunately point-to-point connections are distance sensitive and generally if the distance from site A to the telco to site B is greater than five miles this will not be an effective option.

In general, to conform with other standardized library resources, maximize efficiency of network connections, and minimize complexity of administration, it is recommended that Library Automation Systems use the TCP/IP suite of protocols for connectivity. Systems should at least have a client component that can access resources through the Internet using TCP/IP.

2.1 AUTOMATION OF IN-HOUSE OPERATIONS

A good library automation system (software) will necessarily have following components (modules):

- Acquisition
- Cataloging
- Circulation
- Serials' Control
- OPAC and
- Library Statistics

Required details of each of above modules is given below with the help of an existing popular software SydneyPLUS.

2.1.1 Acquisitions

The Acquisitions module improves purchasing decisions and helps you better manage costs through detailed reporting on the financial operation of your information center.

Managing costs. Keeping an eye on the bottom line. Making better use of purchasing dollars. It's essential to track and control costs, but you can only manage that which can be measured.

The Acquisitions module is a powerful financial tool. It improves purchasing decisions because it lets you efficiently track and manage costs. One of our clients, for example, found cost-savings of 10 per cent after using the Acquisitions module to manage subscriptions. Flexible and adaptable, the system also promotes conscientious spending by tracking purchases at the agency, department, individual and project level.

Key Features and Benefits

- Easily interfaces with back office accounting systems to eliminate duplicate data entry.
- Track budgeted vs. actual expenses on a daily, monthly or annual basis.
- Record what was purchased, by whom and when.

- Adapt the cost control system to meet your needs by spreading costs over multiple cost centers or centralizing cost management.
- Reduce costs by identifying areas that are inefficient and purchasing only those items that offer a return on investment.

2.1.2 Catalog

Powerful and flexible, the Cataloging module improves operational efficiency—enabling the creation of an information resource tailored to the exact needs of your organization. Information alone is not enough—to ensure your organization saves time and money your users need to be able to quickly and accurately find the knowledge resources they need.

The backbone of the SydneyPLUS system, the Cataloging module is an essential tool for managing information resources. With its unprecedented flexibility, this module can catalog any item imaginable. The result is a robust and dynamic storehouse of information resources that improves staff efficiency, helping employees make better decisions, faster.

Key Features and Benefits

- Centralize a variety of information resources under a single search and retrieval system.
- Improve efficiency thanks to a simplified process for importing records in either XML or MARC and by automatically generating authority headings.
- Eliminate redundant data entry by using a single office-wide database.
- Categorize information with robust authority controls so users can easily find what they need.
- Enhance the value of the catalog by linking virtually any digital file to a record.

2.1.3 Circulation

The Circulation module reduces costs by streamlining information sharing across the enterprise. Organizations invest in information—in books, reports, journals and other resources —but this investment only has value if it is used and shared. Information can also be expensive, and organizations need to ensure that their knowledge investment is spread among as many users as possible.

The SydneyPLUS Circulation module addresses these two key business needs. It streamlines the information-sharing process, improving the availability of resources because their exact whereabouts are always known. And it directly contributes to reduced costs. Acquisitions budgets can be stretched as resources are more effectively shared with greater numbers of users.

Key Features and Benefits

- Reduce costs by minimizing information resource "losses" and by identifying and eliminating obsolete or unused resources from your collection.
- Maximize the value of information resources by distributing them fairly and efficiently across the organization.
- Improve workflow by seeing all activity for a borrower or item.
- Develop more relevant information resources by tracking item usage.

2.1.4 Serials

The Serials Module helps organizations maximize the full value of their investment in serials and journals—often one of the highest cost items in a company's knowledge asset budget.

Serials are one of the largest investments a company makes in knowledge resources. Many large firms spend hundreds of thousands (if not millions) annually on paper and digital subscriptions, yet they have no way of tracking whether this is money well spent.

The Serials Module is a powerful tool in simplifying serials management. Clients who use this module invariably report that they have reduced costs and freed up staff time to focus on higher-level tasks. They also report that they have saved money by finally being able to identify and eliminate duplicate, unauthorized or unread subscriptions.

Key Features and Benefits

- Reduce costs by eliminating unnecessary duplication of subscriptions and unauthorized or unread subscriptions.
- Reduce costs by negotiating volume discounts.
- Maintain the completeness of journal subscriptions by ensuring that all expected issues have arrived.
- Improve efficiency thanks to a simplified check-in process and automatic reporting.

- Expand your firm's knowledge resources by incorporating e-journals.
- Streamline the process of claiming undelivered materials.

2.1.5 MARC Record

The MARC Record Interface saves time and costs when creating a high-quality and accurate catalog of your organization's knowledge resources.

Your online catalog is the most visible aspect of your library. The sooner it's up and running the better. The MARC Record Interface ensures fast implementation, high quality records and reduced costs compared to manual data entry. The result is an up to date and accurate catalog.

Key Features and Benefits

- Save time by minimizing data entry.
- Promote consistency using automatically generated authority headings.
- Optimize resource utilization and reduce costs by using outside resources to update catalog data.

2.1.6 Systems Architecture

- **Industry-standard Architecture Reduces Cost of Ownership:** The knowledge management system provides the infrastructure you need to maximize the value of your information resources. SydneyPLUS for MS-SQL & Oracle uses advanced system architecture based on industry-standard SQL technology. The system is easily integrated within your existing network architecture, eliminates the cost of supporting proprietary technology and reduces your total cost of ownership.
- **Minimal Bandwidth Needed:** Thin-client architecture improves efficient information sharing, making the system ideal for global deployment, Internet use and WAN connections. With its small network footprint, there's no need to purchase extra bandwidth. And the system's use of XML further improves its efficiencies, making it easy to integrate data from outside sources.
- **Robust Reporting Tools:** SydneyPLUS's open architecture means that it can use any third-party reports, including Crystal Reports, a standard ODBC-compliant reporting program. This significantly

reduces the cost of ownership while ensuring full functionality and flexibility.

- **Modules Improve Efficiency, Reduce Costs:** SydneyPLUS is a completely integrated and modular system. All data entry functions have been streamlined, and information entered in one area is immediately accessible to all other modules and users. Because separate modules address each function of the library, you can purchase only those modules that you require. Yet you can add additional modules at any time. And by incorporating the modules available through our knowledge management solution, you can transform your library into an opportunity-generating resource.

2.2 LIBRARY CATALOG

A library catalog (or library catalogue) is a register of all bibliographic items found in a particular library or group of libraries, such as those belonging to a university system spread out over several geographic locations. A bibliographic item can be any information entity (*e.g.*, books, computer files, graphics, realia, cartographic materials, etc.), that is considered library material (*e.g.*, a single novel in an anthology), or a group of library materials (*e.g.*, a trilogy), or linked from the catalog (*e.g.*, a webpage) as far as it is relevant to the catalog and to the users (patrons) of the library.

The card catalog was a familiar sight to library patrons for generations, but it has been effectively replaced by the Online Public Access Catalog (OPAC). Some still refer to the online catalog as a "card catalog", but this is incorrect. Some libraries with OPAC access still have card catalogs on site, but these are now strictly a secondary resource and are seldom updated. Many libraries who have retained their physical card catalog will place a sign advising the last year that the card catalog was updated. Some libraries have eliminated their card catalog in lieu of the OPAC for the purpose of saving space for other use, such as additional shelving.

2.2.1 Goal of a library catalog

A library catalog typically serves to answer the following questions:

- Does the library have a specific work?
- Which works by a particular author does the library have?
- Which works about a particular subject does the library have?

Some large libraries possess almost all publications of some period, or some country, and their catalog also serves to answer the following questions:

- Which works have been written by a particular author?
- Which works about a particular subject have been published?

In addition, a catalog may serve internal purposes of the library, for example as an inventory of everything that should be present.

2.2.2 CATALOG TYPES

Traditionally, there are the following types of catalog:

- **Author Catalog:** A formal catalog, sorted alphabetically according to the authors' or editors' names of the entries.
- **Title Catalog:** A formal catalog, sorted alphabetically according to the title of the entries.
- **Keyword Catalog:** A subject catalog, sorted alphabetically according to some system of keywords.
- **Mixed Alphabetic Catalog Forms:** Sometimes, one finds a mixed author / title, or an author / title / keyword catalog.
- **Systematic Catalog:** A subject catalog, sorted according to some systematic subdivision of subjects.
- **Shelf list Catalog:** A formal catalog with entries sorted in the same order as bibliographic items are shelved. Producing formal catalogs is relatively easy, as the cataloger can follow a set of cataloging rules. However, a formal catalog cannot answer question 3 (Which works about some subject does the library have?). A subject catalog just serves that goal, but it is much more difficult to produce, as the cataloguer has to get an accurate impression of the contents of the bibliographic item.

Library catalogs originated as manuscript lists, arranged by format (folio, quarto, etc.) or in a rough alphabetical arrangement by author. Printed catalogs, sometimes called dictionary catalogs enabled scholars outside a library to gain an idea of its contents. These would sometimes be interleaved with blank leaves on which additions could be recorded, or bound as guardbooks in which slips of paper were bound in for new entries. The first card catalogs appeared in the nineteenth century, enabling much more flexibility, and towards the end of the twentieth century the OPAC was developed (see below).

2.2.3 CATALOGING RULES

Cataloging rules have been defined to allow for consistent cataloging of various library materials across several persons of a cataloging team and across time. Users can use them to clarify how to find an entry and how to interpret the data in an entry. Cataloging rules prescribe:

- Which information from a bibliographic item is included in the entry.
- How this information is presented on a catalog card or in a cataloging record;
- How the entries should be sorted in the catalog.

The larger a collection, the more elaborate cataloging rules are needed. Users cannot and do not want to examine hundreds of catalog entries or even dozens of library items to find the one item they need.

Currently, most cataloging rules are similar to, or even based on, the International Standard Bibliographic Description (ISBD), a set of rules produced by the International Federation of Library Associations (IFLA) to describe a wide range of library materials. These rules organize the bibliographic description of an item in the following areas: title and statement of responsibility (author or editor), edition, material-dependent information (for example, the scale of a map), publication and distribution, physical description (for example, number of pages), series, note, and standard number (ISBN). The most commonly used set of cataloging rules in the English speaking world are the Anglo-American Cataloging Rules, 2nd Edition, or AACR2 for short. In the German-speaking world there exists the Regeln fr alphabetische Katalogisierung, abbreviated RAK. AACR2 has been translated into many languages, however, for use around the world. AACR2 provides rules for descriptive cataloging only and does not touch upon subject cataloging.

Library items that are written in a foreign script are, in some cases, transliterated to the script of the catalog.

2.2.4 Sorting

In a title catalog, one can distinguish two sort orders:

- In the grammatic sort order (used mainly in older catalogs), the most important word of the title is the first sort term. The importance of a word is measured by grammatic rules; for example, the first noun may be defined to be the most important word.

- In the mechanic sort order, the first word of the title is the first sort term. Most new catalogs use this scheme, but still include a trace of the grammatic sort order: they neglect an article at the beginning of the title.
- The grammatic sort order has the advantage that often, the most important word of the title is also a good keyword (question 3), and it is the word most users remember first when their memory is incomplete. However, it has the disadvantage that many elaborate grammatic rules are needed, so that only expert users may be able to search the catalog without help from a librarian.

In some catalogs, person's names are standardized, *i. e.*, the name of the person is always (catalogued and) sorted in a standard form, even if it appears differently in the library material. This standardization is achieved by a process called authority control. An advantage of the authority control is that it is easier to answer question 2 (Which works of some author does the library have?). On the other hand, it may be more difficult to answer question 1 (Does the library have some specific material?) if the material spells the author in a peculiar variant. For the cataloguer, it may incur (too) much work to check whether Smith, J. is Smith, John or Smith, Jack.

For some works, even the title can be standardized. The technical term for this is uniform title. For example, translations and re-editions are sometimes sorted under their original title. In many catalogs, parts of the Bible are sorted under the standard name of the book(s) they contain. The plays of William Shakespeare are another frequently cited example of the role played by a uniform title in the library catalog.

Many complications about alphabetic sorting of entries arise. Some examples:

- Some languages know sorting conventions that differ from the language of the catalog. For example, some Dutch catalogs sort IJ as Y. Should an English catalog follow this suit? And should a Dutch catalog sort non-Dutch words the same way?
- Some titles contain numbers, for example 2001: A Space Odyssey. Should they be sorted as numbers, or spelled out as two thousand and one?

In a subject catalog, one has to decide on which classification system to use. The cataloger will select appropriate subject headings for the bibliographic item and a unique classification number (sometimes known as a "call number") which is used not only for identification but also for the purposes of shelving, placing items with similar subjects near one another, which aids in browsing by library users, who are thus often able to take advantage of serendipity in their search process.

2.2.5 Online Catalogs

Online cataloging has greatly enhanced the usability of catalogs, thanks to the rise of Machine Readable Cataloging = MARC standards in the 1960s. Rules governing the creation of catalog MARC records include not only formal cataloging rules like AACR2 but also special rules specific to MARC, available from the Library of Congress and also OCLC. MARC was originally used to automate the creation of physical catalog cards. Now the MARC computer files are accessed directly in the search process. OPACs have enhanced usability over traditional card formats because:

- The online catalog does not need to be sorted statically; the user can choose author, title, keyword, or systematic order dynamically.
- Most online catalogs offer a search facility for any word of the title; the goal of the grammatic word order (provide an entry on the word that most users would look for) is reached even better.
- Many online catalogs allow links between several variants of an authors name. So, authors can be found both under the original and the standardised name (if entered properly by the cataloger).

2.2.6 Online Public Access Catalog (OPAC)

Many libraries have their catalog accessible via Internet; some of them can be queried using a simple browser, other using a special version of browser (with Java Script and CSS features), some others using Z39.50 clients.

If you are looking for a book or another item, you can access to the OPAC of your nearest library and search for it there. If you cannot find a particular book in the catalog, it may be obtainable through interlibrary loan, also known to librarians as I.L.L. Ask the nearest Reference librarian for assistance. If you want to look if a book exists and you have few elements to identify it, you can use a meta-searcher: you can fill the query form once and spread you search over many library catalogues. A service such as MultiOpac does this task for you.

An Online Public Access Catalog or OPAC is a computerized online catalog of the materials held in a library. The library staff and the public can usually access it at several computer terminals within the library, or from home via the Internet. Since the mid-1980s, it has replaced the card catalog in most libraries. Since the mid-1990s, character-based OPAC interfaces are being replaced by Web-based interfaces. OPACs are often part of an integrated library system.

The largest OPAC is the giant OCLC WorldCat database, which is a catalog of materials held at collaborating public and private libraries worldwide. WorldCat only shows whether a given item is owned by a library system. It does not breakdown information by branch library or show which copies are currently on loan to users. WorldCat users must connect to a library's own OPAC to see that information.

3.0 Retrospective Conversion and Weeding

Retrospective Conversion is the process of converting shelf cards or pre-existing online records into machine-readable format. There are several methods for converting your existing shelf-list cards into MARC records. Typically, you search a large database such as OCLC for a given record that matches your shelf list card. Once this record is found, you download the item from OCLC or another remote database.

Finally, you customize this record, adding local information such as call number and your library's holding information. It is important that the shelf list card display adequate information in order to find the appropriate match in the remote database. One way to ensure a good match is to make sure that your shelf list or pre-existing catalog is in a standard MARC format before conversion.

3.1 TIPS FOR RECORD CONVERSION

Establish a source for conversion. Will you use your shelf list, public catalog, or copy the verso (inside title page) for performing your retro-conversion? Identify one source that works best, and use it!

Convert the most popular items first. What circulates the most? What items or subject areas generate the most inquiries? Popular collections will be the most crucial to convert. Work your way down from here.

Examine Your Card Catalogs. Is there enough information available to establish a match in the database of MARC records? Is local data added to your MARC records available? (call number or copy information.

What Type of Match are You Looking For?

Is a general match sufficient; for example, as long as the title is Gone With The Wind and the author is Margaret Mitchell this match will serve your needs? Or, are you looking for more specific information for a record match such as the particular edition of Gone With The Wind? General matches require less specific information-title and publication date. If you want a closer match, you will need more detailed information such as ISBN number or Library of Congress Card Number (LCCN).

3.1.1 Shelf List Preparation

Make sure the following fields are listed on each shelf card:

- Library of Congress Number
- ISBN (or ISSN for periodicals)
- Author
- Title
- Publisher
- Publication date
- Check the shelf list against your physical collection.

What to look for when testing data:

- Call numbers with semicolons, or of atypical origin
- Order records
- Suppressed records should still be suppressed
- Provisional records
- Claim notes and action dates
- Open orders by type and prices
- Diacritics in authority and bibliographic records
- 949 fields of bibliographic record
- Unlinked items or short MARC records
- Copy holding notes and 852, 866, 867, 868, and 899 fields
- Long and/or complex call numbers (government documents, Dewey, Cutter)
- Existing system control numbers
- Weeding

General Criteria for Weeding

- Item has not circulated in five years
- Library owns a later edition

- Library owns multiple copies
- Item is in poor condition

Refer to your library's collection development policy, for guidance in what items or category of items to weed.

3.2 Barcoding

A *barcode* (also bar code) is a machine-readable representation of information in a visual format on a surface. Bar coding is a computer coding system that uses a printed pattern of lines or bars to identify products, mail and packages, customer accounts, and the like.

The industrial use of barcodes can be traced back as far as the 1960s. Some of these early implementations were used to identify rail road cars. Common barcodes started appearing on grocery shelves in the early 1970s as the UPC code to automate the process of identifying grocery items. Today, barcodes are just about everywhere and are used for identification in almost all types of business. When barcodes are implemented in business processes, procedures can be automated to increase productivity and reduce human error. Barcoding should be used whenever there is a need to accurately identify or track something.

Originally barcodes stored data in the widths and spacings of printed parallel lines, but today they also come in patterns of dots, concentric circles, and hidden in images. Barcodes can be read by optical scanners called barcode readers or scanned from an image by special software. Bar codes are read by optically scanning the printed pattern and using a computer program to decode the pattern.

When the code is read by an optical scanner linked to a computer, the computer can provide and record information about the item, such as its price or the quantity sold, from and to databases. In a linear bar code system, the code itself contains no information about the item to which it is assigned but represents a string of identifying numbers or letters.

The original North American Universal Product Code (UPC), which dates to 1971, used a set of two dark (usually black) and two light (usually white) bars of specified thicknesses to represent 12 numbers, but beginning in 2005 the Uniform Code Council, now known as GS1 US, adopted the similar European Article Numbering Code (EAN), which encodes 13 numbers and had become the international standard.

The standards for the international product bar code system are managed by GS1, formerly known as EAN International, which is based in Brussels. The dark bars may be from one to three units wide and the light bars from one to four units. For registration purposes two one-unit dark bars are placed at each end and in the middle. Each item is assigned a unique numeric code, which is printed as a bar code on the item's packaging.

So-called two-dimensional (2D) bar codes permit the encoding of information about an item in addition to an identifying code. In a 2D bar code, two axes, or directions, are used for recording and reading the codes and the bar size is reduced, increasing the space available for data in the way that a column of words improves on a column of letters. Some 2D codes do not use bars at all, such as the United Parcel Service's hexagon-based Maxicode.

3.2.1 Applications

Barcodes (and other machine readable tags like RFID) are used wherever physical objects need to be tagged with information that is to be processed by computers. Instead of typing strings of data into a terminal, the operator only has to display the code to a barcode reader. They also work well in fully automated environments, such as baggage routing at airports.

The data contained in a barcode varies with the application. In the simplest case, an identification number is used as an index in database where the complete information is kept. The EAN-13 and UPC codes commonly found on retail articles work this way.

In other cases the barcode holds the complete information itself, with no need for an external database. This led to the development of barcode symbologies that can express more than decimal digits, ranging from additionally encoding just the upper case alphabet to the complete ASCII character set and beyond. The drive to encode ever more information in combination with the space requirements of simple barcodes led to the development of matrix codes (a type of 2D barcode), which do not consist of bars but rather a grid of square cells. Stacked barcodes are a compromise between true 2D barcodes and linear codes, and are formed by taking a traditional linear symbology and placing it in an envelope that allows multiple rows.

3.2.2 Barcode Types and Use

The type of barcode to use for a particular situation depends upon:

1. Implementation;
2. Data you need to encode in the barcode and
3. How you wish to print the barcode.

There are several different types of barcode standards for different purposes—these are called symbologies. Each type of symbology (or barcode type) is a standard that defines the printed symbol and how a device such as a barcode scanner reads and decodes the printed symbol.

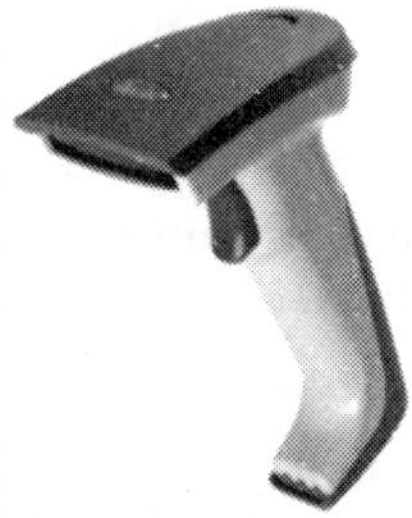

Fig. 6.1: Barcode Reader

If an industry standard has been established for your implementation of barcoding then you most likely do not have a choice in selecting your barcode symbology. Industry standards are usually established when multiple parties or companies are involved in the id process. Note that the standard is not necessarily the same as the barcode symbology. Barcode standards define how to use the barcode symbology in a particular situation. For example, ISBN is a standard for labeling books and periodicals that uses the EAN-13 symbology.

3.2.3 Reading Barcodes

One of the most common methods of reading barcodes is by use of the hand held barcode scanner. When a barcode is scanned in, the data is sent to your PC as if it had been typed by the keyboard.

Most barcode scanners can read common linear symbologies such as code 39, UPC, EAN, Code 128 and Codabar if it is enabled in the scanner.

3.3 BARCODE READER

A barcode reader (or barcode scanner) is a computer peripheral for reading barcodes printed on various surfaces. Like a flatbed scanner, it generally consists of a light source, a lens and a photo conductor translating optical impulses into electrical ones. Additionally, nearly all barcode readers currently produced contain decoder circuitry analyzing the barcode's image data provided by the photo conductor and sending the barcode's content to the scanner's output port.

3.3.1 Types of Barcode reader

Many different types of barcode scanners are available. They can be distinguished as follows:

A typical barcode scanner (photo by Argox)

By light source

- **LED scanners:** Also referred to as CCD scanners — even though the CCD is in fact the photo imaging sensor, not the light source.
- **Laser scanners:** much more expensive than LED scanners but are generally capable of longer maximum scanning distances.
- **Imager scanners:** These scanners take an image of the linear barcode, generally more rugged as they have no moving parts.

By Housing

- **Handheld scanner:** With a handle and typically a trigger button for switching on the light source
- **Pen scanners (or wand scanners):** A pen-shaped scanner that is swiped across a barcode stationary scanners
- Wall—or table-mounted scanners that the barcode is passed under or beside. These are commonly found at the checkout counters of supermarkets and other retailers.
- **Pda scanners:** A personal digital assistant (PDA) with a built-in barcode reader.

Technology and Engineering

Laser Scanners: Typical scanners utilise a 650 nm laser diode, leveraging on the economies of scale of diodes manufactured for CD/DVD drives. The diode is normally housed in a metal casing for heat dissipation. The laser light emitted is focused through a lens and reflected off a rotating polygonal mirror, or an oscillating mirror. This produces a line or raster pattern which may be aimed at a barcode. The light reflected off the barcode is captured onto a photodiode whereby the current is amplified and interpreted by the decoding circuit. The signal obtained is the modulation of the laser spot (whose profile approximates that of a Gaussian beam), over the alternating black and white lines of the barcode.

3.4 LIBRARY BARCODES STANDARDS

14-digit Barcodes: The two primary types used by library automation vendors are Codabar or Code 39 (Code 3 of 9). Either of these two should work with library automation vendors.

Smart and Dumb Barcodes

Smart barcodes display detailed information such as title, author and call number. In addition, smart barcodes come with an online item record for each material in your database. These records can be pre-loaded if your automation vendor performs your retrospective conversion. Dumb barcodes are simply a barcode number. You will have to assign these barcodes to each item in your collection.

3.4.1 Advantages of Smart Barcodes

Save time during the bar-coding process because items are easily identified.

Disadvantages

- Higher costs
- Inaccuracies between what is actually on the shelf and what title is assigned to the barcode.
- Some items may have barcodes while others somehow were skipped

Bar-coding an entire collection can be extremely time consuming. One way to make this process easier is to purchase duplicate barcodes one for the shelf list card and the other for the item. Each time you barcode an item, barcode the shelf list card as well. For duplicate copies, barcode duplicates on the same shelf list card. When you do retrospective conversion, you will have the barcode for the item record as well, thus, creating your item record database.

3.4.2 Place Barcode Numbers

1. Place the barcode where it is easiest to scan.
   - On the cover of the book (upper-left hand corner)
   - On the fly-leaf inside the front cover for books
   - Inside near the book pocket
2. Wherever you place it, make sure it can be scanned by any automation checkout system you may purchase such as a self-check out device.
3. Don't forget to place book tape over the barcode to protect them. Even better, purchase laminated barcodes.

3.4.3 Barcode Accuracy and Misreads

In a recent study at Ohio University, several different bar code symbologies were tested to determine their accuracy and are listed below. Keep in mind that a well-trained data entry operator will usually make a data entry error once every 300 keystrokes. Therefore, implementing the least accurate symbology is a huge step forward to increase production and reduce data entry errors. The most inaccurate symbology (UPC) is used in the retail industry.

Table 1.1

Barcode Type	*Worst Case Accuracy*	*Best Case Accuracy*
DataMatrix	1 error in 10.5 million	1 error in 612.9 million
PDF417	1 error in 10.5 million	1 error in 612.4 million
Code 128	1 error in 2.8 million	1 error in 37 million
Code 39	1 error in 1.7 million	1 error in 4.5 million
UPC	1 error in 394 thousand	1 error in 800 thousand

3.5 RETROCONVERSION

A key part of the librarian's role has always been to guide scholars towards appropriate sources of information, a task made much easier since the advent of networked computerised catalogues. As a result, researchers no longer have to travel to an individual library in order to check its card catalogue but can rather search a library's holdings remotely.

However, the process of replacing or duplicating card catalogues is labour-intensive and still far from complete, particularly when it comes to old and rare material.

3.5.1 Patron Records and Barcodes

Patron Records: Depending on the automation system you purchase, you may be able to load patron data that you have in a local database into another system. Some automation vendors can load Microsoft Access data or a People Soft database into their system. If you can use pre-existing data this will be a big money and time saver.

Just as you create specific subject categories in your collection fiction, non-fiction, and so forth, create specific types in your patron database. Possible categories may include adult, juvenile, student, out of county, etc. The vendor you choose should support multiple types of patrons.

Library cards are another consideration. Depending on budget and maintenance factors, you may want to give each patron their own card, or perhaps it is easier to keep patrons barcode numbers in a Rolodex.

Patron Fields to Consider:

- Name (first, last)
- Street Address
- Mailing Address
- Zip Code
- Telephone Number
- Social Security Number (Not required)
- Grade Level
- Notes
- Birth date

Type of Patron

Adult, Juvenile, Student, Senior Location: Local, out of county, out of state.

3.5.2 Electronic Texts and OCR

Government organizations, research institutions, universities, and private-sector organizations distribute a growing number of electronic documents through their Web sites, with permission to download them. You can access documentation on the sites of professional associations, the press and television networks, although access is not always free. These documentary sources are currently the most commonly used for term extraction in terminology work. It is simply a matter of locating them with the help of search and navigation guides on the Internet, indexing them, and retrieving them with such tools as Isys Desktop 6 or AltaVista Discovery. Some documentation may only be available in hardcopy.

If only a hardcopy of the document is available, you can have the text optically scanned in order to obtain an electronic version. Once the text is available in computer-readable form, you can use computer-assisted term-extraction software or automatic term-extraction software such as Nomino, MultiTrans, or EdiTerm.

Given the current state of the art in OCR, this option is not recommended for very large documents or documents with complicated layouts (graphics, diagrams, tables, multiple languages, etc.

3.6 ISBN

The International Standard Book Number, or ISBN (sometimes pronounced "is-ben"), is a unique1 identifier for books, intended to be used commercially. There is another quite similar system, the International Standard Serial Number (ISSN), for periodical publications such as magazines. The ISBN system was created in the United Kingdom in 1966 by the booksellers and stationers W H Smith and originally called Standard Book Numbering or SBN. It was adopted as international standard ISO 2108 in 1970.

Electronic Texts and OCR

Government organizations, research institutions, universities, and private-sector organizations distribute a growing number of electronic documents through their Web sites, with permission to download them. You can access documentation on the sites of professional associations, the press and television networks, although access is not always free. These documentary sources are currently the most commonly used for term extraction in terminology work. It is simply a matter of locating them with the help of search and navigation guides on the Internet, indexing them, and retrieving them with such tools as Isys Desktop 6 or AltaVista Discovery. Some documentation may only be available in hardcopy.

If only a hardcopy of the document is available, you can have the text optically scanned in order to obtain an electronic version. Once the text is available in computer-readable form, you can use computer-assisted term-extraction software or automatic term-extraction software such as Nomino, MultiTrans, or EdiTerm.

Given the current state of the art in OCR, this option is not recommended for very large documents or documents with complicated layouts (graphics, diagrams, tables, multiple languages, etc.).

4.0 Enumerative Bibliography

A bibliography is a list, either indicative or comprehensive, of works:

- By a particular author
- On a particular subject

- Published in a particular country
- Published in a specified period
- Mentioned in, or relevant to, a particular work (a bibliography of this type, sometimes called a *reference list* should normally appear at the end of any paper in scientific literature)

A bibliography may be arranged by author, date, topic or some other scheme. Annotated bibliographies give descriptions about how each source is useful to an author in constructing a paper or argument. Creating these blurbs, usually a few sentences long, establishes a summary for and expresses the relevance of each source prior to writing.

Bibliographies differ from library catalogues by including all relevant publications rather than items actually found in a particular library. However, some national libraries' catalogues also serve as national bibliographies, as they contain (almost) all their countries' publications.

4.1 ANALYTICAL BIBLIOGRAPHY

The critical study of bibliography is further subdivided into descriptive, historical and textual bibliography. Descriptive bibliography is the close examination of a book as a physical object, recording its size, format, binding, and so on, while historical bibliography takes a broader view of printing and publishing. Textual bibliography is another name for textual criticism.

4.1.1 Constructing and Managing Bibliographies

Bibliographic management system is deemed to be a specialized kind of database management system (DBMS) for keeping a structured database of bibliographic records. All DBMSs have the same basic structure, which is an elaboration of the layer-cake model. Its most important feature is the layer of software between the user and the database, since this allows for flexibility in what the user sees and receives from the database and for more efficient operations.

For bibliographic management, the "input templates" specify what kind of an entity is to be recorded (*e.g.*, book by one or more authors, edited collection of essays, article, interview), and the "output templates" the format in which the bibliographic records are to be presented (*e.g.*, MLA, Chicago). Bibliographic managers also come with a set of tools specific to bibliographic operations. Any bib-manager worthy of use should:

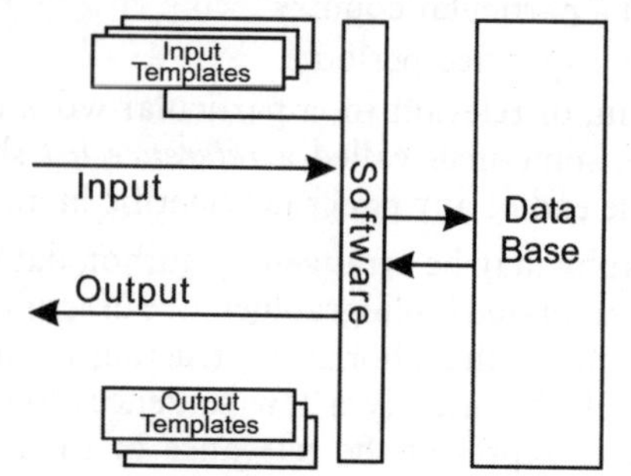

Structure of a Database Management System

Fig. 6.2: DBMS

- Allow the user to select the kind of bibliographic entity about to be recorded, and based on that selection present only the fields relevant to it.
- Provide a way for the user to define a new type of entity by selecting which fields are to be presented in what order, and what they are called.
- Allow the user to choose among a sufficiently large number of pre-defined standard formats for output.
- Provide a way for the user to define new output formats with complete flexibility as to which fields will be presented in what order, and what formatting effects are to be applied to each, if any.
- Allow for one or more commentary fields of relatively unlimited length.
- Provide search-and-replace and other common wordprocessing functions, including a rational, mnemonically adhensive means of entering accented characters.
- Allow the user to define a subset of the current bibliography, *i.e.* a selective "view" of the database rather than a copy of some items.
- Construct a Bibligraphy according to the selected format.
- Export its data in a common format, so that these data can be imported into another manager or other software.

In general the first four of the above are crucial, since they define the software as a manager of bibliographic data rather than a determiner of what kind of data may be managed. Furthermore, they make possible the following (see the provious chapter Network related fig.)

A tool or set of tools for scholarly record-keeping.Bibliographic management is only one kind of scholarly record-keeping. More broadly, research often if not usually treats bibliographic information as part of note-taking. In many kinds of scholarship the end is not to compile a bibliography but to document ideas. In the final, written form these ideas take, they tend to be ordered by the argument that subsumes them rather than by their original source. Most bibliographic managers appear to have been constructed on the model of the annotated bibliography, *i.e.*

What scholars require, however, is software that can accommodate this model and be used for general note-taking, for which a more useful model is the 3-by-5 card or "slip", such as those used by Sir James Murray while composing entries for the Oxford English Dictionary.

4.1.2 Encoding Methodology

Libraries as information dissemination centers have a long history processing multilingual data and serving users who speak different languages. During the past two decades library applications such as integrated library systems and online bibliographic databases have significantly contributed to the global resource sharing and enhanced library services to serve multicultural and multilingual populations.

An essential component of any library application is an encoding methodology that allows computers to process characters and symbols used to represent language information in written form. For years the encoding mechanism was not developed under a unified umbrella nor did it reach various languages equally. The computer industry has not fully supported libraries serving various multilingual populations.

In the past, software developers concentrated on office automation or other computer applications that only dealt with one or two languages at a time. Librarians' pleas for systems that process multilingual data simultaneously receive little support, and libraries have been forced to build their own multiscript applications because commercial computer software has not supported the variety of scripts needed. Additionally, even though various standards for individual scripts were developed, there was a lack of international standards for a unified language character-coding table which would support general-purpose software and operating systems. Library professionals cannot solve such problems by themselves; yet, to many companies, the library is not a high profit market.

The Internet and World Wide Web applications revolutionized the world of information exchange. Web client software and web search engines that handle multilingual data have pushed the demands for operating system support and international standards for a unique and comprehensive character set for all languages to a new height. A significant joint effort of developing international standards to meet such a need can be traced back to 1991 when the Unicode Consortium was formed with members including many major computer and high tech companies such as Apple Computer, Xerox, HP, IBM, etc.

4.2 DATA REPRESENTATION THROUGH DIGITALCOMPUTERS

4.2.1 Binary

Data is basically represented in digital computers through machine language which is in binary form. Binary form is used as the basic computing units use switching which have only two states either "on" or "of" *i.e.*, "Close" or "Open". Binary is the base two number system that computers use to represent data. It consists of only two numbers: "0" and "1". In the table below, binary numbers are shown with their decimal equivalents.

4.2.2 Bit (Binary Digit)

A bit is the smallest unit of data in a computer. A bit has a single binary value, either 0 or 1. Although computers usually provide instructions that can test and manipulate bits, they generally are designed to store data and execute instructions in bit multiples called bytes. In most computer systems, there are eight bits in a byte. The value of a bit is usually stored as either above or below a designated level of electrical charge in a single capacitor within a memory device.

Half a byte (four bits) is called *a nibble*. In some systems, the term octet is used for an eight-bit unit instead of byte. In many systems, four eight-bit bytes or octets form a 32-bit word. In such systems, instruction lengths are sometimes expressed as full-word (32 bits in length) or half-word (16 bits in length).

All digital computers represent data as a collection of bits. A *bit* is the smallest possible unit of information. It can be in one of two states—off or on, 0 or 1. The meaning of the bit, which can represent almost anything, is unimportant at this point. The thing to remember is that all computer data—a text file on disk, a program in memory, a packet on a network—is ultimately a collection of bits.

If one bit has two different states, how many states do two bits have? The answer is four. Likewise, three bits have eight states. For example, if a computer display had eight colors available, and you wished to select one of these to draw a diagram in, three bits would be sufficient to represent this information. Each of the eight colors would be assigned to one of the three-bit combinations. Then, you could picking one of the colors by picking the right three-bit combination.

A common and convenient grouping of bits is the byte or octet, composed of eight bits. If two bits have four combinations, and three bits have eight combinations, how many combinations do eight bits have? If you don't want to write out all the possible byte patterns, just multiply eight twos together—one two for each bit. Two times two is four, so the number of combinations of two bits is four. Two times two times two is eight, so the number of combinations of three bits is eight. Do this eight times—or just compute two to the eighth power— and you discover that a byte has 256 possible states.

Obviously, if a byte has 256 possible states, its exact state can be represented by a number from 1 to 256. However, since zero is a very important number, a byte is more typically represented by a number from 0 to 255 in decimal system. This is very common, and with bit pattern 00000000 representing zero, and bit pattern 11111111 representing 255. The numbers matching these two patterns, and everything in between, can be computed by assigning a weight to each bit, multiplying each bit's value (0 or 1) by its weight, and then adding the totals. For example, here's how 217 is represented as 11011001 in binary: (Go on dividing the number by 217)

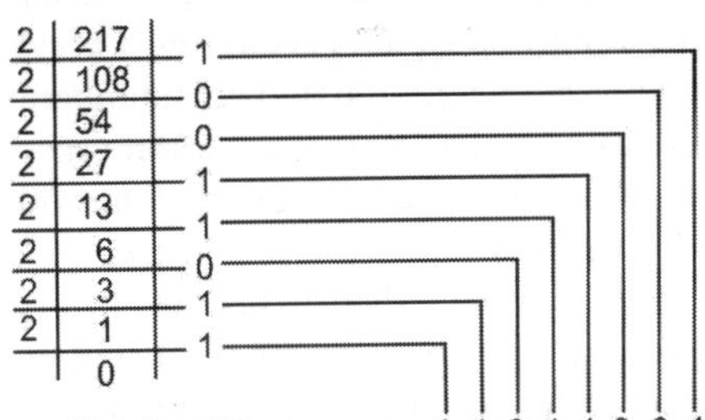

Required Binary ⟶ 1 1 0 1 1 0 0 1

The most common bit patterns in Internet engineering are those with a string of one bits, followed by a string of zero bits. Here are all such bytes, along with their decimal representation, computed just like the example using 217.

4.2.3 Byte

In most computer systems, a byte is a unit of information that is eight bits long. A byte is the unit most computers use to represent a character such as a letter, number, or typographic symbol (for example, "g", "5", or "?"). A byte can also hold a string of bits that need to be used in some larger unit for application purposes (for example, the stream of bits that constitute a visual image for a program that displays images).

4.2.4 Octet

In computers, an octet (from the Latin octo or "eight") is a sequence of eight bits. A octet is thus an eight-bit byte. Since a byte is not eight bits in all computer systems, octet provides a non-ambiguous term. This term should not be confused with octal, a term that describes a base-8 number system.

4.2.5 Octal

Octal (pronounced AHK-tuhl, from Latin octo or "eight") is a term that describes a base-8 number system. An octal number system consists of eight single-digit numbers: 0, 1, 2, 3, 4, 5, 6, and 7. The number after 7 is 10. The number after 17 is 20 and so forth. In computer programming, the octal equivalent of a binary number is sometimes used to represent it because it is shorter.

4.2.6 Data Processing and Transmission

In computers and in data transmission between them, *i.e.* in digital data processing and transfer, data is internally presented as octets, as a rule. An octet is a small unit of data with a numerical value between 0 and 255, inclusively. The numerical values are presented in the normal (decimal) notation here, but notice that other presentations are used too, especially octal (base 8) or hexadecimal (base 16) notation. Octets are often called bytes, but in principle, octet is a more definite concept than byte. Internally, octets consist of eight bits (hence the name, from Latin octo 'eight'), but we need not go into bit level here. However, you might need to know what the phrase "first bit set" or "sign bit set" means, since it is often used. In terms of numerical values of octets, it means that the value is greater than

127. In various contexts, such octets are sometimes interpreted as negative numbers, and this may cause various problems.

Different conventions can be established as regards to how an octet or a sequence of octets presents some data. For instance, four consecutive octets often form a unit that presents a real number according to a specific standard. We are here interested in the presentation of character data (or string data; a string is a sequence of characters) only.

In the simplest case, which is still widely used, one octet corresponds to one character according to some mapping table (encoding). Naturally, this allows at most 256 different characters being represented. There are several different encodings, such as the well-known ASCII encoding and the ISO Latin family of encodings. The correct interpretation and processing of character data of course requires knowledge about the encoding used. For HTML documents, such information should be sent by the Web server along with the document itself, using so-called HTTP headers (cf. to MIME headers).

Previously the ASCII encoding was usually assumed by default (and it is still very common). Nowadays ISO Latin 1, which can be regarded as an extension of ASCII, is often the default. The current trend is to avoid giving such a special position to ISO Latin 1 among the variety of encodings.

4.2.7 Character Repertoire

A set of distinct characters. No specific internal presentation in computers or data transfer is assumed. The repertoire per se does not even define an ordering for the characters; ordering for sorting and other purposes is to be specified separately. A character repertoire is usually defined by specifying names of characters and a sample (or reference) presentation of characters in visible form. Notice that a character repertoire may contain characters which look the same in some presentations but are regarded as logically distinct, such as Latin uppercase A, Cyrillic uppercase A, and Greek uppercase alpha.

4.2.8 Character Code

The code serves the important purpose of standardizing the approach to dealing with text on different computer systems.

The fundamental idea in standardizing character codes is to allow data entry to proceed using the standard QWERTY keyboard. Most word processors and text editors associate the keys on the keyboard with specific ASCII codes and hence can support data entry in any language that assigns the displayable ASCII codes to its alphabet. This is done through the use of fonts, where associated with each ASCII code is the shape of the letter that should be displayed.

Sometime around the 1960s, a few decisions were made about just what ones and zeros (bits) would represent characters. One important choice that most modern computer users give no thought to was the decision to use 8-bit bytes on nearly all computer platforms. In other words, bytes have 256 possible values. Within these 8-bit bytes, a consensus was reached to represent one character in each byte. So at that point, computers needed a particular encoding of characters into byte values; there were 256 "slots" available, but just which character would go in each slot? The most popular encoding developed was Bob Bemers' American Standard Code for Information Interchange (ASCII), which is now specified in exciting standards like ISO-14962-1997 and ANSI-X3.4-1986(R1997). But other options, like IBM's mainframe EBCDIC linger on, even now.

A *mapping*, often presented in tabular form, which defines a one-to-one correspondence between characters in a character repertoire and a set of non-negative integers. That is, it assigns a unique numerical code, a code position, to each character in the repertoire. In addition to being often presented as one or more tables, the code as a whole can be regarded as a single table and the code positions as indexes. As synonyms for "code position", the following terms are also in use: code number, code value, code element, code point, code set value - and just code.

Character code is a method (algorithm) for presenting characters in digital form by mapping sequences of code numbers of characters into sequences of octets. In the simplest case, each character is mapped to an integer in the range 0—255 according to a character code and these are used as such as octets. Naturally, this only works for character repertoires with at most 256 characters. For larger sets, more complicated encodings are needed. Encodings have names, which can be registered.

Notice that a character code assumes or implicitly defines a character repertoire. A character encoding could, in principle, be viewed purely as a method of mapping a sequence of integers to a sequence of octets. However, quite often an encoding is specified in terms of a character code (and the implied character repertoire). The logical structure is still the following:

- A character repertoire specifies a collection of characters, such as "a", "!", and "d".
- A character code defines numeric codes for characters in a repertoire. For example, in the ISO 10646 character code the numeric codes for "a", "!", "d", and " " (per mille sign) are 97, 33, 228, and 8240. (*Note:* Especially the per mille sign, presenting 0/00 as a single character, can be shown incorrectly on display or on paper. That would be an illustration of the symptoms of the problems we are discussing.)
- A character encoding defines how sequences of numeric codes are presented as (*i.e.*, mapped to) sequences of octets. In one possible encoding for ISO 10646, the string a!d is presented as the following sequence of octets (using two octets for each character): 0, 97, 0, 33, 0, 228, 32, 48.
- The phrase character set is used in a variety of meanings. It might denotes just a character repertoire but it may also refer to a character code, and quite often a particular character encoding is implied too.

For our purpose following three codes are most important:

- ASCII (American Standard Code for Information Interchange)
- ISCII (Indian Standard Code for Information Interchange)
- Unicode (Universal Code)

4.3 ASCII

The name ASCII, originally an abbreviation for "American Standard Code for Information Interchange", denotes an old character repertoire, code, and encoding. Most character codes currently in use contain ASCII as their subset in some sense. ASCII is the safest character repertoire to be used in data transfer. However, not even all ASCII characters are "safe"!

ASCII has been used and is used so widely that often the word ASCII refers to "text" or "plain text" in general, even if the character code is something else! The words "ASCII file" quite often mean any text file as opposite to a binary file. As of today, the ASCII code is probably the only code correctly identified on all the computer systems. Information made up of pure ASCII coded text is thus viewable on almost all computers. E-mail is an example of an application that uses pure ASCII coded text in the messages.

In the early days of information processing, the Roman character set served as the basis for interacting with computer systems. European languages which use a slightly different set of character codes manage quite well with the ASCII approach by replacing a few of the special characters in ASCII by symbols specific to each language.

The ASCII code covers a range of 256 characters of which 96 (codes from 32 to 127) are reckoned as standard displayable ASCII. Codes between 128 and 255 are often used for displaying symbols useful for tabular information, graphics etc. Some European languages such as Greek, Russian etc., also support codes for their languages in the range 128 to 255, to allow bilingual information (English as well as the specific language) to be displayed easily. The International Standards Organization has come up with recommendations known as the Latin Character sets, which encode the alphabet of the different European languages. For an excellent review of these the viewer is encouraged to look at Internationalization.

ASCII itself is of somewhat limited extent. Only the values of the lower-order 7-bits of each byte might contain ASCII encoded characters. The top 7-bits worth of positions (128 of them) are "reserved" for other uses (back to this). So, for example, a byte that contains "01000001" might be an ASCII encoding of the letter "A", but a byte containing "11000001" cannot be an ASCII encoding of anything. Of course, a given byte may or may not actually represent a character; if it is part of a text file, it probably does, but if it is part of object code, a compressed archive, or other binary data, ASCII decoding is misleading. It depends on context.

The reserved top 7-bits in common 8-bit bytes have been used for a number of things in a character-encoding context. On traditional textual terminals (and printers, etc.) it has been common to allow switching between codepages on terminals to allow display of a variety of national language characters (and special characters like box-drawing borders), depending on the needs of a user. In the world of Internet communications, something very similar to the codepage system exists with the various ISO-8859- encodings. What all these systems do is assign a set of characters to the 128 slots that ASCII reserves for other uses. These might be accented roman characters (used in many Western European languages), they might be non-roman character sets like Greek, Cyrillic, Hebrew, or Arabic (or in the future, Thai and Hindi). By using the right codepage, 8-bit bytes can be made quite suitable for encoding reasonable sized (phonetic) alphabets.

The definition of ASCII also specifies a set of control codes ("control characters") such as linefeed (LF) and escape (ESC). But the character repertoire proper, consisting of the printable characters of ASCII, is the following (where the first item is the blank, or space, character):

```
  ! " # $ % & ' ( ) * + , - . /
0 1 2 3 4 5 6 7 8 9 : ; < = > ?
@ A B C D E F G H I J K L M N O
P Q R S T U V W X Y Z [ \ ] ^ _
` a b c d e f g h i j k l m n o
p q r s t u v w x y z { | } ~
```

The appearance of characters varies, of course, especially for some special characters. Some of the variation and other details are explained in The ISO Latin 1 character repertoire - a description with usage notes.

A Formal View on ASCII

The character code defined by the ASCII standard is the following: code values are assigned to characters consecutively in the order in which the characters are listed above (rowwise), starting from 32 (assigned to the blank) and ending up with 126 (assigned to the tilde character ~). Positions 0 through 31 and 127 are reserved for control codes. They have standardized names and descriptions, but in fact their usage varies a lot.

The character encoding specified by the ASCII standard is very simple, and the most obvious one for any character code where the code numbers do not exceed 255: each code number is presented as an octet with the same value.

Octets 128 – 255 are not used in ASCII. (This allows programs to use the first, most significant bit of an octet as a parity bit, for example.)

The Misnomer "8-bit ASCII"

Sometimes the phrase "8-bit ASCII" is used. It follows from the discussion above that in reality ASCII is strictly and unambiguously a 7-bit code in the sense that all code positions are in the range 0–127.

It is a misnomer used to refer to various character codes which are extensions of ASCII in the following sense: the character repertoire contains ASCII as a subset, the code numbers are in the range 0 - 255, and the code numbers of ASCII characters equal their ASCII codes.

4.4 UNICODE

Since there are many languages and the ASCII encoding supports only 96 regular letters for data entry and display, some special mechanism is needed to associate the codes with the letters of different languages, if multilingual information is to be displayed. This mechanism has been provided through Unicode, which is essentially derived from ASCII but provides some means of identifying the script associated with the characters.

Unicode caters to a very large set of before we see what Unicode is, it makes sense to step back slightly to think about just what it means to store "characters" in digital files. Anyone who uses a tool like a text editor usually just thinks of what they are doing as entering some characters—numbers, letters, punctuation, etc. But behind the scene a little bit more is going on. "Characters" that are stored on digital media must be stored as sequences of ones and zeros, and some encoding and decoding must happen to make these ones and zeros into characters we see on a screen or type in with a keyboard.

ISO/IEC 10646 is a relatively new character set standard, published in 1993 by the International Organization for Standardization (ISO). Its name is "Universal Multiple-Octet Coded Character Set". Throughout this overview I use its acronym, UCS.

Unicode Standard Version 3.0 is the software specification that assures a single, universal way to represent text worldwide. Version 3.0 supports 49,194 characters, including 31per cent more ideographs for Japanese, Chinese and Korean markets. Implementation support is greatly expanded, with double the character property data, and four times as many technical specifications for supporting implementations.

The Unicode Standard is a major component in the globalization of e-business, as the marketplace continues to demand technologies that enhance seamless data interchange throughout companies' extended — and often international — network of suppliers, customers and partners. This new version reaffirms the broad, cross-industry commitment to the standard among leading IT vendors, enabling reliable transmission and storage of text data anywhere in the world. Unicode is the default text representation in XML, an important open standard being rapidly adopted throughout e-business technology.

The Unicode Standard assigns every character a unique number, ensuring the same representation for text regardless of country, language, or operating system. As a result, computer programs written to its specifications can be used around the world without modification. Unicode-enabled programs—client, server, operating system, or middleware—can share textual data worldwide. Text can be transmitted freely, without suffering the data loss that occurs with older systems.

"The adoption of Unicode enables computer users to read and write documents in their own native languages, and this new version extends this ability to virtually all corners of the globe", said Mark Davis, President of the Unicode Consortium. "This universal representation for text data is a crucial building block in the evolution of the Web and e-business".

Unicode is enabled in all modern Web browsers, almost all operating systems, and Internet standards such as HTML, Java, ECMAScript, XML, and LDAP. The Unicode Standard, published by Addison-Wesley Longman, is widely available in bookstores or may be obtained directly from the Unicode Consortium.

The development of the Unicode Standard is a milestone in international computing because it supports the creation of global software that can be easily adapted for local needs. It brings good news for library professionals. However, as observed by the authors, the implementation of the Unicode Standard has not received full attention or strong support from some library communities, such as the Chinese library communities in Asia (Mainland, Hongkong, Taiwan, as well as other multilingual regions and countries in Asia), that present special and unique issues to multiscript information processing. It is the purpose of this paper to analyze and explain those issues to both the librarians and the Unicode developers in order to encourage the shifting from diverse character code sets to the Unicode Standard in library applications.

The Unicode Worldwide Character Standard is a character coding system designed to support the interchange, processing, and display of the written texts of the diverse languages of the modern world. In addition, it supports classical and historical texts of many written languages. It covers the principal written languages of the world, as well as technical symbols in common use. The Unicode Standard is the international standard used to encode text for computer processing. It is a subset of the International Standard ISO/IEC 10646, Universal Multiple-Octet Coded Character Set.

This development is a milestone in international computing because it supports the creation of global software that can be easily adapted for local needs. It brings good news for library professionals. However, as observed by the authors, the implementation of the Unicode Standard has not received full attention or strong support from some library communities, such as the Chinese library communities in Asia (Mainland, Hongkong, Taiwan, as well as other multilingual regions and countries in Asia), that present special and unique issues to multiscript information processing. It is the purpose of this paper to analyze and explain those issues to both the librarians and the Unicode developers in order to encourage the shifting from diverse character code sets to the Unicode Standard in *library* applications.

Unicode solves the problems of previous character encoding schemes by providing a unique code-number for every character needed, worldwide and across languages. Over time, more characters are being added, but the allocation of available ranges for future uses has already been planned out, so the room exists). In Unicode encoded documents, no ambiguity exists about how a given character should display (for example, should byte value 0x89 appear as e-umlaut, as in codepage 850, or as the per-mil mark, as in codepage 1004?). Furthermore, by giving each character its own code, there is no problem or ambiguity in creating multi-lingual documents that utilize multiple character sets at the same time. Or rather, these documents actually utilize the single (very large) character set of Unicode itself.

Unicode is managed by the Unicode Consortium, a non-profit group with corporate, institutional, and individual members. Originally, Unicode was planned as a 16-bit specification. However, this original plan failed to leave enough room for national variations on related (but distinct) ideographs across East Asian Languages (Chinese, Japanese, and Korean, nor for specialized alphabets used in mathematics and the scholarship of historical languages. As a result, the code space of Unicode is currently 32-bits (and anticipated to remain fairly sparsely populated, given the 4 billion allowed characters).

UCS is the first officially standardized coded character set with the purpose to eventually include all characters used in all the written languages in the world (and, in addition, all mathematical and other symbols). This is certainly a very ambitious goal, but the current first edition at least covers all major languages and all commercially important languages.

To be able to give every character of this grand repertoire a unique coded representation, the designers of UCS chose a uniform encoding, using bit sequences consisting of 16 or 31 bits (in the two coding forms, UCS-2 and UCS-4). This is the reason for the phrase "multi-octet" in the name of the standard.

Unicode is a coded character set specified by a consortium of major American computer manufacturers, primarily to overcome the chaos of different coded character sets in use when creating multilingual programs and internationalizing software. From version 1.1 on, Unicode is scrupulously kept compatible with ISO/IEC 10646 and its extensions. The consortium is also an important contributor to the ISO work to further develop ISO/IEC 10646.

In short, Unicode can be characterized as the (restricted) 2-octet form of UCS on (the most general) implementation level 3, with addition of a more precise specification of the bi-directional behavior of characters, when used in the Arabic and Hebrew scripts. Unicode is presently at version 1.1. Extensions in the soon forthcoming version 2.0 will make it possible to access also the wider coding space of UCS-4, within this 16-bit encoding.

UCS is intended to be usable both for internal data representation in computer systems and in data communication. UCS is already employed in commercial products from Microsoft, Novell, Apple and others. It is implemented in free software like Linux, and is proposed for inclusion in advanced data communication standards like HTML.

Strong but in my opinion ill-founded criticism has met UCS from programmer groups in Japan. It has, however, recently been adopted as a Japanese national standard.

ISO/IEC 10646 is a fundamental standard, potentially affecting almost all parts of information technology. But it specifies only a coded character set, not a complete system for text representation. It provides the basis for internationalization, but does not in itself give a complete solution of the problems in this field.

The simple kind of text for whose representation a coded character set standard is sufficient, plain text, is essentially only a linear sequence of graphic characters, with a fixed division into lines and possibly pages.

4.4.1 Encodings

A full 32-bits of encoding space leaves plenty of room for every character we might want to represent, but it has its own problems. If we need to use 4 bytes for every character we want to encode, that makes for rather verbose files (or strings, or streams). Furthermore, these verbose files are likely to cause a variety of problems for legacy tools. As a solution to this, Unicode is itself often encoded using "Unicode Transformation Formats" (appreviated as UTF-*). The encodings UTF-8 and UTF-16 use rather clever techniques to encode characters in a variable number of bytes, but with the most common situation being the use of just the number of bits indicated in the encoding name. In addition, the use of specific byte value ranges in multi-byte characters is designed in such a way as to be friendly to existing tools. UTF-32 is also an available encoding, one that simply uses all four bytes in a fixed-width encoding.

The design of UTF-8 is such that US-ASCII characters are simply encoded as themselves. For example, the English letter "e" is encoded as the single byte 0x65 in both ASCII and in UTF-8. However, the non-English "e-umlaut" diacritic, which is Unicode character 0x00EB is encoded with the two bytes 0xC3 0xAB. In contrast, the UTF-16 representation of every character is always at least 2 bytes (and sometimes 4 bytes). UTF-16 has the rather straightforward representations of the letters "e" and "e-umlaut" as 0x65 0x00 and 0xEB 0x00, respectively. So where does the odd value for the e-umlaut in UTF-8 come from. Here is the trick: no multi-byte encoded UTF-8 character is allowed to be in the 7-bit range used by ASCII, to avoid confusion. So the UTF-8 scheme uses some bit shifting, and encodes every Unicode character using up to 6 bytes. But the byte values allowed in each position are arranged in such a manner as not to allow confusion of byte positions (for example, if you read a file non-sequentially).

4.4.2 Characters Representing Several Scripts of the World

Unicode is yet to be recognized by many word processing and data entry software running on different computers of the world (as of Jan. 2005). Microsoft Windows (Windows 2000/XP), Linux and Java support Unicode but bulk of the systems continue the plain ASCII approach. Unicode is an international standard that should necessarily be understood by persons developing multilingual software. Web browsers provide support for displaying Unicode text on different computer systems.

- Before taking up the issue of coding Indian language characters, we make some important observations are made.
- Eight bit character codes are entirely adequate for languages whose alphabet is a small set and the written text consists of only the individual letters themselves and possibly some punctuation.
- Data entry in any language can be effected with ease by encoding the letters of the alphabet along the lines of the ASCII code and using appropriate fonts with the Word Processing software. If the codes are also assigned in the range 128–255, then data entry is not straightforward and will require special input mechanisms.
- A document prepared on the basis of Unicode encoding, will be truly Multilingual. However, data entry on word processors supporting Unicode (*e.g.*, Microsoft Word) still remains cumbersome. More about this in the section on Unicode for Indian Languages.

4.4.3 Encoding Forms of Unicode

In order to understand how Unicode is encoded and stored in computer files, a short excurse into some of the technical details can not be avoided. This section is especially intended for encoders, who run into trouble with the default mechanism of their favorite software platform, which is usually designed to hide these details.

Unicode allows the encoding of about 1 million characters. This is the theoretical upper limit, but at present less than 10 per cent of this code space is actually used. The code space is arranged in 17 planes of 65536 code points each, with Plane 0, the Basic Multilingual Plane (BMP) being the one where most characters are defined.

In order to store the numeric values of the code points in a computer, they have to be serialized. Unicode defines two encoding forms for serialization: UTF-8 and UTF-16.

UTF-16 simple stores the numerical value as a 16 bit integer, while characters with higher numerical values are expressed using two UTF-16 values from a range of the BMP set aside for this purpose, they are called Surrogate Pairs. Since most computers store and retrieve numeric values in bundles of 8 bit (bytes), the 16 bits of one UTF-16 value have to be stored in two separate bytes. Conventions of whether the byte with the higher value (Big-Endian) or the lower value (Little-Endian) differ in the same way and for the same reasons as the egg openers in Gulliver's Travels, for those reasons, there are two storage forms of UTF-16: UTF-16-LE and UTF-16-BE. If UTF-16 is used without any further specification, it is usually UTF-16-BE, which is the default for example on Microsoft Windows platforms.

UTF-8 avoids the whole issue of endianess by serializing the numbers in chunks of single bytes. To achieve this, it uses sequences of multiple single bytes to encode a Unicode numeric value. The length of such sequences depends on the value of the Unicode character; values less than 128 (the range of the ASCII or ISO 646 characters) are just one byte in length, which means they are identical to ASCII. This means that English text and also the tags used for markup do not differ in UTF-8 and ASCII. This is one of the reasons why UTF-8 is rather popular. It is also the default encoding for XML files in the absence of a specific encoding declaration and the recommended encoding to use. In UTF-8, most accented characters require a sequence of two bytes, East-Asian character need three, and the characters beyond the BMP need four or more bytes.

In most cases, there is no need to worry about the specific encoding form, except to make sure that the encoding declaration, which is optionally included in the first line of an XML file in the form `<?xml version ="1.0" encoding="utf-8"?>` does indeed faithfully reflect the actual encoding. Problems do occasionally arise with UTF-16 files read with the wrong endianess. The TEI-Emacs bundle, which is distributed on the CD-ROM accompanying this book, makes all attempts to act according to the encoding declaration. If this fails, because of a mismatch between the encoding used and the encoding declared, it is still possible to force Emacs into opening a file with the desired encoding and correct the mismatch.

4.5 INDIAN SCRIPT CODE FOR INFORMATION INTERCHANGE (ISCII)

ISCII was proposed in the eighties and a suitable standard was evolved by 1991. Here are the salient aspects of the ISCII representation:

- It is a single representation for all the Indian scripts.
- Codes have been assigned in the upper ASCII region (160–255) for the *aksharas* of the language.

- The scheme also assigns codes for the *matras* (vowel extensions).
- Special characters have been included to specify how a consonant in a syllable should be rendered. Rendering of Devanagari has been kept in mind.
- A special attribute character has been included to identify the script to be used in rendering specific sections of the text.

4.6 OBSTACLES IN USE OF THE UNICODE STANDARD IN LIBRARY APPLICATIONS

While examples and observations are from current CJK (Chinese, Japanese, Korean) information processing practice, librarians from other regions, especially third-world countries, may find them interesting.

Major reasons for lukewarm reception to Unicode in part of East Asia can be explained from the following technological aspects, while the unfamiliarity with the content of the Unicode Standard among the majority of librarians might be a major cause behind various of these obstacles:

(1) No Bibliographic Records or MARC Database that use Unicode

Currently major MARC databases are either using ASCII (including extended ASCII) or a local character set. For example, when dealing with CJK scripts, USMARC uses EACC and CNMARC uses GB (Chinese National Standards). There is no national or regional bibliographic utility willing to convert MARC databases into an Unicode-based database or to generate records using Unicode. For instance, both RLIN and OCLC bibliographic files contain over 30 million titles in over 360 languages. Over 1.5 million records in the RLIN database contain CJK, Cyrillic, Hebrew and/or Arabic scripts. In the OCLC's OLUC (Online Library Union Catalog), records in 45 languages had over 14,000 items. But, a plan to convert any of these languages into Unicode appears impossible due to the immense workload involved.

Bibliographic databases using Unicode can be created today, as demonstrated by products from DRA, VTLS and CGI (to name but three companies) and Carl/UnCover's experimental practice. However, what is still missing is probably the ability to exchange MARC records containing Unicode values. To exchange, the internal Unicode data has to be converted to records complying with the 8-bit USMARC or UNIMARC formats.

(2) No Bibliographic Standards have Adopted Unicode until very Recently

We cannot blame bibliographic utilities for not creating data in Unicode because no MARC standards support Unicode currently. The most popular USMARC only defined some conventional language characters sets—for example EACC—as part of its standards. Apparently now both U.S. and European organizations are working on updating the specifications. For USMARC, MARBI has approved the first proposal submitted by one of its subcommittees to add Unicode; and for UNIMARC, the CHASE group has published its recommendations on format change to UNIMARC to include Unicode. However, it is noticed by the authors that there is no report on how Unicode data should be entered into those MARC formats. We would highly encourage librarians in the East Asian regions to participate in the discussions regarding this issue.

(3) Few Library Systems in the Market Fully Support Unicode

Some of the library system vendors riding on the technology development are currently developing Unicode versions of their systems (*e.g.*, DRA, VTLS, CGI), while some others are experimenting with such an approach (*e.g.*, Carl/UnCover, and Sirsi Corp). However it is also true that a few other system vendors who developed CJK systems based on EACC/CCCII have attempted to sell their existing software. In fact, none of the current installation of CJK systems is using the Unicode. This has led to confusion among librarians in East Asia about whether it is feasible to switch to the Unicode CJK and has added hazy understandings about the relationship between Unicode CJK subset and EACC.

(4) Requirement for an Ideographic Repertoire that Meets Library Nneeds

The biggest concern that librarians in Asia have regarding Unicode is that the number of Han characters is far from enough for processing library materials. This concern exists to all the CJK-standards that need to expand the number of characters to the maximum number found in an authoritative dictionary such as Kang-Xi Zi Dian (Kang Xi Dictionary). The Unified Repertoire and Ordering (URO), also know as "Unified Han", which is in both the Unicode Standard and ISO/IEC 10646 includes about 20 thousand ideographs. In comparison, some Chinese software packages already have over 60 thousand characters. A small-sized set which contains the most commonly used characters may be good enough for office automation software, but it is far from the number of characters needed by libraries to deal with rare books and full text documents. For many librarians, it looks as if making a revision of a local standard is much easier than requesting extension of an international standard such as the Unicode Standard. In fact, ISO/IEC JTC1/SC2/WG2 is in the process of adding six thousand more characters to ISO/IEC 10646, and these will also be part of the Unicode Standard; however, many librarians are not aware of it.

Furthermore, in Asian countries, librarians' opinions on their needs are not strong enough to pass important messages to the ISO SC2/WG2 Ideographic Rapporteur Group (IRG) which has successfully developed Unified Han and carries an ongoing task of identifying and collating additional ideographic characters. Librarians' needs can be demonstrated by two examples.

First is the number of characters: many Chinese personal names contain characters that are rarely used in texts; therefore, they are not included in "Unified Han." Second is the various forms of the same character: for rare books cataloging, to record the information on a title page information exactly, a librarian often needs to use a variation of a character that is not considered as a standard abstract format and therefore not included in the Unicode Standard or ISO/IEC 10646 because of the rules governing the content of the Unified Repertoire and Ordering. (Here we are not talking about glyph/font issue.) It is the authors' wish that a "character thesaurus" be made available so that a standard character can be used to retrieve all the materials that have used various forms of this character. Meanwhile, the character can be displayed in desired forms that match its appearance on title pages or rare book full texts.

Special script problems like this can be solved only when CJK library communities become actively involved in the study and development of the Unified Han repertoire. Version 2.0 of the Unicode Standard provides an extension mechanism using surrogate code pairs to encode extremely rare characters, and other encoding strategies may be developed. "The task of identifying, collating, and verifying ideographs becomes progressively more difficult as more focus is given to historical and rarely used characters" (footnote 7). Librarians should play an important role in this task. Librarians can take various approaches such as to work with the appropriate group within their national standards body, or to communicate via ISO TC 46 or directly to ISO/IEC JTC1/SC2/WG2 and the IRG to make their needs known.

(5) Ordering of CJK characters for displays remains a problem

The Unicode Standard points out that relying on the order of characters in a character set for sorting is usually inadequate for a culturally expected result. For the Chinese characters (which occur in C, J and K), sorting is a more difficult job because the set combines several different national/regional standards and their different versions. In most cases, a sort table would be needed. In Unicode, the Unified Ideographs are predominantly ordered according to each character's radical and stroke count. If a different order is required, then the application (or system) must maintain a separate "order weight" table which permits mapping each unified ideograph to its "order weight". Library system vendors would be required to develop methods that employ a sort table to sort the display based on Pinyin- or Radical/stroke-order for Chinese, for example. This also means the systems using CJK ideographs need a special process for sorting.

(6) Lack of awareness of Computer Software and Operating Systems that Support Unicode

One of the important rationales for CJK librarians to consider using Unicode is web applications and the support to the applications from Microsoft Windows and popular Web browsers. Currently, both Netscape and Internet Explorer support Unicode/UTF8 display, as does MS Office 97 and Windows NT. By using Unicode, all the information being displayed on a library Web site can be seen by all users in CJK speaking regions without requiring installation of specific software, as long as a font or fonts with the appropriate character repertoire are installed. Without Unicode support, one needs a specific font for each coding set; for instance, not long ago, one would have needed Twinbridge to read big 5-coded Web pages, and Richwin for Web pages that use GB code. The text generated by one software package would not be read by another; conversion would often fail or be incomplete; and format and font style would be lost during the conversion. It has to be noted that some of the multi-language add-on enabler software have a high crash rate on Windows operating systems or web browsers due to their own quality problems, such as incorrectly using control characters in code sequence and lacking of compatibility.

There is no doubt that Unicode-support features in operating systems, databases, and general applications provided by the computer industry will make it easier for companies to build systems for libraries which require multiscript support.

(7) No Word Segmentation Implemented in Library Systems

This problem applies to text in a number of languages, e.g. Thai and Chinese, where words are not written separately. For Chinese data, there is no natural way for computers to tell where the end phrase or word is. Multilingual processing systems need to implement an intelligent way to automatically segment a word from a paragraph so that a keyword index can be generated and a better search can be performed. So far, there is no CJK library system meets this very basic requirement.

In conclusion, without Unicode, users may have to use different software and terminals to display or enter data in different languages, particularly when dealing with more than a few scripts, especially non-Roman scripts. This may be okay for some computer applications, but it is certainly not acceptable for library users. We could go on, adding scripts one at a time, but a better way is to build global software capable of handling all scripts. Information and human knowledge should not be separated because of languages; library systems should support users to retrieve information across languages, and users should be able to access such information online from anywhere in the world.

4.6.1 Other Codes

East Asian Character Code (EACC) is an American national standard that RLG developed in conjunction with the Library of Congress. It covers traditional and simplified Chinese characters as well as Japanese variants, the Japanese *hiragana* and *katakana* characters, and the Korean *hangul* characters. The base standard from which RLG worked—CCCII, Chinese Character Code for Information Interchange, is a 3 byte standard published in Taiwan. RLG adapted CCCII and added characters from other standards:

- CCCGSII, Code of Chinese Character Graphics Set for Information Interchange, a national standard in China (GB2312-80),
- JIS, the Japanese Industrial Standard, and
- KIPS, the Korean Information Processing System
- RLIN East Asian Charcter Code.

To create the RLIN East Asian Character Code (renamed East Asian Character Code when it was adopted as an American National Standard in 1988). This is the only East Asian character set that incorporates all character graphics lists in the four major East Asian character set noted above and internally links all their common variant forms. These links enable users of diverse linguistic backgrounds to search for one character from and retrieve all its related forms. RLIN CJK was released in September 1983. OCLC adopted REACC and its character set, and released its implementation in 1986. In 1988, the RLIN East Asian Character Code was approved by the National Information Standard Organization as an American National Standard Z39.64-1989.

Eight files organized by bibliographic material type are accessible through the RLIN library and archival support system, and are available for searching as one combined "BIB" file through Eureka and Zephyr. The files contain millions of records in more than 365 languages. RLIN is the only online catalog to support all the scripts used in the LC-designated "JACKPHY" languages (Japanese, Arabic, Chinese, Korean, Persian, Hebrew, and Yiddish) plus Cyrillic.

OCLC's WorldCat offers more than 36 million bibliographic records, representing 370 languages. As of June 30, 1997, records in 45 languages in the OCLC's OLUC (Online Library Union Cataloge) had over 14,000 items. Records in three languages (French, German, and Spanish) already exceeded 1.5 million; records in 12 languages were counted to have 100,000—680,000; records in other 24 languages were counted to be 14,000—90,000. (**Source:** OCLC Annual Report, 1997:10)

The Unified Repertoire and Ordering (URO 2.0) is a result of unified Han characters from disparate character standards. It successfully represents all of the standard Han ideographs with just over 21,000 unique characters instead of 121,000 code points if simply combining existing ideographic-character standards set up by different countries and regions. The Unicode Standard uses this character set to represent Han characters in Unicode codespace. The URO also appears in ISO/IEC 10646-1:1993 as the set of Unified CJK Ideograph characters.

ISO SC2/WG2 Ideographic Rapporteur Group (IRG)— formerly called the Chinese/Japanese/Korean Joint Research Group (CJK-JRG) is an international body of experts set up under the International Organization for Standardization. It is formed with representatives from China, Japan, Korea, the United States, Vietnam, Hong Kong, and Taiwan who have worked together to identify, categorize, and set the order of the Han based ideographs and developed the Unified Repertoire and Ordering (URO 2.0) which is used in both the Unicode Standard and ISO/IEC 10646-1:1993.

4.7 BIBLIOGRAPHIC REFERENCES

Traditional documentary search tools and methods include:

- Catalogues produced by libraries and publishing houses.
- Directories of published titles, such as Books in Print.
- Consultation of documentary services by telephone or in person.
- Interlibrary loans.
- Bibliographies published in specialized periodicals and other works.

The terminologist's toolbox also includes computerized tools. The following paragraphs provide additional information on other bibliographic reference sources.

- Documentary Data Banks
- Computerized Catalogues
- Online Search Services
- Reference Portals
- Documentary Data Banks

Among these data banks, DIALOG is considered to be the largest in the world. It gives access to more than 500 databases covering different subject fields. By selecting appropriate options from a menu, it is possible to obtain exact references, summaries, descriptors, or the complete text of periodical or journal articles. PASCAL and DELPHES are among the many databases available through DIALOG. This data bank is accessible via Datapac (using a modem) or the Internet; payment for access is in American dollars.

4.8 COMPUTERIZED CATALOGUES

The computerized catalogues of large libraries facilitate tasks such as referencing, verification of bibliographic information, interlibrary loans, or location of a book or a periodical. For example, AMICUS, the National Library of Canada's information system, includes more than 40 search indexes (author, program, association, department, subject heading, title, subject, ISBN, ISSN, publisher, series, etc.). It is accessible via the Internet. The AG-Canada database is similar to

AMICUS as far as access and types of possible documentary searches are concerned (via Internet, payment required). The largest legal database in Canada is QUICKLAW, which gives access to several databases on precedents and administrative decisions made by government organizations, and to legislative databases containing federal and provincial laws and regulations.

4.9 ONLINE SEARCH SERVICES

These provide terminologists and other users with bibliographic information by giving access to computerized search services in commercial data banks, in the databases mentioned above, or in the catalogues of Canadian libraries registered with GEAC.

4.10 REFERENCE PORTALS

These give access to, among other things, multilingual and multidisciplinary terminology data banks. The names of the data banks are presented in alphabetical order. Selecting a name and clicking on it calls up the querying menu of the desired bank. It is possible to switch from one data bank to another while keeping the same screen, so that search results can be compared, and pertinent information can be retained for terminology research by using a word processor's copy-and-paste function.

4.11 MULTILINGUAL DATABASE MANAGEMENT SYSTEMS

A system such as Termbase is intended for use by translators to manage multilingual terminological data. It allows you to.

- Create records that include English, French, Spanish, German and Italian terms.
- Extract, exchange and update records.
- Statistically manage the database's contents.
- Output in RTF format.
- Define access restrictions per user.
- Tools such as TermStar and MultiTerm are used to manage databases that can be configured by the user, who can define and group numerous databases for consultation purposes. These databases can hold a large number of entries and a wide variety of languages, and can be integrated with word-processing software such as Microsoft Word and Corel WordPerfect. They allow you to:

- Define and store filters for restricting searches.
- Protect certain parts of the database contents.
- Perform global changes on groups of records.
- Link images to records.

4.12 MULTILLINGUAL TECHNOLOGIES (GIST)

The two groups of C-DAC involved in the development of multilingual technologies are:

Graphics and Intelligence Based Script Technology (GIST)

The GIST Group of C-DAC facilitates use of Indian languages in IT. In its endeavor to stay abreast with technologies worldwide, GIST has been adopting the latest concepts to be able to stay tuned with the Internet enabled world.

Applied Artificial Intelligence (AAI)

The AAI Group of C-DAC is involved in a number of activities such as knowledge-based understanding systems, Web-based solutions include Chat Server, and many other strategic software developments and solutions.

5.0 SOUL

University libraries are complex entities, having large collections and serving a huge clientele. To carry out various operations in a library effectively, there is a need for automation. Computer and communication technologies have brought revolutionary changes in the information acquisition, processing, storage, retrieval and dissemination. Keeping in view the latest trends in Information Technology (IT), INFLIBNET Center has developed a Windows based Library Management Software "SOUL", which provides total solution for Library Automation. SOUL is designed using Client-Server Architecture which imparts extra strength to storage capacity, multiple access to single database, various levels of security, back up and storage facilities etc. This software has been designed after a comprehensive study of different library related functions practised in university libraries. It has MS-SQL Server 6.5 RDBMS as the back end. This user friendly software is quite easy to work with.

This software is intended to be provided to all the university libraries, who have been covered under INFLIBNET for which a separate agreement will have to be signed by each library. An extensive training in SOUL will be provided to the library staff on-site upon its installation by the INFLIBNET staff. For others who are interested in using this software, particularly public funded organizations, separate modalities are being worked out including pricing, support, training etc. In due course different versions to work on different platforms suiting the needs of other libraries will be brought out.

The in-built network feature of the software will allow multiple libraries of the same university to function together as well as access to the distributed databases installed at various university libraries and union catalogue mounted at INFLIBNET using VSAT network.

5.1 SOFTWARE FOR UNIVERSITY LIBRARIES (SOUL)

The SOUL is a state-of-the-art library automation software designed and developed by the INFLIBNET. It is user-friendly software developed to work under client-server environment. Although looking at the name of the software, one may think that it is meant for University libraries only, but in fact it is flexible enough to be used for automating any type or size of library in India. While designing this software, the international standards, bibliographic formats, networking protocols, and typical functions of all types and sizes of libraries, particularly at university level, have been taken into account.

The functions have been grouped into six categories, looking into the functional divisions of Indian University libraries. At present SOUL uses RDBMS on Windows N.T. operating system as back end to store and retrieve the data. However, keeping in view the trends in IT towards Linux operating system, efforts are under way also to provide SOUL to work on Linux platform. The inputs received from expert team consisting of practising librarians and the feed backs received from users of our earlier software, ILMS, have given a strong base for designing this software. SOUL is near total solution offered by INFLIBNET to Indian libraries. It puts library staff at ease in exploring all the functions to their advantages with the help of professionally prepared manual.

5.2 HARDWARE AND SOFTWARE REQUIRED

The minimum hardware and software configuration required to use the SOUL is given below:

Client:

Pentium @233 MHz with 64 MB RAM

1.2 GB HDD
32 x CDROM Drive
1.44" Floppy Drive
Color Monitor (SVGA)
Ethernet card 10/100 Mbps
Windows-NT Operating System

Server:

MS-SQL Server 6.5 Pentium @233 MHz with 32 MB RAM
1.2GB HDD with 10MB Free space
1.44" Floppy Drive
Color Monitor (SVGA)
Ethernet card 10/100 Mbps
Windows-95 Operating System

5.3 STRONG FEATURES OF SOUL

Following are few of the strong features of SOUL, which should induce librarians to use SOUL in their libraries.

- Windows-based user-friendly software.
- Well-designed screens, logically arranged functions with extensive help messages make the software user-friendly.
- It is based on client-server architecture allowing scalability to the users.
- It uses RDBMS to organise and query the data.
- SOUL does not need an extensive training. With very little familiarity, one can begin using it.
- It is specially designed to work in the large academic libraries, capable of handling large number of records.
- It is a multi-user software and there is no limit on simultaneous accesses.
- Supports internationally known standards such as CCF and AACR II. etc.
- Provides export and import facility and adheres to ISO 2709 format.
- Incorporates all required features to work in a networked environment *i.e.*, LAN and WAN.

- OPAC is versatile and very user-friendly with all options in- built.
- OPAC is accessible over the Web using any GUI based browsers.
- Provides comprehensive list of reports, master databases and authority files.
- Provides facility to create, view and print records in regional languages.
- Functionally it covers every conceivable operation of University library.
- Available at affordable cost.

SOUL has been fully tested at a number of University libraries and critically evaluated by team of experts and practising librarians.

5.4 MODULES

The SOUL has been divided into following six broad modules as shown in the following figure:

Fig. 6.3: Modules of Soul Interface

These modules have further been divided into sub-modules looking at the nature of functions handled by various functional divisions in University libraries. Brief description of the same along with first screens have been given in the following pages.

5.4.1 Acquisition Module

This particular module provides facilities to handle work relating to acquisition of reading materials of all types except serials, starting from suggestion / recommendation by faculty till accessioning, invoice processing. Acquisition module comprises following six broad sub-modules as given in the following figure:

The above module has got following sub-modules:

- Suggestion
- Master Databases
- Payment
- Order processing
- Receiving/Accessioning
- Reports

Note: sub-module enables creation / updation of databases of new items suggested by a faculty member along with bibliographical details and also suggestor's details such as name, department, user code, budget code etc. Books received through placement of purchase orders as well as those received as gratis are handled from this module. Selecting items for approval, incorporating further details like approval date, budget head, approver's name, rejected items, its date and reasons for rejection are also handled here. Number of required reports having combination of several fields can be easily generated.

Order Processing: Selecting the items for ordering, vendor/publisher selection, assigning order and reference number, setting deadlines for supply and other details can be done under this sub-module. Facilities for sending reminders, cancelling orders, generating certain reports specific to these modules, are provided.

Receiving / Accessioning: This function supports cross checking with order, receiving partial/ full supply of items listed in order, duplicate checking with an existing title, assigning accession numbers along with further necessary details required and merging the records with existing database, if items are already in the collections.

Payments: Payments sub-module allows processing of the regular invoices, advance payments, and recording of every information pertaining to each invoice including conversion rates, handling charges discounts etc. Facility for generating reminders to supplier or publisher. Searching the status of payment and generation of reports are other strong features added in this sub-module.

Master Databases: Publisher, vendor, Currency and budget codes (both source wise and department wise) are few major master databases that are frequently handled/created here. Updation, deletion of all this is possible from here.

Reports: There are 13 major reports relating to acquisition that can be created with number of parameters.

Acquisition module is capable of handling almost every function that is being carried out in acquisition division of university libraries.

5.4.2 Cataloguing Module

Catalogue module function begins with selecting the items that have already been accessioned in the previous module and furnishing rest of the information as per AACR-II rules. Providing user services such as recent addition services, CAS, creation and updation of authority files *etc.* are other major functions supported by SOUL. A comprehensive worksheet covering almost every field facilitates data entry of all types of books, conference proceedings, theses etc. Facility to create database in the regional languages, using respective scripts is also provided. This module allows the library staff to conduct comprehensive searches for existing items before cataloguing new items and has provision of import and export of records and retrospective conversion. Functionally this module has been organised into following sub-modules as given in following figure:

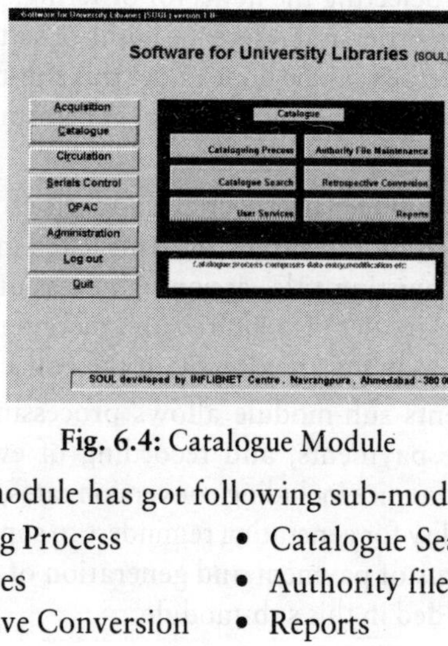

Fig. 6.4: Catalogue Module

The above module has got following sub-modules:

- Cataloguing Process
- Catalogue Search
- User Services
- Authority file Maintenance
- Retrospective Conversion
- Reports

Catalogue Process function allows to pick-up the accessioned item, under process, for the cataloguing purpose. Here one can add remaining information as per specified standards, such as additional bibliographical information, subject headings, classification number etc. Editing of existing records for maintaining consistency can also be done here.

Catalogue Search enables search of the existing items, its status, identifying duplication etc. for the purpose of day-to-day cataloguing. This is similar to OPAC. User Services sub-module has three major functions *viz.*, generating current awareness list (by date, subject etc.), compiling of bibliographies with various combinations and alert services to individual users.

Authority File Maintenance includes creation, updation and use of major authority files for names such as publisher, languages, corporate bodies, meetings, authors, physical media, types of material and also for subject descriptors. This is a unique feature added to this software taking into account the consistency that each library needs to maintain while creating records.

Retrospective Conversion has two major functions *viz.* data entry of old collection with minimum information without going to first sub-module and import and export of data from and to external sources. By using this function libraries can download the matched records from INFLIBNET union catalogue or other sources and export the records for contributing to union catalogue etc. A versatile ISO2709 interface developed at INFLIBNET, which is built-in to this module, enables to carry out the job.

Reports module allows generation of catalogue cards as per AACR-II, generation of recent editions reports subject and class number wise and other related reports. The catalogue module basically supports all major functions relating to technical processing and has been designed as per the international standards.

5.4.3 Circulation Module

Circulation being vital front-end function of any library, sufficient care has been taken in designing this module to achieve transactions within minimum possible time. This module has provision for all possible function handled in a typical academic library, *i.e.* membership, issues, returns, ILL, reminders, overdues, reservations, recall etc. All these functions have been organised into following eight logical sub-modules as given in following figure:

The above module has got following sub-modules:

- Membership
- Inter-Library-Loan
- Overdue Collection
- Status Search
- Transactions
- Reports
- Reminder
- Maintenance

Membership Sub-module provides the facility to create all types of member records, assigning unique membership code, borrowing privileges, renewal, issue of no-due certificates, master databases for

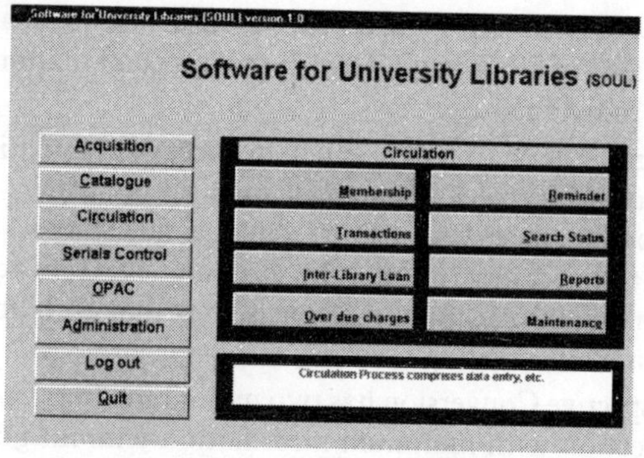

Fig. 6.5: Circulation Module

codes etc. searching the status of membership or an item, suspending the membership and generating related reports.

Transactions handles all major functions such as issue, return, renewal, reservation, recall or reminder of an item etc. This module also handles cancellation of reserved books, lost/missing books, searching the member status and searching catalogue module etc. Transaction is based on Accession Number and Member Code. This module supports generating and reading of barcode labels. A unique facility in this module allows one to see simultaneously the details of members, items borrowed, dues etc. while the transaction process is on. This enables issue counter staff as well end-user to know the exact status of the members borrowing.

Inter-Library-Loan allows lending of items to specified member library and also borrowing items from other libraries, issues, reminders etc. This sub-module has been developed comprehensively to take care of all the details of user libraries, individuals and items loaned.

Overdue collection facilitates collection of overdue charges in full or in part, providing receipts, keeping up-to-date accounting and tallying totals, etc. Using this function one can generate daily, weekly, monthly reports to find out as to how much overdue charges have been collected.

Reminder module handles individual and group reminder generation for all overdue materials. Comprehensive listing of materials that are overdue can also be generated within a specific period giving from and to dates.

Search status enables the library circulation desk staff to check the status of a member or items borrowed by a user and overdue items. Maintenance is yet another comprehensive sub-module, which covers binding, lost and cost recovery of books, damaged books, withdrawn books etc.

Reports sub-module allows the generation of as many as 16 major reports and with many combinations. All possible reports that a large library expects are provided for. Above all this module is capable of handling large transactions. If various functions built in this module are effectively put to use the library staff will save lot of time and help to avoid repetitive jobs.

5.4.4 Serials Module

This module allows one to create an exclusive database for different serials. Handling serials is most complicated job for the library staff. This module has been designed to deal with all these complicated jobs quite effectively. It helps in keeping track of every title received in the library. All functions starting from suggestions, master databases, subscriptions, checking, payment, reminder, binding title history export/import etc., have been covered. For the convenience of users, these functions have been grouped under following logical sub-modules as given in following figure:

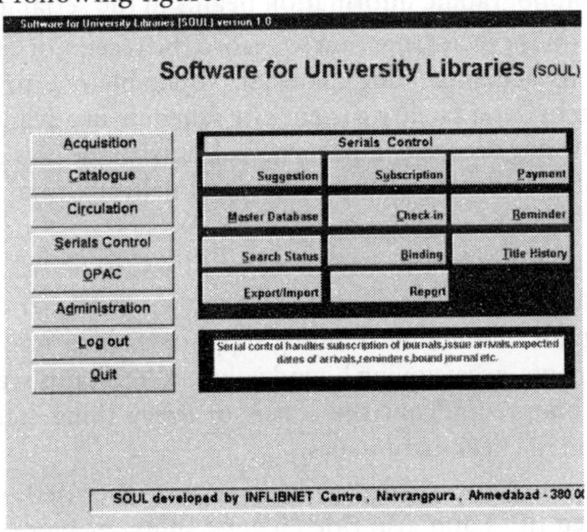

Fig. 6.6: Serials Module

The above module has got following sub-modules:

- Suggestions
- Payment
- Subscription
- Master Databases

- Check-in
- Binding
- Title History
- Report Generation
- Reminder
- Search Status
- Export / Import

Suggestion sub-module enables one to record and keep a track of all the suggestions received for subscribing to a serials. Selection of these titles for approval, preparing budget estimates and generation of related reports are covered under this sub-module.

Subscriptions module takes care of ordering/renewal of serials, follow-up relating to the same, sending reminder, if invoices are not received, generating orders by supplier or publisher are included under this option. Payment function supports processing and recording of all details relating to each invoice, including supplementary invoice such as invoice processing, credit notes processing, reports generation etc.

Master Databases option allows creation of large number of frequently used master databases viz. title entry, language, class number, publisher, binder, country, department, currency, frequency, budget heads, binding type, delivery modes, reports etc. Of these, title entry is main. It is here that the creation of database for each title with bibliographic information begins in the serial module.

Check-in is crucial function to record the receipt of each issue of serial and its accompanying material. To enable one to record the issues, system has a facility to generate schedule in advance for each title by providing necessary inputs viz. Vol. no(s), Issue no(s), frequency, date of publication of first issue, mode of delivery, total number of issues etc.

Sending reminders for non-receipt of issues or issues that are overdue etc. for single or all titles by supplier, publisher etc., can be done using this sub-module. Binding supports making sets, generating order, payments, accessioning bound volumes etc. Status search option facilitates one to find out the status of every thing starting from subscription to check-in of issues.

Title history is provided to keep record of ceased, suspended, discontinued titles and also title change, splits, mergers along with holdings information for each and every title in the database. Export/ Import of data in ISO2709 format is also provided to enable library to transfer the existing records in to SOUL and also contribute data to INFLIBNET union database. Reports is a comprehensive function which has more than 15 built-in reports of all types with different combinations. This adds to the strength of serial module. Serial module is designed to handle large number of titles, with many options giving maximum flexibility to user libraries.

5.4.5 Online Public Access Catalogue Module (OPAC)

One of the major attractions of SOUL is that it has a powerful Online Public Access Catalogue as given in with a choice of search options and variety of display formats. OPAC is a dynamic information desk that allows library staff to post library calendar, library rules and regulations, announcements, or any other information of user interest. SOUL increases the functionality of library's OPAC terminals by allowing the users to access the internal as well as external resources. This enables the users to access various databases developed at INFLIBNET. Library can keep entire collection available at users fingertips. This powerful, yet easy-to-use and user-friendly searching tool allows user to quickly find the materials in the library. Some of the major features of OPAC are:

- SOUL includes Boolean operation when more than one search option is to be used.
- Search results can be sorted according to the preference of search item.
- User has option to select variety of display formats.
- Display of records according to AACR-II format.
- Easy and quick searches with options.
- Status of each book starting from acquisition module is reflected.
- Search key fields, such as, author, title, keywords, class number, accession number, etc.
- Accessible through the GUI- based Web browsers like Netscape Communicator, Internet Explorer etc.,
- User can see the status of currently borrowed items by entering his/her borrower number.
- Search results can be saved and printed.
- Selection of databases can be made according to the choice of users.

5.4.6 Administration Module

Administration module as given in following figure has been to authorise users *i.e.*, the library staff to use various modules.

Assigning login and password to use each module of the system is done by the system administrator. The security function, backups, recovery of data and other utility functions are some of the features added under this module. Users have been categorized into three levels looking into nature of functions handled by the staff at different levels.

5.5 BENEFITS OF USING SOUL

Available at nominal cost to university libraries.

- Available at nominal cost to university libraries.
- Software designed and developed exclusively to work under university envrionment.
- Network feature of the software will allow multiple libraries of university to function together.
- Exhaustive training at INFLIBNET supported by comprehensive manual.
- On-site training.
- Free updation/modification.
- Free technical assistance.
- Follows INFLIBNETs recommended standards and formats such as CCF, AACR2, LCSH.

Conversion

But history has also left us with some four centuries of books, the vast majority of which aren't currently in print. It is suspended that much of the public believes that all of these works are, or will soon, be available in digital form—availability is waiting only on the deployment of high-bandwidth delivery systems, perhaps, and more scanning. This misconception has been fed by a great deal of hype about the evolving network information environment .

If and when these older books are converted, almost all of them will become extremely literal translations of their printed forms. This kind of scanning-based conversion (which actually captures images of the print pages, rather than meaningful digital representations of the characters and words on those pages) is a fairly inexpensive manual process.

More sophisticated conversions that capture the meaning of the content in computer-manipulable ways (at various levels of complexity) are much more expensive and involve extensive human intervention in the conversion process (with the extent of the intervention increasing with the richness of meaning being captured). We may see technological progress over time, which reduces these costs, and allows computational processes to be automatically applied to the page images to capture deeper meaning. Optical character recognition (OCR) technology is one important tool that has been slowly improving in quality over the past two decades, but accurate OCR still requires human review and editing. But all of these considerations speak solely to the mechanics of converting books to digital form.

The legalities of such conversions are a much more serious barrier, and one about which the public remains unaware. Roughly speaking, at least in the United States, any book published before the early 1920s is in the public domain (the details of precisely what is in the public domain are very complicated, and aren't crucial here). If you can find a copy, you can scan it, or, if you are willing to pay the labor costs, you can even re-keyboard it with added structural markup into a more sophisticated digital representation. Whether you obtain a new copyright for your converted digital version of the work seems to be legally murky [15], and seems to depend significantly on how much value you add in doing the conversion [16]. This is important because it has implications for the availability of investment capital to convert public domain materials, and for how these materials need to be protected as they are made available, if they need to generate a revenue stream.

Note also that for nearly all readers, really old works are more appealing in modern editions (with modern spelling, typography, and the like). While the original editions may be in the public domain, commercially acceptable ones may still be under copyright and cannot be converted without negotiating permissions from the copyright holder. Octavo's work with some of the great classics of publishing and intellectual history provides an excellent case study here; they had to go beyond simply imaging pages to also linking a parallel modern translation of the works to make them truly accessible and valuable to today's general readers.

After the early 1920s matters get much more complex. Some of the books published since then are now in the public domain [17]; the vast majority of them remain under copyright, however, and can't be converted to digital form without the copyright holder's permission. But who holds the copyright, and how do you find out? At least in mass-market publishing, it's common for contracts between publishers and authors to stipulate that the rights to the book revert to the author a few years after the publisher takes the work out of print. When the author dies, these rights pass to his or her heirs as part of the author's estate. In other areas, such as scholarly journal publishing, it's more common to see all rights permanently transferred to the publisher as a condition of publication, which has greatly facilitated projects like JSTOR, which are performing massive conversions of back runs of scholarly journals—all the rights for large amounts of material can be negotiated with a single owner.

The U.S. Congress has made this problem worse with the passage of the Sonny Bono Copyright Extension Act of 1998, which extended the term of copyright from *life of the author plus 50 years to life of the author plus* 70 *years*. This law was motivated mainly by the desire of a few media companies such as Disney to extend protection of a relative handful of commercially very profitable works. But it has had the effect of establishing a 20-year moratorium on the entrance of new works into the public domain, and greatly increasing the cost of converting a huge body of commercially fallow books into digital form. The cost of clearing rights for these works is likely to be hundreds of times greater than the costs of actually digitizing the works. For the vast majority, it's not clear that anyone will bother. While it might be possible to generate some return on the investment in digitizing, it's unlikely that the return will cover the additional costs of rights clearance.

The key point here is that for the vast majority of the enormous number of books published since 1920 or so conversion to and availability in digital form (even as direct translations of printed works) is far from guaranteed. In some relatively few cases, publishers hold the rights and can convert them—mainly in cases where the work is still in print on the publisher's backlist. But here too there is controversy and uncertainty. At least in the United States (the situation varies in other nations), publisher contracts with authors prior to the mid-1980s use language like "to print, publish and sell in book form", and it is unclear whether book form includes digital books.

Currently there is litigation involving the publishing industry, authors, and third parties (notably Rosetta Ebooks) who want to license e-book rights from authors covered under such contracts with the print publishing industry. This litigation will determine whether print publishers also hold e-book rights for the books that are part of their active backlist published prior to the time when contracts became explicit about e-book rights [18]. Note that the in-print backlist is of particular importance for conversion to digital format because these are works where a marketplace demand has already been established this is why the publishers are keeping them in print.

In the other and most commonplace situation where rights have reverted to the author when the book went out of print, e-publishing rights can at least in theory be cleared to permit works to be digitized, but this is going to be a case-by-case decision by one of the myriad of organizations (both commercial and nonprofit) and even individuals interested in making books available in digital form.

Converting our literary heritage is going to be slow, incremental, expensive, and somewhat haphazard. And in terms of making a really substantial corpus of material of high interest to the general public available quickly, the publishing industry, with its current in-print works and the out-of-print works it already holds rights to, is the only game in town. Whatever the shape of the future of the digital book, what happens with the availability of the current in-print corpus of material from the publishing industry is going to be important. Perhaps even important enough to keep the new genres marginalized, at least outside of the scholarly world, long enough for the traditional publishing industry to firmly establish new ground rules for digital books—even to define entirely the idea of what constitutes a digital book in the public mind. And this corpus will look very much like literal translations of printed works.

7
UNIT
Digital Libraries

1.1 TRADITIONAL LIBRARY

The collection of the traditional libraries is mostly print media, manuscripts etc. and are not well organized. The documents are deteriorating at a rapid rate, the collection information is not easy to locate and so does not easily reach to user. Again the traditional libraries are confined itself within a physical boundary.

1.2 INTRODUCTION AND BACKGROUND

The term "Digital Library" has a variety of potential meanings, ranging from a digitized collection of material that one might find in a traditional library through the collection of all digital information along with the services that make that information useful to all possible users.

We are in the age of digital environment where I.T. has been used extensively to record, store, and disseminate the information in the digital form. The IT has almost converted the world into a global village. The revolution in the IT sector is influencing the information industry also. The libraries are also changing to meet the demand put on it. The new generation whose demand for information is never meet is always demanding that traditional libraries should be developed as a well equipped and interconnected digital libraries. If we consider that the information is for used and for all then why not libraries for all? Why not libraries become universal for all those who seek for information.

A number of large organizations in advanced countries have already more or less switched over to digital mode *i.e.* they are making the digitized images of periodical which are not available in electronic form and there after no new procurement are being made on print media. Examples of such organization are University Micrographic Inc. (UMI of USA), National Institute of Scientific and Technical Information (INIST of France) etc.

According to Arms a digital library is a managed collection of information with associated services where the information is stored in digital format and accessible over a network. The digital library federation in the USA define the digital library as Digital libraries are organizations that provide the resources, including the specialized staff, to select, structure, offer intellectual access to, interpret, distribute, preserve the integrity of, and ensure the persistence over time of collections of digital works so that they are readily and economically available for use by a defined community or set of communities. A digital library is an organized collection of digitized material or it's holding in the digital form, which can be accessible by a computer on the network by using TCP/IP or other protocol.

1.3 DEFINITION OF DIGITAL LIBRARY

The Digital Library is:

- Organized collection of multimedia and other types of resources. Resources are available in computer processable form.
- The function of acquisition, storage, preservation, retrieval is carried out through the use of digital technology.
- Access to the entire collection is globally available directly or indirectly across a network.
- Support users in dealing with information objects.
- Helps in the organization and presentation of the above objects via electronic/digital means etc.

The "digital library" is not merely equivalent to a digitized collection with information management tools. It is also a series of activities that brings together collections, services and people in support of the full life cycle of creation, dissemination, use and preservation of data, information and knowledge. The challenges and opportunities that motivate an advanced digital library research initiative are associated with this broad view of digital library environment.

Much of the question about the scope of the term is how broad a view should be taken of the digital library. Does it encompass all of information management or is a more tightly constrained view appropriate?

A digital library is much more than just the collection of material in its repositories. It provides a variety of services to all of its users (both humans and machines, and producers, managers, and consumers of information). Thus we start our definition with the notion of the collection of services that the digital library represents. There are a large and varied set of such services, including services to support management of collections, services to provide replicated and reliable storage, services to aid in query formulation and execution, services to assist in name resolution and location, etc.

The Collection of Information Objects:

The basis for a digital library, however, must be the information objects that provide the content. A basic characteristic of the digital library is that the information objects are found in collections with associated management and support functions. The types of information objects vary from traditional "documents" through to live objects (*e.g.* sensor readings) or dynamic query results.

Supporting Users Deal with Information Objects

The goal of the digital library is to assist users by satisfying their needs and requirements for management, access, storage, and manipulation of the variety of information stored in the collection of material that represents the "holdings" of the library. Users may be humans or they may be automated processes acting on behalf of or in support of human needs. Users also vary and include those who are "end" users (those not involved in the management and operation of the library but rather are the customers), library operators, and information "producers" who want their material available through the library.

The Organization and Presentation of those Objects:

The key to effective collections management is to implement simple structural organizations and be able to present those organizations in a way that library users find useful and can understand easily. In traditional libraries, books are primarily stored by subject, title, author, and date, and accessed by following signs to the appropriate floor, room, bookcase, shelf, and spine-labeled book. The size and relative celebration of each portion of the collection gives patrons information about the collection and can reveal the library's collection management objectives as well.

A library is created to serve a community of users. Users who participate in the digital library should be aware of its design and be able collectively to refine that design to better serve their own information needs. Therefore, the ongoing human usability of a digital library depends on the clear and unobtrusive exposure of the library's design, its near-term goals, and its overall objectives.

Furthermore, digital libraries should continue the ongoing tradition of coupling utility with aesthetics in the organization and presentation of materials.

Available Directly or Indirectly

These information objects may be digital objects or they may be in other media (*e.g.* paper) but represented in the library via digital means (*e.g.* metadata). They may be available directly over the network (*e.g.*, using a query service of the library to find and then retrieve electronically the information object) or indirectly (*e.g.* the result of the query may give instructions on how to obtain the object, but that is done outside the scope of the library itself).

Electronic/Digital Availability

Although the objects may not even be electronic, and although the objects themselves may not be available directly over the network, the objects must be represented electronically in some manner through, *e.g.* metadata or catalogs. Otherwise, we would not consider the objects to be part of the digital library.

1.4 OBJECTIVES

A digital library should be a seamless extension of the library that provides scholars with access to information in any format that has been evaluated, organized, archived, and preserved. Access to this evolving collection of digital information is provided through personalized systems as well as through the services of information professionals. The digital library adds value and saves time while shifting the times of access. It reduces need for proximity to information resources, but still emphasizes the quality of those resources. It is a library that can be individually customized and, ultimately, will be easy to use.

One goal of the digital library is to improve teaching and assignments through the incorporation of library materials. The information resources that are important to a class should be format independent, whether the class is taught in a campus classroom or at a distance to a distributed and asynchronous class. To accommodate both classroom teaching and distance education (for neither will completely replace the other in the immediate future), students will need access to information resources and not all are, can, or will be digital. Because a resource is not digital, does not mean that it cannot be used in distance education. Digitizing an article may be one practical solution; another solution, however, may be linking to an existing article database. In some situations, sending a library book to a student's home may be the best way to get the information resource to the student, and making that possible through online requests should be a component of the DL, as it is the library. In some situations, the digital format may have restrictions placed on it (*e.g.* broadcasting, simultaneous number of users, etc.).

Libraries are more than their information resources, their collections, the buildings that house them, the systems that they run on, or the services they provide. Libraries have information professionals that make judgments and interpret user needs; they provide services and resources to people (students, faculty, and others, and organizations). One of our goals is to expand the computing functions from efficient, timely, and quantity based systems. Some traditional library services can be replicated in the DL, partially or wholly, but some cannot be replicated. Some works will not be digitized but are still of use to DL users. Online instruction is important, but sometimes meeting face-to-face, or having a telephone conversation, between student and information professional is the best method. Information seekers should not be denied any library resource or service because it is not available online. The DL and the library should be complimentary, intertwining systems that exist to serve the user community on the highest order.

1.5 SCOPE

Digital libraries provide for collection, development, organisation, access, annotation and preservation, and deal both with information in digital form as well as digital management of information residing on physical media. However, additional emphasis should be made on the need to consider users and usage as part of any analytical framework used to study digital libraries.

Digitied libraries/literature presents the matter in Multimedia format that uses several different media to convey information (*text, audio, graphics, animation, video*, and *underactivity*).

As the information is presented in various formats, multimedia enhances user experience and makes it easier and faster to grasp information. Presenting information in various formats is nothing new, but multimedia generally implies presenting information in various digital formats. It is also used in visual arts to describe works created using more than one medium.

Digital libraries and archives can enable worldwide access to a never-ending supply of distributed information and knowledge that is constantly available, easily updated, and convenient to use. In a manner never quite achieved before on the same scale, digital library issues are bringing together cooperative groupings of libraries with participation from and funding by public, private, and governmental bodies to discuss common standards, collaborative management, intellectual property rights, electronic publishing, and document delivery. And all this is happening on a global level!

There are several reasons why information experts are choosing to create digital libraries. There is not only the need to preserve and conserve materials, but also the desire to increase the customer base by making older and rarer or more-heavily-used documents available to a much wider audience. In addition, there is the fact that the level of electronic publication is increasing rapidly in key areas for instance, STM journals and patents, where pundits anticipate that all patents will be digitally produced by the end of this year. This has implications on a library's acquisition, storage, preservation, and service policies.

1.6 HISTORY OF DIGITAL LIBRARIES

Over the past 20 years, the computer has become the main tool for accessing specialized knowledge and the favoured means of transmitting information. The process began with the computerization of catalogues held by libraries and large publishing houses, and with online access to documentary databases such as PASCAL.

Computerized catalogues were followed by bibliographic CD-ROMs and database-querying terminals such as Electre, and electronic text media such as Frantext, a textual databank.

The computerized library led to the electronic library, in which a digitized copy of the holdings is made available to readers from computer-aided reading stations installed in reading rooms. You will find these in the national libraries of London, Paris and Washington for instance. The digitization of holdings involves the transfer of texts, images, sound and films to electronic media. You may then consult the holdings in networks using a set of tools for querying, selecting, annotating, and on-screen editing.

1.7 TYPES OF AUTOMATED LIBRARIES

Automated library: A library with machine-readable catalogue, computerized acquisition, circulation and OPAC are called as automated library. The holding of this type of libraries are same as that of traditional libraries.

Electronics library: When automated libraries goes for LAN (Local Area Networking) and CD-ROM networking and started procuring Ejournal and other similar kind of publication then it is known as electronic library. The resources of the electronic libraries are in both print and electronic form. The electronic media are used for storage retrieval and delivery of information.

Digital library: It is a later stage of electronic library. In digital library high speed optical fibers are used for LAN and the access is over WAN and provide a wide range of Internet based services *i.e.* audio and video conferencing and like other. The majority of the holding of a digital library is in the computer readable form and also acts as a point of access to other online sources.

Hybrid library: The libraries, which are working both in electronic or digital and print environment, are known as hybrid library. Actually it is a transitional state between print and digital environment. It is estimated that in near future libraries will be of hybrid nature, some of the very strong point in favor of this view are centuries old reading habit of paper, convenience of handling and reading a paper document then the digitized one (in case of digitized some equipment are must needed to read the document), incompatible standard of electronic product, different display standard of digital product and its associated problem etc.

1.8 VIRTUAL LIBRARY/DOCUMENTARY DATABASES

Virtual libraries are delocalized and universally accessible; instead of having a physical address, they are constructed on Internet sites and bear evocative names such as Alexandrie, Bibiliotheca Universalis and Project Gutenberg. You can access them via a computer connected to the Internet and may also download documents for later consultation.

Searchable text databases can come in very handy when looking for documents or researching a term. For example, the New York Times (www) (en) and the Wired (www) (en) sites allow you to search their archives online.

1.9 A PARADIGM SHIFT FROM TRADITIONAL LIBRARY TO VIRTUAL LIBRARY

The development is already taking place. The traditional closed access libraries are shifting towards open access library. The open access libraries are shifting towards automated library, the automated one towards the electronics, the electronics to digital and finally end in digital library and its different aspect/badan barman virtual library. Is it really true? The truth is that nobody knows what will be the future of libraries. In the following para an attempt has been made to categorized the different types of libraries based on the technology used. It's the best time to mention that there is no strict line of demarcation between the latter four types of libraries.

1.10 NEED FOR A DIGITAL LIBRARY

Time is a major factor for each modern user of the library and digitization is the only solution to the problem. Digital libraries are needed to provide quality-based service at the user doorstep. In general digital libraries are needed for the following reasons:

- **Easy to understand:** The visual or graphical information system of digital libraries is more popular as compared to text-based information system.
- **Shifting of the environment:** The new generation user becomes only happy when they will be able to read from the computer screen. The use of digital libraries is the cleanest technologies to fulfil the slogan "Burn a CD-ROM save a tree".
- **Multiple function of same information:** In case of digital libraries by using hypertext it is possible to structure and organized the same digital information in a variety of ways, which serve multiple functions.

- **Information explosion:** Digital library is expected to be able to handle the problem of information explosion somehow. It will be able to handle and manage large amount of digital content by simply providing link, without actually procuring the document.
- **Information retrieval:** By using digital library one will be able to retrieve information specifically *e.g.* a particular image, photo, a definition etc.
- **Distance learning:** Learning from home, office or other places, which are convenient to user.
- **To procure online publication:** More and more information are going to be published over internet, digital library is needed to procure the online publication and to provide link to important sources of information.
- Searching time problem in traditional libraries.
- **Low cost of technology:** When we consider the storage capacity of digital document and its maintained then it can be easily realize that the cost of technologies is much more less than that of traditional libraries.

1.11 ADVANTAGES OF HAVING LIBRARY DIGITIED

A major advantage of digital libraries is that people from all over the world can gain access to the information at any time, as long as an Internet connection is available. This also eliminates the logistical problems inherent in organizing and lending print materials.

Whereas *traditional libraries* are limited by storage space, *digital libraries* have the potential to store much more information, simply because digital information require very little physical space to contain them. As such, the cost of maintaining a digital library is much lower than that of a traditional library. A traditional library must spend large sums of money paying for staff, book maintenance, rent, and additional books. Digital libraries do away with these fees.

Digital libraries can immediately adopt innovations in technology providing users with improvements in electronic and audio book technology as well as presenting new forms of communication such as wikis and blogs.

A digital library is not confined to a particular location or so called building, it is virtually distributed all over the world. The user can get his/ her information on his own computer screen by using the Internet. Actually it is a network of multimedia system, which provides fingertip access. The spoken words or the graphical display of a digital library is again having a different impact from the words that are printed. In the new environment owing a document will not be problem for the library because the user will pay for its uses.

- **No physical boundary:** The user of a digital library need not to go to the library physically, people from all over the world could gain access to the same information, as long as an Internet connection is available.
- **Round the clock availability:** Digital libraries can be accessed at any time, 24 hours a day and 365 days of the year.
- **Multiple accesses:** The same resources can be used at the same time by a number of users.
- **Structured approach:** Digital library provides access to much richer content in a more structured manner *i.e.* we can easily move from the catalogue to the particular book then to a particular chapter and so on.
- **Information retrieval:** The user is able to use any search term bellowing to the word or phrase of the entire collection. Digital library will provide very user-friendly interfaces, giving click able access to its resources.
- **Preservation and conservation:** An exact copy of the original can be made any number of times without any degradation in quality.
- **Space:** Whereas traditional libraries are limited by storage space, digital libraries have the potential to store much more information, simply because digital information requires very little physical space to contain them. When the library had no space for extension digitization is the only solution.
- **Networking:** A particular digital library can provide the link to any other resources of other digital library very easily thus a seamlessly integrated resource sharing can be achieved.
- **Cost:** The cost of maintaining a digital library is much lower than that of a traditional library. A traditional library must spend large sums of money paying for staff, book maintains, rent, and additional books. Digital libraries do away with these fees.

1.12 DRAWBACKS OF HAVING LIBRARY DIGITIED

Some people have criticized that digital libraries are hampered by copyright law, because works cannot be shared over different periods of time in the manner of a traditional library. The content is, in many cases, public domain or self-generated content only. Some digital libraries, such as Project Gutenberg, work to digitize out-of-copyright works and make them freely available to the public.

Digital libraries cannot reproduce the environment of a traditional library. Many people also find reading printed material to be easier than reading material on a computer screen. Also, due to technological developments, a digital library can rapidly become out-of-date and its data may become unaccessible.

The computer viruses, lack of standardization for digitized information, quick degrading properties of digitized material, different display standard of digital product and its associated problem, health hazard, nature of the radiation from monitor etc. makes digital libraries at times handicap.

- **Copyright:** Digitization violates the copyright law as the thought content of one author can be freely transfer by other without his acknowledgement. So one difficulty to overcome for digital libraries is the way to distribute information. How does a digital library distribute information at will while protecting the copyright of the author?
- **Speed of access:** As more and more computer are connected to the Internet its speed of access reasonably decreasing. If new technology will not evolve to solve the problem then in near future Internet will be full of error messages.
- **Initial cost is high:** The infrastructure cost of digital library *i.e.* the cost of hardware, software; leasing communication circuit is generally very high.
- **Bandwidth:** Digital library will need high band for transfer of multimedia resources but the bandwidth is decreasing day-by day due to its over utilization.
- **Efficiency:** With the much larger volume of digital information, finding the right material for a specific task becomes increasingly difficult.
- **Environment:** Digital libraries cannot reproduce the environment of a traditional library. Many people also find reading printed material to be easier than reading material on a computer screen.
- **Preservation:** Due to technological developments, a digital library can rapidly become out-of-date and its data may become inaccessible.

1.13 REQUIREMENT FOR DIGITAL LIBRARIES

The Internet and World Wide Web provide the impetus and technological environment for the development and operation of a digital library. The Internet provides the TCP/IP and or its associated protocol for accessing the information and web provide tools and technique for publishing the information over Internet. In the digital environment it is reasonable to say that a central back up or archive should be created at the national level, which will store information out put of the region as well as information from out side the country. Some of the requirement for a digital libraries are:

- **Audio visual:** Color T.V., V.C.R., D.V.D., Sound box, Telephone etc.
- **Computer:** Server, P.C. with multimedia, U.PS. etc.
- **Network:** LAN, MAN, WAN, Internet etc.
- **Printer:** Laser printer, Dot matrix, Barcode printer, Digital graphic printer etc.
- **Scanner:** H.P. Scan jet, flatbed, Sheet feeder, Drum scanner, Slide scanner, Microfilming scanner, Digital camera, Barcode scanner etc.
- **Storage devices:** Optical storage device, CD-ROM, Jukebox etc.
- **Software:** Any suitable software, which is interconnected and suitable for LAN and WAN connection.

1.14 RESOURCES OF A DIGITAL LIBRARY

The resources of a digital library are those, which the computer can store, organize, transmit and display without any intervening conversion process. It includes both print and electronic or digital material. The digital material may be of multimedia types or any other *i.e.* only digital audio, video, full text information, photograph, drawing, digitized sound, e-book, v-book, electronic tax, map, image, 3D representation etc. The collection may also include structured/unstructured text, scanned images, graphic audios, video recording etc.

Online Resources

- Local database of traditional books in machine-readable form.
- E-book, v-book, electronic tax, map, image, sound, video, and multimedia etc.
- E-journal.
- LAN, MAN, WAN for Web browsing, e-mail etc.
- Well trained manpower for online help.

Off line resources

- C.D-ROM, Jukebox etc.
- Audio visual aid etc.

1.15 ROLE OF LIBRARIAN IN DIGITAL ENVIRONMENT

Though the digital environment is built as a system, which can be used by its ultimate end-user directly from their desktop pc, but the role of librarian cannot be overlooked. In digital environment also the librarian and information scientist will be needed for packaging and repackaging of information, for electronic publishing, for reference purpose, to advice the user about the strategy to identify relevant electronic sources etc. Thus the librarian will be more or less a hypertext engineer. In the new environment it will be very difficult for the librarian to decide what should be organized? How to give citation? How to organize the collection? Etc. because the new environment will be really challenging one for the librarian to decide who are author who are publisher and who are user? In the new environment any one who has access to the network of digital libraries can be a publisher by merely posting messages to an online discussion group or by other means. A virtual environment is really open for all.

1.16 DEFINING DIGITAL BOOKS AND E-BOOK READERS

Imprecise and inconsistent terminology has been a major source of confusion in the hype over e-books, and an obstacle to disentangling the issues involved. It is essential to distinguish between the idea of a digital book and a book-reading appliance.

A digital book is just a large structured collection of bits that can be transported on CD-ROM or other storage media or delivered over a network connection, and which is designed to be viewed on some combination of hardware and software ranging from dumb terminals to Web browsers on personal computers to the new book reading appliances.

Digital books cover a wide spectrum of material, ranging from literal translations of printed books, created by scanning pages or generating a PDF file, to complex digital works that are the intellectual successors of certain genres of book-length works, but which cannot be reasonably converted back into printed form. To a large extent, digital books exist (or at least should exist) independent of the devices that may be used to access, render and view them. A key role of standards (to be discussed later) is to make this independence formal, and to ensure that a digital book can be used with a wide range of viewing environments that may change over time.

Not every digital book can be viewed using every viewing technology. Some are highly targeted to specific viewing technologies, while others are versatile and can be easily delivered to many diverse viewing environments. Also, recognize that while it may be technically straightforward to deliver a book to a wide range of viewing environments, the publisher may deliberately choose to limit the environments a digital book can be delivered to. And of course, viewing technologies can be thought of as defining markets. Authors may choose to author for markets that they believe are large or easily reached or profitable, and as a consequence may choose to create works that deliver well to particular viewing technologies.

1.17 INTER-OPERABILITY OF DIGITIED RESOURCES

A fundamental issue for digital libraries is interpretability. The capability of digital libraries to exchange and share documents, queries, and services. Interoperability also applies to exchanging these within various components of a single digital library. From a slightly different perspective, interoperability in digital libraries is the ability to generate a single (virtual) view on many different library components without sacrificing autonomy. An important aspect of this is to coordinate changes of such autonomous components while providing guarantees on quality of data and services.

In the world of printed materials, libraries have sophisticated and well-regulated mechanisms for sharing their holdings. This enables scholars from the U.S., for example, to gain access to materials in Europe without having to travel to Europe. It is necessary for digital libraries to develop mechanisms that achieve the same effect. Interoperability for digital libraries, however, is more complex than for traditional libraries for several reasons.

First, there are myriad technical and engineering issues associated with connecting together networks, databases, and other computer-based systems; while the World-Wide Web is a good start at linking the world's information sources, it is neither comprehensive nor organized well enough to provide real library service.

Second, digital libraries will provide a greater array of services than do traditional libraries; for example, they already have sophisticated search engines, the ability to browse massive amounts of material quickly, and to reformat documents on-the-fly. We expect that digital libraries will provide even greater services, such as negotiating rights to works (acting as proxies for human users), reformat queries into various natural languages, create composite works, and on and on.

Third, the types of information available in digital libraries, and the format of this information, will be in much greater variety than is typically the case for traditional libraries. Consider, for example, that the World Wide Web distributes video (in many forms), audio (also in many forms), (marked-up) text, computer software, and other forms of information. The trend is for this variety to become even greater.

Fourth, and most importantly, digital libraries will be composed of a large number of loosely connected components. This is because the demand for both services and information can be expected to grow so large that it is not possible for a single corporation, professional organization, or government to provide all that is necessary for a digital library; the problem is exacerbated when digital libraries are considered on an international scale. Additionally, a large percentage of the information and services that will be made available through digital libraries will be privately held intellectual property, and it is unlikely that control over this is property will be yielded to third parties. Thus, a large number of groups, some small, some large, will provide digital library content and services. Compounding this, interoperability is naturally an international issue, since users will demand content and services from anywhere in the world.

An important observation is that interoperability may occur at different levels of abstraction, which are described subsequently. We refer to the different levels as interoperability layers. From this perspective, referred to as vertical view, interoperability is the definition of languages, with requisite syntax, semantics, and pragmatics, to connect digital libraries. These languages will be defined at various levels of abstraction, and will be specialized for particular uses. We do not expect to have a single language for interoperability, but a variety of them. This perspective, however, is not optimal to capture the dynamic aspects of interoperability in digital libraries, such as query processing and coordination issues. To this end, we have come up with a reference architecture of digital libraries that allows for another perspective to interoperability issues, the horizontal perspective.

Interoperability Layers: Figure 1 depicts the vertical view. It builds from the transport of bits along a network, to the commonly called 'middleware', where data structures, distributed objects and various protocols (*e.g.*, http, CORBA, Z39.50) are used to format and deliver bits to the proper place. The concerns at this point are reliable delivery of packets, and are not specialized to digital libraries. Hence, they will not be considered any further subsequently.

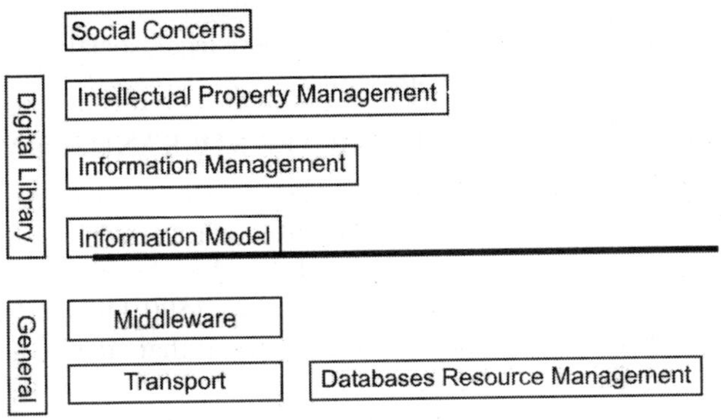

Fig. 7.1: Middleware

1.16.1 Interoperability Framework

At the next level, the framework is concerned with transforming packets into the fundamental elements of digital libraries: documents of all types, queries, and other related items, and thus is distinguished from other network applications. This includes formatting, searching, and displaying. At the information model level, there are definitions of document and query types. This includes concepts as SGML and XML as document descriptions, CORBA Query Service and InfoBus as example mechanisms for delivery of digital library content and services.

Building upon the information-model level is the document-management level. At this level, the concern is about properly managing information, *e.g*, documents, and not on composing or transporting it. Thus, in a sense, we treat documents and services as atomic.

The intellectual-property-rights management level follows the document-management level. At this level, the concerns of property rights are paramount and the actual documents and services can be safely abstracted. Examples of the activities at this level include negotiation, contracts, and various economic issues.

Finally, the social-information-management level is concerned with broader information and service-use issues. This level, like the intellectual-property-rights level, is concerned with various legal, social, and cultural issues. An example of interoperability at this level is the PICS standard, where Websites managers rate their sites on various content categories, such as violence, to allow possible users to be aware of what the site contains. This is particularly important in international settings, where legal, social, and cultural issues are more keenly expressed.

2.1 IMAGE FORMATS CHARACTERISTICS AND FORMATS

Pixels: There is a delicate balance between the crispness of a picture and the number of pixels needed to display it. Let's say you have two images, each is 5 inches across and 3 inches down. One uses 300 pixels to span that five inches, the other uses 1500. Obviously, the one with 1500 uses smaller pixels. It is also the one that offers a more crisp, detailed look. The more pixels, the more detailed the image will be. Of course, the more pixels the more bytes the image will take up.

So, how much is enough? That depends on whom you are speaking to, and right now you're speaking to me. I always go with 100 pixels per inch. That creates a ten-thousand pixel square inch. I've found that allows for a pretty crisp image without going overboard on the bytes. It also allows some leeway to increase or decrease the size of the image and not mess it up too much.

The lowest I'd go is 72 pixels per inch, the agreed upon low end of the image scale. In terms of pixels per square inch, it's a whale of a drop to 5184. Try that. See if you like it, but I think you'll find that lower definition monitors really play havoc with the image.

2.1.1 Graphics Types

Computer graphics are anything that can be displayed on the screen except the text and sometimes even text falls into the graphics category if it is save in a graphics format. There are basically two types of computer graphic:

- Raster (bitmapped)
- Vector/structured.

Raster (Bitmapped) graphics are images that are mapped to the monitor or screen. The screen is made up of tiny dots called pixels. These dots can display various colours depending on the type of computer hardware and software you have. Using shades of red, green and blue (RGB) an image can be displayed on the screen by mapping different colors to the screen in different sequences.

Vector graphics use objects created from mathematical formulae to represent things like lines, curves, fills, line thickness, etc. to create the image.

Each type of graphic has it's own advantages and disadvantages. Older versions of HTML were only able to recognizes bitmapped graphics so most graphics created for the Internet, using standard HTML, are created or converted to a bitmap format. The newest version of HTML or XHTML is able to display vector graphics but not all browsers are able to display these graphics. Within each of the two main types there are dozens of different formats.

2.1.2 Graphics Filename Extensions

The three main bitmapped format graphics used on the Internet are .gif, .jpeg (.jpg) and .png. There are many others including .bmp, .tiff (.tif), .pcx, .ppm, .tga and a host of others. Some of the structured formats are .ai, .cmx, .eps, .wpg, .cgm and a host of others.

Bitmapped graphics can be created and modified in a paint program and vector or structured graphics can be created and modified in a draw program.

The main tools in a graphics program allow you to select a section of a picture, erase part of a picture, fill a defined area, select a color, magnify a section, draw free hand, draw with various tools such as a straight line; a curved line; a rectangle; an oval; and a polygon. You can also modify a drawing by changing the size, color, placement, and depending on the program, hundreds of other modification.

2.2 RASTER OF BITMAP IMAGES

The most common and comprehensive form of storage for images on a computer is as a raster or bitmap image. The terms raster and bitmap can be used interchangeably as they refer to images that are created as two-dimensional sets of points on a computer display.

Bitmap images use a grid (the bitmap or raster) of small squares known as pixels to represent images. Each pixel is assigned a specific location and color value.

Raster graphics is the TV-like method that uses dots to display an image on screen. Raster graphics images are created by scanners and cameras and are also generated by paint packages. A picture frame is divided into hundreds of horizontal rows, with each row containing hundreds of dots, called pixels.

Unlike TV, which uses one standard (NTSC) for the country, there are dozens of raster graphics standards. Also, unlike TV, which records and displays the dots as infinitely variable shades and colors (analog), computer graphics have a finite number of shades and colors (digital). When you scan an image or paint an object into the computer, the image is created in a reserved area of memory called a bitmap, with some number of bits corresponding to each dot (pixel).

Photoshop and other paint and image-editing programs generate bitmap images. When working with bitmap images, you edit pixels rather than objects or shapes. Depending on the hardware capabilities, each point can display from two to millions of colors. The use of the word comprehensive means that an image looks as much as possible like the real-world or the original product. This means that the proportion, size, color, and texture must be as accurate as possible.

Bitmap images are the most common electronic medium for continuous-tone images, such as photographs or digital paintings, because they can represent subtle gradations of shades and color. Bitmap images are resolution-dependent—that is, they contain a fixed number of pixels. As a result, they can lose detail and appear jagged if they are scaled on-screen or if they are printed at a lower resolution than they were created for.

Fig. 7.2: Bitmapping

Consider above figure, where a piece of the image has been enlarged over 3 and 24 times to show to show the individual points that make up only a small portion of the total image.

Bitmap images are good for reproducing subtle gradations of color, as in photographs. They can have jagged edges when printed at too large a size or displayed at too high a magnification.

A bitmap image is resolution-dependent that is, it contains a fixed number of pixels to represent its image data. As a result, a bitmap image can lose detail and appear jagged if viewed at a high magnification on-screen or printed at too low a resolution. Bitmap images are the best choice for representing subtle gradations of shades and color, for example, in photographs or painted images.

You have three different ways to make a bitmap:

- Make a bitmap from scratch with a paint program MS Paint, Adobe Photoshop.
- Grab a bitmap from an active computer screen with a screen capture program, and then paste it into a Paint program or your application.
- Capture a bitmap from a photo, artwork, or a television image using a scanner or video capture device that digitizes the image.

2.2.1 Bitmap Formats

Common bitmapped file formats include PICT, BMP, GIF, TIFF, Quick Time movies, AVI movies, and JPEG, with PICT and QuickTime being the standards for cross-platform development. Programs that work with bitmapped images are often called Paint programs. The three main Internet formats, GIF, JPEG, and Bitmap, are all Raster formats.

Some other Raster formats include the following:

Extension name	*Type of Image/Graphics*
• CLP	Windows Clipart
• DCX ZOFT	Paintbrush
• DIB OS/2	Warp format
• FPX	Kodak's FlashPic
• IMG GEM	Paint format
• JIF JPEG	Related Image format
• MAC	MacPaint
• MSP	MacPaint New Version
• PCT	Macintosh PICT format
• PCX Z	Soft Paintbrush
• PPM	Portable Pixel Map (UNIX)
• PSP	Paint Shop Pro format
• RAW	Unencoded image format
• RLE	Run-Length Encoding (Used to lower image bit rates)
• TIFF	Aldus Corporation format
• WPG	WordPerfect image format

GIF and *JPEG* are the two most popular graphics formats that have become standards for graphical use particularly on the Web, and are universally read by most browsers without plug-ins. Both are bitmap formats, images consisting of individual pixels. As a general rule, bitmaps are fixed in their size and resolution. So as you manipulate and scale them, they begin to lose their original quality.

PNG is a relatively new bitmap graphics format. It combines the best of GIF and JPG into one file format and there is no burden of royalty.

Graphics and bandwidth are constantly at odds on the Web, and the designer must mediate in this struggle, always trying to strike the balance between high-impact visual appeal and low-impact delivery. To do this successfully, you need to know the difference between formats used on the Web, and how and when they are best used.

2.2.3 GIF: Better For Designed Graphics

GIF stands for Graphics Interchange Format, a bitmap graphics format defined in 1987 by CompuServe, using a lossless compression system that employs the Unisys LZW algorithm. GIF images are 8-bit, 256 colors, can be indexed to a fixed palette, and provide support for transparency. CompuServe originally defined the GIF format for its online graphic images. It's an 8-bit format, which supports a maximum of 256 colors.

Raster image formats (RIFs) should be the most familiar to Internet users. A Raster format breaks the image into a series of colored dots called pixels. The number of ones and zeros (bits) used to create each pixel denotes the depth of color you can put into your images.

If your pixel is denoted with only one bit-per-pixel then that pixel must be black or white. Why? Because that pixel can only be a one or a zero, on or off, black or white. Bump that up to 4 bits-per-pixel and you're able to set that colored dot to one of 16 colors. If you go even higher to 8 bits-per-pixel, you can save that colored dot at up to 256 different colors.

Does that number, 256 sound familiar to anyone? That's the upper color level of a GIF image. Sure, you can go with less than 256 colors, but you cannot have over 256. That's why a GIF image doesn't work overly well for photographs and larger images. There are a whole lot more than 256 colors in the world. Images can carry millions. But if you want smaller icon images, GIFs are the way to go.

Raster image formats can also save at 16, 24, and 32 bits-per-pixel. At the two highest levels, the pixels themselves can carry up to

16,777,216 different colors. The image looks great! Bitmaps saved at 24 bits-per-pixel are great quality images, but of course they also run about a megabyte per picture. There's always a trade-off, isn't there?

One of the GIF big advantages is images can be indexed to a set palette, whereas JPEG images can't. This is important to Web designers who can use indexing to more easily optimize images for browser-safe palettes (browser-safe palettes display 216 colors, which includes all the colors in the Windows and Macintosh system palette). Designing for a browser-safe palette ensures consistent, cross-platform, device-independent display. GIF images use a lossless compression scheme, meaning no data is lost in the compression or decompression process.

The GIF format was invented (and is owned) by CompuServe, the online service providers. Like the windows. BMP format, it stores the image as a rectangular grid of pixels, specifying the color of each pixel in the image. A GIF file is much smaller than the equivalent BMP file because it uses special compression to group together patches of continuous color. GIF files are good for drawn graphics and text, because they produce a small file and bold, clear lines. They can only have a maximum of 256 colours so photographs don't come out as well. You can produce a GIF file with a transparent background, which looks nice if your page has a background image.

If you find that your GIF files are too large, consider using the new PNG format. GIF files perform compression by analysing horizontal bands of the same color: PNG files, on the other hand, can analyse both horizontal and vertical areas. Consequently, they can achieve compression levels up to 30 per cent better than GIFs.

The oldest version of the format is called GIF 87a. This is the "basic" version of the format. The more advanced version is called GIF89a. This extension allows you to specify one color in the GIF to be transparent, to attach comments to a file, to produce multi-image ("animated") files, and to produce an image in what's referred to as "interlaced" format.

A normal GIF file specifies the image one line at a time, starting at the top of the image and working downwards. An interlaced GIF also specifies one line at a time, but puts the lines in a different order in the file, so that the whole image has at least some information from the beginning.

The effect of this becomes clear when you are downloading the image over the Internet. A non-interlaced GIF files from the topdown, as you would expect. An interlaced GIF appears initially out of focus-a bit like those bits on Crimewatch where they're disguising someone's face—and the image then gradually comes into focus.

2.2.4 JPEG : Better For Photographs

JPEG stands for Joint Photographic Experts Group, the committee that originally created this bitmap graphics format. JPEG uses a lossy compression system. JPEG images are 24-bit, 16.7 million colors, provide the ability to specify the degree of compression, and can be displayed in stages, called a "progressive JPEG".

The JPEG format is based on a lossy compression algorithm, which means your image will lose some quality when it's compressed. However, depending on the quality of your original and the amount of compression you apply to the image, the data loss is often imperceptible to the human eye. Whereas GIF is only 8-bit, JPEG images are 24-bit, capable of displaying 16 million colors.

The JPEG format was designed to render photographs, and other images intended to look realistic to humans. It uses various psychological tricks to suggest detail which is not actually there the savings from this process can result in very small images indeed.

When you convert a file to JPEG format, most applications will ask for an "accuracy" value (some describe this as a "compression ratio"). If you choose a high accuracy, then you will retain all the detail in your image, but the resulting file will be very big—were you to choose 100 per cent accuracy the image would be identical to an equivalent TIFF file, and would be just as big. If you are willing to go for a low accuracy, you can get a very small image size (images less than 100 bytes in size are possible), but you lose detail until eventually the image is almost unrecognisable.

Losing detail from an image doesn't sound like a very good idea, but for photographs it often doesn't matter—some people even say that a photo or a painting looks nicer on a lower accuracy setting since you are less aware of the pixels of the screen.

A GIF file can only have a maximum of 256 colors, which is fine for line art, but can make a photo look like a painting-by-numbers picture. All JPEG files are saved off in 24-bit format, allowing 16 million colors. Of course, unless you have a 24-bit graphics card, your PC can't display all of them at once, but reducing the number of colors at the graphics-card stage will produce a more realistic image. And, of course those of your readers who do have 24-bit colors will see your pictures in all their glory.

2.2.5 PNG: Using The Best of the Two Others

PNG (Portable Network Graphics) is a bitmap graphics format that has been approved as a standard by the W3C. PNG offers up to 256 indexed colors, support for 24-and 48-bit true color, full alpha-channel support for transparency, and typically 10 per cent to 30 per cent better lossless compression than GIF.

2.2.6 TIFF

Tagged-Image File Format (TIFF) is used to exchange files between applications and computer platforms. TIFF is a flexible bitmap image format supported by virtually all paint, image-editing, and page-layout applications. Also, virtually all desktop scanners can produce TIFF images.

TIFF format supports CMYK, RGB, Lab, indexed-color, and grayscale images with alpha-channels and Bitmap-mode images without alpha-channels. Photoshop can save layers in a TIFF file; however, if you open the file in another application, only the flattened image is visible. Photoshop can also save annotations, transparency, and multiresolution pyramid data in TIFF format.

2.2.7 PDF

Portable Document Format (PDF) is a flexible, cross-platform, cross-application file format. Based on the Postscript imaging model, PDF files accurately display and preserve fonts, page layouts, and both vector and bitmap graphics. In addition, PDF files can contain electronic document search and navigation features such as electronic links.

Generic PDF files are created using applications other than Photoshop, such as Adobe Acrobat and Adobe Illustrator, and can contain multiple pages and images. When you open a Generic PDF file, Photoshop rasterizes the image. Photoshop and Image Ready recognize two types of PDF files: Photoshop PDF files and Generic PDF files. You can open both types of PDF files; however, you can only save images to Photoshop PDF format.

Photoshop PDF files are created using the Photoshop Save As command. Photoshop PDF files can contain only a single image.

Photoshop PDF format supports all of the color modes and features that are supported in standard Photoshop format. Photoshop PDF also supports JPEG and ZIP compression, except for Bitmap-mode images, which use CCITT Group 4 compression.

2.2.8 PSD Photoshop format

Photoshop format (PSD) is the default file format for newly created images—and the only format supporting all available image modes (Bitmap, Grayscale, Duotone, Indexed Color, RGB, CMYK, Lab, and Multichannel), guides, alpha-channels, spot channels, and layers (including adjustment layers, type layers, and layer effects).

2.3 VECTOR GRAPHICS

Vector graphics (also known as object-oriented graphics), maintains the image as a series of points, lines, arcs and other geometric shapes. Vector graphics are made up of lines and curves defined by mathematical objects called vectors. Vectors describe an image according to its geometric characteristics. For example, a bicycle tire in a vector graphic is made up of a mathematical definition of a circle drawn with a certain radius, set at a specific location, and filled with a specific color. You can move, resize, or change the color of the tire without losing the quality of the graphic.

Vector graphics are resolution-independent—that is, they can be scaled to any size and printed at any resolution without losing detail or clarity. As a result, vector graphics are the best choice for representing bold graphics that must retain crisp lines when scaled to various sizes—for example, logos.

Fig. 7.3: Bicycle Vector

Vector graphics are good for reproducing crisp outlines, as in logos or illustrations. They can be printed or displayed at any resolution without losing detail.

Because computer monitors represent images by displaying them on a grid, both vector and bitmap images are displayed as pixels on-screen.

Vector graphics images are typically produced by programs such as Abode Illustrator or Corel's CorelDraw! are created with curves and other mathematical shapes instead of pixels. Because a mathematical curve always stays smooth no matter how closely you look at it, vector graphics always point at the highest smoothness level, and the graphic's the size remains small. For print, the most common type of vector graphic language is Postscripts, which is a cross-platform standard for manipulating, storing, and printing vector graphics, including fonts. Programs that work with vector graphics are usually referred to as illustration or draw programs.

Drawing programs such as Adobe Illustrator create vector graphics, made of lines and curves defined by mathematical objects called vectors. Vectors describe graphics according to their geometric characteristics. For example, a bicycle tire in a vector graphic is made up of a mathematical definition of a circle drawn with a certain radius, set at a specific location, and filled with a specific color. You can move, resize, or change the color of the tire without losing the quality of the graphic.

Most multimedia authoring systems provide for use of vector-drawn objects such as lines, rectangles, ovals, polygons, and text.

- Computer-aided design (CAD) programs have traditionally used vector-drawn object systems for creating the highly complex and geometric renderings needed by architects and engineers.
- Graphic artists designing for print media use vector-drawn objects because the same mathematics that put a rectangle on your screen can also place that rectangle (or the fancy Bezier curves of a good line-art illustration) on paper without jaggies. This requires the highest resolution of the printer, using a page description languages such as Postscript.
- Programs for three-dimensional (3-D) animation also use vector-drawn graphics. For example, the various changes of position, rotation, and shading of light required to spin the extruded corporate logo shown in following figure must be calculated mathematically,

A few of the many still images required to "spin" a logo.

2.3.1 Vector Formats: PGML, VML, SWF

PGML, SWF and VML are all vector formats, which describe graphics in terms of mathematical formulas rather than pixels. Vectors are typically much smaller than bitmap images, and provide the ability to scale and zoom in on images without any loss in resolution or clarity. Both PGML and VML have been submitted for consideration by the W3C. Both standards are based on XML (Extensible Markup Language (XML).

In the past few months, several vector standards for the Web have been proposed. Under consideration by the World Wide Web Consortium (W3C), the main standards body for the Web, are the vying formats of PGML and VML. Macromedia, which backs the VML proposal, has recently opened up its vector Flash format to other developers.

3.1 DIGITAL AUDIO FORMATS

Digital audio comes in many different formats, and multiple formats will be a fact of life for the foreseeable future. Groups like MPEG have created open standards, but even formats based on the same MPEG standard may not compatible with each other because of proprietary components.

Fortunately for consumers, many hardware and software players are able to support multiple formats—so if you purchase digital music in any of the major formats (MP3, WMA, etc.) you will be in a good shape. If a format does become obsolete, plenty of tools are available for converting digital audio to different formats.

3.2 DIGITAL AUDIO FILES

An audio file has two main parts: a header and the audio data. The header is used to store information about the file, including the resolution, sampling rate and type of compression. Often a "wrapper" is used to add features, such as license management information or streaming capability, to a digital audio file.

The format of a digital audio file refers to the type of audio data within the file. The file type refers to the structure of the data within the file. It is common for the same format to be used by more than one file type. For example, the PCM format is found in both WAV and AIFF files.

Table 7.1: Common Digital Audio Formats

Type	*Extensions*	*Codec*
AIFF (Mac)	.aif, .aiff	PCM
AU (Sun/Next)	.au	u-law
CD audio	(CDDA) N/A	PCM
MP3 .mp3	MPEG	Audio Layer-III
Windows Media Audio	.wma Proprietary	(Microsoft)
QuickTime	.qt	Proprietary (Apple Computer)
RealAudio	.ra, ram	Proprietary (Real Networks)
WAV	.wav	*PCM

• Can be used with other codecs.

3.3 WAV

WAV (or WAVE), short for WAVE form audio format, is a Microsoft and IBM audio file format standard for storing audio on PCs. It is a variant of the RIFF bitstream format method for storing data in "chunks", and thus also close to the IFF and the AIFF format used on Macintosh computers. It takes into account some peculiarities of the Intel CPU such as little-endian byte order. The RIFF format acts as a "wrapper" for various audio compression codecs. It is the main format used on Windows systems for raw audio.

The Microsoft Windows WAV format is used in the Windows 3.1, 95, 98 and NT operating systems for recording and playback of recorded sound. The advantage of placing a "WAV" file on your computer is the ease of making the file on your part—a recorder/player for this type of file is already located on your users machine (if it is a Windows computer). The disadvantages are that many people on the Internet will be unable to play the file if they are using another operating system and the fact that WAV files are typically larger, taking longer for the end-user to download.

This file format is widely supported by windows application. It was developed by Microsoft as a subset of the RIFF format. It is capable of sampling rates of 8 and 16 bits (mono and stereo). Though a WAV file can hold audio compressed with any codec, by far the most common format is pulse-code modulation audio data. Since PCM uses an uncompressed, lossless storage method which keeps all the samples of an audio track, professional users or audio experts may use the WAV format for maximum audio quality. WAV audio can also be edited and manipulated with relative ease using software.

WAV files can be used as an intermediary storage type for ripping songs from cassette tapes. People using Windows' Sound Recorder can redirect the output from their old Walkman-type cassette players' ear phone jacks to the PC audio input jacks via an audio cable. Although Sound Recorder can only record for 60 seconds, there are easy ways to lengthen the file for long songs or even the entire tape (record something, choose decrease speed from the effects menu, repeat until file is sufficiently long, and then record over it from the beginning). The WAV file is then converted via freeware packages to other, more compact formats.

As file sharing over the Internet has become popular, the WAV format has declined in popularity, primarily because uncompressed WAV files are quite large. More frequently, compressed but lossy formats such as MP3, Ogg Vorbis and Advanced Audio Coding are used to store and transfer audio, since their smaller file sizes allow for faster transfers over the Internet, and large collections of files consume only a conservative amount of disk space. There are also more efficient, lossless codecs available, such as Monkey's Audio, TTA, WavPack, FLAC, Shorten, Apple Lossless and WMA Lossless.

The WAV format is limited to files that are less than 2 GiB in size, due to the way its 32-bit file size header is read by most programs. Although this is equivalent to more than 3 hours of CD-quality audio (44.1 kHz, 16-bit stereo), it is sometimes necessary to go over this limit. The W64 format was created for use in Sound Forge. This format can be converted using the LGPL libsndfile library.

Contrary to the popular misconception, audio CDs do not use WAV as their storage format. The commonality is that both audio CDs and WAV files have the audio data encoded in PCM, but WAV is a data file format for computer use; if you rip an audio CD to WAV files and burn them onto a CD-R as a data disc (ISO) and insert it into a player that can handle only audio CDs, the CD will not play.

3.4 MUSICAL INSTRUMENT DIGITAL INTERFACE (MIDI)

MIDI is a communications protocol—which is a standard set of rules—for communication between two or more musical instruments or musical instruments and computers. Yes, you read that last sentence right—you can hook instruments to computers (more on this in a minute). MIDI is a software program, or sequencer, enables you to make the recording and then view and edit the score of the performance on your computer. The beauty of a MIDI recording is that you can use it to correct individual notes, change the sound, or even transpose the piece. These editing capabilities alleviate the tedium of recording and rerecording until you hit a perfect performance.

MIDI represents a protocol that sends musical data from one MIDI component (such as a keyboard) to another, or to a computer that records the performance. MIDI information only consists of performance data, which includes the instructions for creating music. It communicates things such as turning "on" and "off" notes, a note's velocity, program changes, damper pedal use, and event timing.

Fig. 7.4: Synthesizer

If you only need to see the notes or the score of your composition, a MIDI recording is all you need. With it, you can hear a playback of your performance through your sound card or instrument, but no actual recording of the sound occurs. In order to record the music so that others can hear it on CD, you must make an audio recording, as well. The audio recording consists of the actual analog or digital audio signal. These signals do not pass through the MIDI cable that carries MIDI data from one component to another or through the universal sound card MIDI cable that carries data from the instrument to the computer.

Some Microsoft Windows programs contain MIDI files. (Microsoft Corp's Microsoft Windows is a graphical user interface, or GUI. It allows users to choose icons—pictures—that represent commands and to choose commands from pull-down menus instead of having to give them manually.) According to Gideon Rosenblatt, Microsoft Corp. multimedia publishing product manager, "We use MIDI files in Microsoft Bookshelf to allow us to include the national anthems of over 160 countries".

In order to play MIDI files, you don't even need a keyboard as long as you have a sound card (an add-in circuit board that you install inside your computer's case to increase its sound capabilities) that supports MIDI. This is because sound cards contain small synthesizer chips on them.

Windows actually includes a MIDI player. So if you have your MIDI files and a sound card, you've got what you need to play MIDI files.

The keyboard may be one of the most expensive components of your home recording studio, so you'll want to give careful consideration to your purchase. If you don't own a keyboard synthesizer and aren't sure if you're ready to spend the money for one, consider purchasing a keyboard controller instead. Unlike a keyboard that has an onboard synthesizer, the keyboard controller has no internal sounds. It provides only the keyboard that you connect to your computer and play to transmit the MIDI data of the performance. Expect to pay roughly $ 200 for a controller as opposed to more than $ 500 for a keyboard that has its own internal sound synthesizer.

Advantages and Disadvantages of MIDI over Digital Sound

MIDI has several advantages over digital audio and two huge disadvantages. First, the advantages:

- MIDI files are much more compact than digital audio files, and the size of a MIDI file is completely independent of playback quality. In general, MIDI files will be 200 to 1,000 times smaller than CD-quality digital audio files. Because MIDI files are small, they don't take up as much RAM, disk space, and CPU resources.
- In some cases, MIDI files may sound better than digital audio files at the MIDI sound source you are using is of high quality.
- You can change the length of a MIDI file (by varying its tempo) without changing the pitch of the music or degrading the audio quality. MIDI data is completely editable-right down to the level of an individual note. You can manipulate the smallest detail of a MIDI composition (often with submillisecond accuracy) in ways that are impossible with digital audio.

Disadvantages

- Because MIDI does isn't sound, you can be certain that playback will be accurate only if the MIDI playback device is identical to the device used for production. Even with the General MIDI standard the sound of a MIDI instrument varies according to the electronics of the playback device and the sound generation method it uses.
- MIDI cannot easily be used to play back spoken dialog, although expensive and technically tricky digital samplers are available.

3.5 MP3

MP3 is a popular digital audio encoding and lossy compression format invented and standardised in 1991 by a team of engineers working in the framework of the ISO/IEC MPEG audio committee under the chairmanship of Professor Hans Musmann (University of Hannover - Germany). It was designed to greatly reduce the amount of data required to represent audio, yet still sound like a faithful reproduction of the original uncompressed audio to most listeners. In popular usage, MP3 also refers to files of sound or music recordings stored in the MP3 format on computers.

MP3 (MPEG Audio) — MPEG audio is a standard for high-quality audio and video files that has gained widespread use. MPEG (Motion Picture Experts Group) is a family of compression formats and audio/video storage formats developed by a cooperative effort under the joint direction of the International Standards Organization (ISO) and the International Electro-technical Commission (IEC). There are many companies that offer free players capable of playing or creating MPEG files. It offers the advantage of high quality with a smaller file size.

CD-quality MP3 sound files can still be large, however, for use on a Web page that may be downloaded through a typical modem connection. The end-user will have to have an audio player on their computer capable of handling the format. MP3 has found a receptive audience among musicians on the Net, who offer samples of their works available at sites such as the Internet Underground Music Archives. It has been used, however, to encode and transmit copyrighted music extensively by individual Internet users and the music industry has been lobbying for standards in the format that will prevent illegal copying of copyrighted works.

The name is derived from "MPEG-1 Audio Layer 3" more formally known as ISO/IEC 11172-3 Layer 3. The files recorded in this format are saved with the .mp3 filename extension. This extension is also sometimes shared by audio files encoded using the newer MPEG-2 Audio Layer 3 standard. The invention of this file extension name (.mp3) would be due to the FhG institute.

MP3 is a lossy compression format. It provides a representation of pulse-code modulation-encoded (PCM) audio data in a much smaller size by discarding portions that are considered less important to human hearing (similar to JPEG, a lossy compression for images).

A number of techniques are employed in MP3 to determine which portions of the audio can be discarded, including psychoacoustics. MP3 audio can be compressed with different bit rates, providing a range of tradeoffs between data size and sound quality.

The MP3 format uses, at its heart, a hybrid transform to transform a time domain signal into a frequency domain signal:

- 32 band polyphase quadrature filter
- 36 or 12 tap MDCT; size can be selected independent for sub-band 0...1 and 2...31
- aliasing reduction postprocessing.

MP3 Surround, a version of the format supporting 5.1 channels for surround sound, was introduced in December 2004. MP3 Surround is backward compatible with standard stereo MP3, and file sizes are similar.

In terms of the MPEG specifications, AAC (Advanced Audio Coding) from MPEG-4 is to be the successor of the MP3 format, although there has been a significant movement to create and popularize other audio formats. Nevertheless, any succession is not likely to happen for a significant amount of time due to MP3's overwhelming popularity (MP3 enjoys extremely wide popularity and support, not just by end-users and software but by hardware such as DVD and CD players).

3.5.1 Quality of MP3 Audio

Because MP3 is a lossy format, it is able to provide a number of different options for its "bit rate" — that is, the number of bits of encoded data that are used to represent each second of audio. Typically rates chosen are between 128 and 256 kilobit per second. By contrast, uncompressed compact disc audio has a bit rate of 1378 kbit/s or 1411 kbit/s.

MP3 files encoded with a lower bit rate will generally play back at a lower quality. With too low a bit rate, "compression artifacts" (*i.e.*, sounds that were not present in the original recording) may appear in the reproduction. A good demonstration of compression artifacts is provided by the sound of applause: it is hard to compress because it is random, therefore the failings of the encoder are more obvious, and are audible as ringing.

As well as the bit rate of the encoded file, the quality of MP3 files depend on the quality of the encoder and the difficulty of the signal being encoded. For average signals with good encoders, many listeners accept the MP3 bit rate of 128 kibit/s as near enough to compact disc quality for them, providing a compression ratio of approximately 11:1. However, listening tests show that with a bit of practice many listeners can reliably distinguish 128 kibit/s MP3s from CD originals; in many cases reaching the point where they consider the MP3 audio to be of unacceptably low quality. Yet other listeners, and the same listeners in other environments (such as in a noisy moving vehicle or at a party) will consider the quality acceptable.

3.5.2 Bit Rate

The bit rate is variable for MP3 files. The general rule is that more information is included from the original sound file when a higher bit rate is used, and thus the higher the quality during playback. In the early days of MP3 encoding, a fixed bit rate was used for the entire file.

Bit rates available in MPEG-1 Layer 3 are 32, 40, 48, 56, 64, 80, 96, 112, 128, 160, 192, 224, 256 and 320 kibit/s and the available sample frequencies are 32, 44.1 and 48 kHz. 44.1 kHz is almost always used (coincides with the sampling rate of compact discs), and 128 Kbit/s has become the de facto "good enough" standard, although 192 Kibit/s is becoming increasingly popular over peer-to-peer file sharing networks. MPEG-2 and [the non-official] MPEG-2.5 includes some additional bit rates: 8, 16, 24, 32, 40, 48, 56, 64, 80, 96, 112, 128, 144, 160 Kibit/s

Variable bit rates (VBR) are also possible. Audio in MP3 files are divided into frames (which have their own bit rate) so it is possible to change the bit rate dynamically as the file is encoded (although not originally implemented, VBR is in extensive use today). This technique makes it possible to use more bits for parts of the sound with higher dynamics (more sound movement) and fewer bits for parts with lower dynamics, further increasing quality and decreasing storage space. This method compares to a sound activated tape recorder which reduces tape consumption by not recording silence. Some encoders utilize this technique to a great extent.

3.5.3 Design limitations of MP3

There are several limitations inherent to the MP3 format that cannot be overcome by using a better encoder.

Newer audio compression formats such as Vorbis and AAC no longer have these limitations.

In technical terms, MP3 is limited in the following ways:

- Bit rate is limited to a maximum of 320 kibit/s
- Time resolution can be too low for highly transient signals
- Encoder/decoder overall delay is not defined
- No scale factor band for frequencies above 15.5/15.8 kHz
- Joint stereo is done on a frame-to-frame basis
- Lack of gapless playback; tracks intendended to flow into another (such as in live performances) do not seamlessly flow: a short gap of silence is heard between the two tracks.
- Nevertheless, a well tuned MP3 encoder can perform competitively even with these restrictions.

3.5.4 Decoding of MP3 audio

Decoding, on the other hand, is carefully defined in the standard. Most decoders are "bitstream compliant", meaning that the uncompressed output they produce from a given MP3 file will be the same (within a specified degree of rounding tolerance) as the output specified mathematically in the standard document. The MP3 file has a standard format which is a frame consisting of 576 samples and all the frames have associated header information (32 bits) and side information(17 bytes). The header and side information help the decoder to decode the associated huffman encoded data correctly.

Therefore, for the most part, comparison of decoders is almost exclusively based on how computationally efficient they are (*i.e.*, how much memory or CPU time they use in the decoding process).

Though proponents of newer codecs such as WMA and RealAudio have asserted that their respective algorithms can achieve CD quality at 64 kbit/s, listening tests have shown otherwise; however, the quality of these codecs at 64 kbit/s is definitely superior to MP3 at the same bandwidth. The developers of the patent-free Ogg Vorbis codec claim that their algorithm surpasses MP3, RealAudio and WMA sound quality, and the listening tests mentioned above support that claim. Thomson claims that its mp3PRO codec achieves CD quality at 64 kbit/s, but listeners have reported that a 64 kbit/s mp3PRO file compares in quality to a 112 kbit/s MP3 file and does not come reasonably close to CD quality until about 80 kbit/s.

3.5 AIFF AND AU

AIFF is the default audio format for the Macintosh, and AU is the default format for SUN systems. Both of these formats are supported on most other platforms and by most audio applications. Each of these formats can be compressed, but compression sometimes creates compatibility problems with other platforms.

3.6 STREAMING AUDIO

Streaming audio avoids many of the problems of large audio files. Instead of having to wait for the entire file to download, you can listen to the sound as the data arrives at your computer. Streaming audio players store several seconds worth of data in a buffer before beginning playback. The buffer absorbs the bursts of data as they are delivered by the Internet and releases it at a constant rate for smooth playback. Many digital audio formats can be streamed by wrapping them in a streaming format, such as Microsoft 's ASF (Active Streaming Format), which can be used to stream MS Audio, MP3 and other formats.

Table 7.2: Streaming Audio Systems

Type	*Primary Format*	*Developer*
Windows Media Technologies	Windows Media Audio	Active Streaming Format (ASF)
Icecast (open source)	MP3	The Icecast Team
QuickTime	QuickTime	Apple Computer
RealSystem	RealAudio	RealNetworks
SHOUTcast	MP3	Nullsoft

4.1 ANIMATION, VIDEO AND DIGITAL MOVIES

Animations and digital video movies are sequences of bitmapped graphic scenes (frames), rapidly played back. But animations can also be made within the authoring system by rapidly changing the location of objects or sprites to generate an appearance of motion. Most authoring tools adopt either a frame—or object-oriented approach to animation, but rarely both.

Movie-making tools take advantage of QuickTike (Macintosh) and Microsoft Video for Windows (also known as AVI, or Audio Video Interleaved) technology and let you create, edit, and present digitized motion video segments, usually in a small window in your project.

To make movies form video you need special hardware to convert the analog video signal to digital data. Movie-making tools such as Premiere, VideoShop, and MediaStudio Pro let you edit and assemble video clips captured from camera, tape, other digitized movie segments, animation, scanned images, and from digitized audio or MIDI files. The completed clip, often with added transition and visual effects, can then be played back-either stand-alone or windowed within your project.

Morphing is an animation technique that allows you do dynamically blend two still images, creating a sequence of in-between pictures that, when played back rapidly in QuickTime, metaporphoses the first image into the second. A racing car transforms itself into a tiger; a mother's face becomes her daughter's.

4.2 DIGITAL VIDEO: VIDEO RECORDING IN DIGITAL FORM

In order to edit video in the computer or to embed video clips into multimedia documents, a video source must originate as digital (digital camera) or be converted to digital. Frames from analog video cameras and VCRs are converted into digital frames (bitmaps) using frame grabbers or similar devices attached to a personal computer. Uncompressed digital video signals require huge amounts of storage, and high-ratio realtime compression schemes, such as MPEG, are essential for handling digital video in today's computers.

4.3 FULL-MOTION VIDEO

For the first few years of its existence, the PC was a text-only device. As users' needs increased and the technology improved, graphics became commonplace on the computer screen. Graphics are now viewed as a prerequisite for virtually all PCs, and our insatiable appetite for PCs that can do more and more has led us into the age of so-called "multimedia". Most people use PCs today that can show not just graphics but sound, animation and full-motion video as well.

The term "multimedia" is, unfortunately, one of those buzzwords that means different things depending on who says it and what they are talking about. It generally refers to the use of software that incorporates elements of sound, graphics and moving video (animation or recorded). Graphics have been around for years and so has sound. What is really new on the PC, relatively speaking, is full-motion video. This section briefly takes a look at the topic, concentrating on the new technologies involved in delivering full motion video to the PC desktop.

Full motion video is being seen increasingly on PCs as they expand their horizons to fill roles they never have before. Some of the areas where full-motion video is increasingly being seen:

Games: Many games today use video to add realism to the gaming experience. In particular, many games use video to make them more "cinematic"; these games are often a cross between a movie and a classical computer game. They are sometimes called "interactive movies" in fact.

4.4 VIDEO PERFORMANCE ISSUES

The idea of playing a movie or TV program on the PC is a relatively new one, and these applications make demands on the performance of a PC system that few others do. The main problem is the sheer volume of data that is involved.

Let's suppose that we wanted to display a 2-hour movie in high-quality, high-resolution graphics. To keep things from getting out of hand, let's choose just 640 × 480 resolution in high color (16 bits per pixel) instead of going "all out" to say, 1024 × 768 in true color. Full-motion video typically means 30 frames per second. To store the data required for a full-motion, two-hour movie using these parameters, would require approximately 133 GB. (That's not a typo—133 billion bytes, and that doesn't include sound.) Also, the data to support this video stream would need to be pumped to the video card at the rate of about 150 million bits per second. In ten years these numbers won't seem like that big of a deal but of course, right now they sure are.

There is a solution to this dilemma: data compression. Due to the nature of the data, video can be easily compressed by a large factor. There are several reasons for this:

Every image that is displayed has large areas of redundancy that can be represented by a smaller amount of data using an encoding algorithm. For example, if the bottom half of the screen in a particular frame is black, that can be represented using a small number of bytes; we don't need to have 150,000 2-byte pixels all filled with zeroes.

The human eye is not sensitive to certain details in a video, which can be removed without appreciable loss to the perceived signal.

If you take any two consecutive frames, the changes from one to the next are usually rather small. For example, if you have a scene centered on someone's face, the background images are probably going to remain static for hundreds or thousands of consecutive frames. There are special algorithms that can describe a frame only by the changes that it represents from the one before, which dramatically cuts down both on the amount of storage required and the time to transmit each frame of data.

4.5 MPEG

The most popular compressed video format is MPEG, which stands for the Motion Pictures Expert Group, the body that defined the standard. MPEG can typically compress as high as 100 to 1, bringing out 2-hour movie down to a much more manageable 1.33 GB. To use MPEG the video must be encoded using an MPEG encoder, and then viewed using an MPEG decoder. Not surprisingly, MPEG has its own problems. The chief one is that the process of encoding and decoding takes a lot of processing power, which is why dedicated MPEG encoding and decoding hardware is used for high-performance applications.

There are several ways that MPEG (or other video) files can be played on a PC. In every case, the process involves reading the file, decompressing the video data, and then feeding it to the video card for display on the monitor. Since the data is compressed to reduce storage, it must be uncompressed to restore it to a visible image. This takes a great deal of processing power. There are software programs that can be used to decode these files, allowing you to play an MPEG file on your computer. As with every other case where an application takes a lot of processing power, it is more efficient (but more costly) to let dedicated hardware do the function. This is the purpose of an MPEG Decoder.

Much as is the case with 3D accelerators, it is possible to get a dedicated card for MPEG decoding, or to get a regular video card that has an MPEG decoder built into it. Dedicated MPEG decoders come with different capabilities. They often use the video feature connector <if_Feature.htm> on the regular video card, to supply video information directly to the card without tying up the system but with an enormous amount of video data.

Warning: Beware of some video cards that advertise that they "support MPEG decoding" when all they mean is that they will work with a software MPEG decoder. Many of these cards aren't doing hardware decoding at all. They still may provide acceptable performance, but make sure you know what you are buying.

The quality of MPEG decoders varies, with generally the more expensive ones providing better quality. A high-quality decoder will allow the display of full-screen, smooth video animation while leaving the system CPU free to do other tasks. A cheaper hardware decoding solution, or a software package, may have problems keeping the supply of data to the video card which can result in choppy performance or lost frames. In some cases, this is compensated for by running the video in a small window instead of full-screen, and in others by slowing the frame rate down to less than 30 frames per second. Others reduce the resolution of the screen.

4.6 CODECS

The basic engine that drives multimedia transmission, particularly in the videoconferencing industry, is the codec (coder-decoder), which is the video counterpart of a modem. Codecs take analog audio and video signals from video cameras and other sources, digitize them, and compress the digital values for transmission at lower bandwidth. At the receiving end codecs decompress the digitized signals, convert them back to analog, and transmit them to audio and video output devices. Current codecs operate with 384-Kbps lines, which represent a compression rate of 230:1, and some codecs also operate with 112-Kbps circuits, which means compression rates of 800:1.

The highest-quality codecs offer more than 175,000 pixels-per-frame resolution, which is a measure of video quality that is so critical in multimedia video applications. This is close to top-quality-broadcast TV video quality, which ranges from 190,000 to 200,000 pixels per frame.

The quality of video images also depends on how often the screen is refreshed; the standard TV video is 30 fps. Codecs reduce the frame rate to 15, 10 or even 5 fps depending on the amount of motion, but most codec vendors claim a maximum potential of 30 fps. In general users cannot perceive a difference in video quality until the frame rate falls to about 18 to 16 fps. At rates typically offered over the 112—and 128-Kbps circuits, the frame rates are in the order of 10 to

15 fps, and some deterioration of video quality is noticeable when fast and continuous motion is involved. Pre—and post-processing of the data and motion compensation techniques further reduce the amount of data handled by codecs, and these methods are key in providing acceptable video quality in 112 and 128-Kbps facilities, which are also available at lower tariffs than those with higher bandwith.9. What can be accomplished using a photoediting program?

4.7 MOVIE PLAYERS

With QuickTime Movie Player and with the Windows Media Player (with AVI installed), you can view and edit movies. You can play a movie forward or backward, and resize it. You can also cut and copy frames from one movie and paste them into another.

4.8 STANDARD FORMATS

Standard formats make it easier for software developers and equipment manufacturers to produce products that are less costly and more compatible with each other. The compatibility provided by standard formats helps assure consumers that their music and equipment won't become obsolete. Cassette tapes, compact discs and PCM are examples of standard audio formats that benefit both consumers and manufacturers.

Following are the three most popular video— cum—audio formats:

- MPEG
- AVI
- Quick time

4.9 MPEG AUDIO

MPEG Audio is a family of open standards for compressed audio that includes MP2, MP3 and AAC.

4.9.1 MPEG-based Proprietary Formats

Several proprietary formats are based on MPEG audio. Some of these are used in special applications, such as voice mail systems, high definition TV and satellite radio. Others compete directly with MP3 and are based on AAC or MP3, with proprietary wrappers. The sound quality of some of these is very good, but their proprietary nature makes them incompatible with many programs and portable players.

4.9.2 Proprietary Formats

Even though MPEG Audio formats are based on open standards and widely used, many companies continue to develop proprietary digital audio formats. Proprietary formats can generate enormous profits for the developer if the format becomes a de facto standard. The marketplace, which tends to favor open standards like MPEG, will ultimately decide which formats will prevail.

4.9.3 RealAudio

RealAudio was the first widely used system for streaming audio and video over the Internet. It is a proprietary format, but it is used by many online music stores for sample clips of songs. The RealPlayer also provides support for MP3.

4.9.4 A Concise History of MP3

The technical term is MPEG-1, Audio Layer 3. First developed by the Fraunhofer Institute fur Integrated Schaltungen in Germany back in 1987, MP3 was originally patented in 1989 and incorporated into the Moving Picture Experts Group (MPEG) specifications in 1992. The MP3 standard, or codec, enabled high—quality yet low bit rate compression of digital audio files. In other words, audio files, which were typically large (such as WAV files from CDs), could be compressed to a fraction of their original size without a significantly noticeable loss in quality to the human ear.

However, the power of MP3 did not begin to be harnessed until developers created encoding, decoding, and playing tools for the standard. The original patent holder, Fraunhofer IIS-A, made the technology free for use and development. Therefore, a number of independent developers and companies were working simultaneously to provide MP3 tools. Timing was everything. Just as home PC processing power was exploding and the Internet was taking off in 1997, researcher Tomislav Uzelac introduced the first desktop MP3 player known as AMP. AMP became the precursor to Winamp and later MacAmp, the first two widely available MP3 player applications for consumers. Soon, companies such as RealNetworks and MusicMatch jumped on the bandwagon. In May of 1999, RealNetworks introduced RealJukebox, its CD-ripping, encoding, file-management, and player application.

With the advent of faster computers comes animation. Though it has been around for years the modern computer has made it possible to include animation in programs without causing them to slow down (much). As with every multimedia format there are a number of types.

You may have seen .gif animations on this website. A GIF animation is a series of separate images or frames that display one after the other to give the impression of movement. Other formats are Audio Visual Interleave's .avi, the aforementioned mpg, Microsoft's Media Player .wmv, Apple's Quick Time .qt, .aif(*) & .mov, RealNetwork's RealVideo .rm(*), Macromedia's Flash creates Shockwave .swf, and JavaScript as well as various others.

There are various animation or multimedia players available for a free download off the Internet.

To create animations, sounds or graphics you will need a program that has the capabilities you want. Visit the various multimedia company websites to read up on their product to see if they can do what you want. (**Hint:** to find a company website type "www.' replace this with the companies name'.com"). Most companies offer free trials that you can download from their website.

You should also be aware that most media content placed on the Internet is considered published material and therefore copyright unless explicitly stated otherwise.

4.10 AUDIO VIDEO INTERLEAVED (AVI)

AVI is a multimedia container format introduced by Microsoft in November 1992, as part of the Video for Windows technology. .avi files contain both audio and video data in a standard container that allows simultaneous playback. Most .avi files also use the file format extensions developed by the Matrox OpenDML group in February 1996. These files are supported by Microsoft, and are known unofficially as "AVI 2.0".

Audio Video Interleaved (AVI) is Microsoft-developed software that plays full-motion interleaved video and audio sequences in Windows, without specialized hardware, at about 15 frames per second in a small window. With acceleration hardware, you can run AVI video sequences at 30 frames per second. Video data are interleaved with audio within the file that contains the motion sequence, so the audio portion of the movie remains synchronized to the video portion.

Video for Windows consists of a set of tools based on the AVI file format, together with an ad-on module for Windows 3.1 and a CD-

ROM. It is designed to enable developers to capture and digitise video sequences using an adapter board, edit them and use OLE to incorporate them in a wide range of applications. Users can play the same sequences back, using hardware codecs or runtime software decompression, on all versions of Windows, although the quality will depend upon the hardware. Three software codecs are included in the package: Microsoft's Video and RLE together with Indeo, Intel's video recording and playback technology based on DVI.

Like Apple's QuickTime, AVI provides the following features:

- Playback from hard disk or CD-ROM.
- Playback on computers with limited memory; data are streamed from the hard disk or CD-ROM player without using great amounts of memory.
- Quick loading and playing, because only a few frames of video and a portion of audio are accessed at a time.
- Video compression to boost the quality of your video sequence and reduce their size.

AVI includes two tools to capture and edit video sequences and play them back: VidCap and VidEdit, respectively. AVI also includes data preparation rools (BitEdit, and WaveEdit); MCIAVI.DRV, the MCI driver for AVI; MediaPlayer; and sample video sequences.

It is a special case of the Resource Interchange File Format (RIFF), which divides the file's data up into data blocks called "chunks". Each "chunk" is identified by a FourCC tag. An .avi file takes the form of a single chunk in an RIFF formatted file, which is then subdivided into two mandatory "chunks" and one optional "chunk".

The first sub-chunk is identified by the "hdrl" tag. This chunk is the file header and contains metadata about the video such as the width, height and the number of frames. The second sub-chunk is identified by the "movi" tag. This chunk contains the actual audio/visual data that makes up the .avi movie. The third optional sub-chunk is identified by the "idxl" tag and indexes the location of the data chunks within the file.

By way of the RIFF format, the audio/visual data contained in the "movi" chunk can be encoded or decoded by a software module called a codec. The codec translates between raw data and the data format inside the chunk. An .avi file may therefore carry audio/visual data inside the chunks in almost any compression scheme, including:

Full Frames (Uncompressed), Intel Real Time Video, Indeo, Cinepak, Motion JPEG, Editable MPEG, VDOWave, ClearVideo/ RealVideo, QPEG, MPEG-4, XviD, DivX and others.

4.11 QUICKTIME

QuickTime is a multimedia technology developed by Apple Computer, capable of handling various formats of digital video, sound, text, animation, music, and immersive virtual reality panoramic images.

It is currently at version 7.0.2 on the Macintosh and Windows platforms.

The QuickTime technology has three major components:

- The QuickTime file format itself openly documented and available for anyone to use royalty-free
- A media player which Apple makes available for free download on its Website and bundles with each of its computers.
- Software development kits available for the Macintosh and Windows platforms. These kits allow people to develop their own software to manipulate QuickTime and other media files.
- Apple released the first version of QuickTime on December 2, 1991 as a multimedia add-on for System 7. The first version of QuickTime laid down the basic architecture which survives essentially unchanged today, including multiple movie tracks, extensible media type support, an open-ended file format, and a full complement of editing functions. The original video codecs included:
- The Apple Video codec (also known as "Rocky Road"), suited to normal live-action video.
- The Animation codec, which used simple run-length encoding and better suited cartoon-type images with large areas of flat color.
- The Graphics codec, optimized for 8-bit-per-pixel images, including ones which had undergone dithering.

Apple released QuickTime 1.5 for Mac OS in the latter part of 1992. This added the SuperMac-developed Cinepak vector-quantization video codec (initially known as Compact Video), which managed the unheard-of feat of playing back video at 320 • 240 resolution at 30 frames per second on a 25 MHz 68040 CPU. It also added text tracks, which allowed for things like captioning, lyrics etc. at very little addition to the size of a movie.

In an effort to increase the adoption of QuickTime, Apple contracted an outside company, San Francisco Canyon Company, to port QuickTime to the Windows platform. Version 1.0 of QuickTime for Windows provided only a subset of the full QuickTime API, including only movie-playback functions driven through the standard movie controller.

QuickTime 7 was released on April 29, 2005 with Mac OS X v10.4 featuring complete MPEG-4 compliance, H.264 codec, live resizing, multi-channel audio, full-screen overlay, and full support for interactive animations created with Apple's new tool Quartz Composer.

4.11.1 QuickTime File Format

A QuickTime file functions as a multimedia container file that contains one or more tracks, each of which store a particular type of data, such as audio, video, effects, or text (for subtitles, for example). Each track in turn contains track media, either the digitally encoded media stream (using a specific codec such as Cinepak, Sorenson codec, MP3, JPEG, DivX, or PNG) or a data reference to the media stored in another file or elsewhere on a network. It also has an "edit list" that indicates what parts of the media to use.

Internally, QuickTime files maintain this format as a tree-structure of "atoms", each of which uses a 4-byte OSType identifier to determine its structure. An atom can be a parent to other atoms or it can contain data, but it cannot do both.

Apple's plans for HyperCard 3.0 illustrate the versatility of QuickTime's file format. The designers of Hyper Card 3.0 originally intended to store an entire HyperCard stack (similar in structure to a complete Website, with graphics, buttons and scripts) as a QuickTime file.

The ability to contain abstract data references for the media data, and the separation of the media data from the media offsets and the track edit lists means that QuickTime is particularly suited for editing, as it is capable of importing and editing in place (without data copying) other formats such as AIFF DV, MP3, MPEG-1, and AVI. Other later-developed media container formats such as Microsoft's Advanced Streaming Format or by the open source Ogg and Matroska containers lack this abstraction, and require all media data to be rewritten after editing.

4.11.2 QuickTime and MPEG-4

On February 11, 1998 the ISO approved the QuickTime file format as the basis of the MPEG-4 •.mp4 container standard. Supporters of the move noted that QuickTime provided a good "life-cycle" format, well suited to capture, editing, archiving, distribution, and playback (as opposed to the simple file-as-stream approach of MPEG-1 and MPEG-2, which does not mesh well with editing). Developers added MPEG-4 compatibility to QuickTime 6 in 2002. However, Apple delayed the release of this version for months in a dispute with the MPEG-4 licensing body, claiming that proposed license fees would constrain many users and content providers. Following a compromise, Apple released QuickTime 6 on 15 July 2002.

4.11.3 QuickTime players

Apple releases official media player software for Mac OS and Windows for free under the brand QuickTime Player. (Earlier versions simply used the name "MoviePlayer.") The player also comes with a number of media-editing and media-creation features, but users have to unlock these by purchasing a key from Apple, turning the media player into QuickTime Pro.

5.1 STORAGE MEDIA FILE FORMATS

The Yellow Book does not specify how data files are to be stored and retrieved. To access any file the computer operating system needs to know where that file is. For convenience, all files are given a unique name and are assigned to one of a number of hierarchically organised directories. Files also have associated attributes (to define read and write permissions for example).

A file system facilitates the storage and retrieval of perhaps many hundreds or even thousands of files. Each computer operating system uses a different file system and therefore there are a number of CD-ROM file systems in use to suit a range of platforms including Windows, Macintosh, Unix etc.

The most common file system for CD-ROM is ISO 9660 which is the international standard version of the High Sierra Group file system and is purpose designed for the PC and MSDOS. The JOLIET extensions provide long filenames.

With the advent of DVD, the Micro-UDF file system has been added to the list. This is suitable for read-only and re-writable discs and allows long filenames etc.

5.2 ISO 9660

Following the completion of the Yellow Book, a group of interested parties formed the High Sierra Group and agreed on a proposal for a file system for CD-ROMs. This was then ratified, in a slightly modified form, by the International Standards Organisation as recommendation ISO 9660.

The ISO 9660 data starts at track time 00:02:16 or sector 166 (logical sector 16) of track one. For a multisession disc (write-once or pressed) the ISO 9660 data will be present in the first data track of each session containing CD-ROM tracks.

The ISO 9660 specification defines two levels:

5.2.1 Level Features

1. Compatible with an MS-DOS file system. Filenames in UPPER case and up to 8 + 3 characters. Subdirectories are allowed to nest up to eight levels deep.
2. Longer filenames, up to 32 characters. Not usable for MS-DOS systems. Most restrictions of Level 1 remain.

5.2.2 JOLIET Extensions to ISO 9660

The Joliet specification was designed to resolve a number of deficiencies in the original ISO 9660 file system (Level 1) particularly when used with Windows95 and later. These include:

- Character Set limitations to upper case characters, numbers and underscore.
- File Name Length limited to 8 characters plus three character extension
- Directory Tree Depth limitations
- Directory Name Format limitations

5.3 DVD

DVD is a family of optical discs of the same size as compact discs (CD) but of a much greater capacity due to a higher recording density.

The developers aimed at a medium which could be used equally in audio and video equipment, in computer systems, in game consoles etc. It would bring together different spheres of electronics.

First DVD stood for Digital Video Disc. But later when the DVD functions extended it meant Digital Versatile Disk.

The development of the DVD format officially started in September 1995 by a group of 10 companies: Hitachi, JVC, Matsushita, Mitsubishi, Philips, Pioneer, Sony, Thomson, Time Warner and Toshiba. A DVD forum based on this consortium was founded in May 1997. The DVD forum is an open for entry organization which today includes more than 200 members.

The overall goal of this organization is to develop and promote the DVD format, work out compatible specifications and give licences to companies in the sphere of the DVD technology. There are special workgroups working on different aspects of the DVD technology. For some specifications there are international standards established.

5.3.1 Important Advantages of the DVD Technology

Today's DVD is a widely spread, time-proved and dynamically developing technology with a great potential. Recording and reproduction of high-quality video and audio in a real-time mode, an effective operation of computer multimedia information, an effective random access to data kept in a great number of small files; up to 4.7 GBytes disc size (around two hours of MPEG-2) per side for one-layer recording and 8.5 GBytes per side for two-layer recording; a possibility of two-level recording on each side; a universal file system UDF; recording and multiple rewriting of DVD discs; back compatibility with current CD discs—their geometrical dimensions are identical, and all DVD equipment can read CD-Audio and CD-ROM discs (MultyRead specification).

5.3.2 The First DVD Formats

Initially the DVD technology was primarily based on 3 formats to meet the requirements of different spheres of DVD application:

- DVD-ROM is used for data recording, including multimedia which is used in computer technologies;
- DVD-Video is used for recording of video materials for their further reproduction on video equipment or with a DVD-ROM drive connected to a computer. The format provides copy protection;
- DVD-Audio is used for recording of high-quality multichannel sound. Besides, the DVD forum recommends an additional support of video, graphics and other information.

Those formats specified the discs only for reading. The information could be recorded only once—in course of production. Later new specifications allowed users to record and rewrite discs. But the major Forum participants couldn't reach an agreement on a single specification for such discs since they wanted to keep control over their own technologies. As a result, we got several competing specifications (DVD-RAM, DVD-RW, DVD+RW formats). Let's study the DVD disc formats which are developed today.

5.3.3 Developing DVD Formats

The developers couldn't reach an agreement on a single format of recordable discs. Because of the competition there are almost no any devices that support several formats. That is why the discs recorded in one format cannot be read in devices of other recording formats. Panasonic was first to develop a drive which could work with DVD-RAM and DVD-R(G) formats (in April 2001).

Some drives can't sometimes understand a format developed after they were released. Home equipment can be designed for a particular segment of the consumer market (DVD-Audio, DVD-Video, both formats), and it is not necessary that it ensures reading of computer discs (this is what the forum has defined). At the same time computer drives work excellently with video, audio, multimedia and other computer discs.

5.3.4 UDF File System

The universal MicroUDF file system adopted in 2000 was a major achievement in providing compatibility of DVD technologies. The MicroUDF file system is a DVD adapted version of the UDF system (Universal Disk Format) which is based on the ISO-13346 standard. This file system is now gradually replacing an aging ISO9660 created for CDs. But until computer devices and discs of the ISO9660 format are not done away with, there is a UDF Bridge system which is a combination of the MicroUDF and the ISO9660. Recording of Audio/ Video DVD discs can be done only with the MicroUDF.

Features of the MicroUDF file System

- Independence from a used soft hardware platform (in this aspect the UDF is optimal for archive systems).
- **High Capacity:** The whole disc can be viewed as a single partition.

- **Optimal data rate:** Reading and recording speed in the UDF data can be higher than the performance of its own file systems when large files are transferred (for example, in multimedia systems)

Maximum Possible File Sizes

- Usage of UNICODE font format which provides an effective international support.
- Extended file attributes support which is used in some operating systems.
- Support of long file names with extension of OS limitations. The maximum length of a file name is 255 symbols;
- Interchangeability of DVD discs in home equipment and computer systems.

With the MicroUDF one DVD disc can contain video films, audio records, photos and computer files. This provides platform compatibility, *i.e.* a DVD disc is a universal medium for Macintosh, DOS/Windows, OS/2, and UNIX.

5.3.5 DVD Read Only Formats

DVD-ROM (Digital Versatile Disc Read Only Memory)

DVD-ROM discs are meant for computer equipment. Information can be recorded only once, at the factory.

The development of DVD is very similar to that of CD and is mainly aimed at improvement of speed characteristics and at inclusion of a recording function. The first generation DVD-ROM devices used the CLV mode and read discs at 1.38 MBytes/s (it is 1x for DVD). The 2G devices could read discs at 2X (2.8 MBytes/s). Modern DVD-ROM devices (3G) uses the CAV mode (rotation management mode) and read data at 4X-6X (5.5–8.3 MBytes/s) and at higher speeds. Modern DVD-ROM drives support reading of almost all formats including CD.

5.3.6 DVD-Video

The DVD-Video format is designed for storage and reproduction of video. This format makes possible only to read data with video players (video recorders). This specification is based on the DVD-ROM one but it provides a special data storing method different from a bitwise disc copying. Video data are recorded in course of production in an encoded form. DVD-video reproduction is possible only on consumer video players (video recorder) or on DVD drives connected to a computer. In a computer data are decoded either on a software or on a hardware level. The modern specification ensures recording of high-quality video (up to 2 hours in MPEG-2), multichannel sound accompaniment in 8 languages, different screen formats, captions in 32 languages, interactive management through a screen menu, up to 9 angular directions of viewing films, copy protection, division of viewing video materials among regions, children access control.

5.3.7 DVD-Audio

This is a new generation of an audio format after CD. The specification defines a high-quality multichannel sound, supports of wide range of sound quality (16, 20, 24 bit sampling at 44.1 to 192 kHz), CD playing by DVD players, additional information (including video, text, menu, screensavers, navigation system), connection with Web-sites which provide information support, extension of features in case of new technologies.

There are two versions of the DVD-Audio format: DVD-Audio which is only for audio contents and DVD-AudioV which is for sound and additional information. Special measures have been established to protect from illegal copying.

5.3.8 DVD Formats for Multiple Recording

Multiple Recording

All known specifications of rewritable DVD discs use a multiple recording technology based on the phase changing (crystalline/amorphous) of an information layer implemented by a laser of 650 (635) nm wavelength (phase-change recording). During reading a leam is reflected from an information layer to define its optical characteristics in its different phase states (the same as in recording).

Material for Multiple Recording

DVD discs are made of AVIST, a material developed by TDK in 1995. Its characteristics meet the requirements of the rewriting technology:

- High reflection power - up to 25–35 per cent, which is sufficient for compatibility of DVD-discs in reproduction.

- Easy to change a phase both at high and at low speeds, which is especially important when working with different applications. Applications working with rewritable CDs (for example, CD-E) implement recording at lower than 3 m/s. DVD-RAM data require a higher speed— from 3 to 6 m/s. In case of compressed video data the speed must exceed 6 m/s.
- Excellent signal/noise ratio and characteristics of phase changing allowed TDK achieve very small dimensions of a marker (less than 0.66 mm).
- AVIST can endure at least 1000 cycles od rewriting even at the speed less than 3 m/s. The higher the speed, the more the number of cycles.

Each format has its own advantages and disadvantages which determine their ranges of application. The most wide spread format is DVD-RAM because of cheaper drives and discs which can work with it.

5.3.9 DVD-RAM (Digital Versatile Disc Random Access Memory)

This is a rewritable format developed by Panasonic, Hitachi and Toshiba.

The format was approved by the DVD forum in July 1997. More than 20 computer companies tested the equipment and discs of this format during 3 months. More than 160 forum participants voted for this specification. Today this format is the most popular.

DVD-RAM drives can read all DVD-ROM drives, while DVD-ROM drives can read only third generation DVD-RAM discs which appeared in the middle of 1999.

The G1 DVD-RAM discs could contain 2.6 GBytes per side. The modern (2G) discs carry 4.7 GBytes per side or 9.4 GBytes on for a double-sided modification.

There are two types of single-sided DVD-RAM discs—in cartridge and without it. The former ones are mainly intended for home video equipment where it is important to avoid external factors in case of intensive manual use of discs. The cartridges can be openable and non-openable.

The best advantages of DVD-RAM discs is a possibility to rewrite up to 100,000 times and a error correction mechanism.

The greatest number of rewriting cycles among all DVD formats, an error correction mechanism and random disc access both in reading and in recording made this format the most effective in secondary data storage—devices. An overwhelming number of devices of mass data storage—robotic DVD libraries—use exactly this technology.

DVD-RAM discs can be used for recording and reading of stream video on equipment supporting the DVD-VR specification (see below).

5.3.10 DVD+RW (Digital Versatile Disc Re-Writable)

The DVD+RW format is promoted only by its developers—Hewlett-Packard, Mitsubishi Chemical, Philips, Ricoh, Sony and Yamaha (the DVD forum doesn't support it).

You can record on DVD+RW discs both video or sound and computer data. These discs can be rewritten about 1000 times.

The DVD+RW served a base for the DVD+RW Video Format. Equipment and discs working in this format are positioned as fully compatible with DVD-Video equipment. It means that DVD+RW discs containing video data can be played on consumer DVD equipment which was produced earlier.

Philips announced its DVD video recorder in September 2001. DVD+RW discs recorded on this equipment can also be read by usual DVD-Video players. This solution was made in response to the DVD-VR specification (see below).

5.3.11 DVD-RW (Digital Versatile Disc Re-Recordable)

It is also known as DVD-R/W and DVD-ER.

The DVD-RW is a multiple recording format developed by Pioneer. The discs can house 4.7 GBytes per side, are produced in single-sided and double-sided modifications and can be used for storing video, audio and other data.

DVD-RW discs can be rewritten up to 1000 times. Unlike DVD+RW and DVD-RAM, DVD-RW discs can be played by DVD-ROM drives of the first generation.

- TDK states that the service life of DVD-RW discs is 100 years.
- DVD disc recordable formats
- DVD-R (Digital Versatile Disc Recordable).

DVD-R is developed by Pioneer. Such devices were first to record DVD discs. The recording technology is similar to the CD-R one and is based on irrevisable change with a laser of spectral characteristics of an information layer covered with special organic matter.

DVD-R discs can contain both computer data, multimedia programs and video/audio information. Given to the type of recorded data such discs can be played by other devices compatible with the format, including DVD-Video players and a majority of DVD-ROM drives. Single-sided DVD-R discs can contain 4.7 or 3.95 GBytes per side. Double-sided discs carry 9.4 GBytes (4.7 GBytes per side). Currently the format doesn't support double-sided recording technology.

Their service life is estimated more than 100 years.

There are two specifications for copy protection: DVD-R(A) and DVD-R(G). These versions use a different laser wavelength in course of recording. It means that such discs can be recorded only on the equipment which corresponds to their specification. Reproduction can be implemented on any equipment with DVD-R support.

DVD-R(A) (DVD-R for Authoring) is used in professional applications. In particular, the Cutting Master Format allows using these discs for recording of original replica (pre-mastering) instead of usual DLT types.

DVD-R(G) (DVD-R for General) is used in a wider range of fields. These discs are protected from bitwise copying of information on them from other discs. This format is supported in devices of mass data storage (for example, in robotic DVD libraries developed by Pioneer).

5.3.12 DVD-VR

The DVD-VR format is based on the DVD-RAM and is supported by the DVD forum. The DVD-VR format allows recording up to 2 hours of high-quality video in a real-time mode in MPEG-2 on a single-sided DVD-RAM disc of 4.7 GB capacity and provides for such functions as editing of recorded video data, recording of different types of static images etc. Electronic devices based on this format are produced by Panasonic, Toshiba, Samsung, Hitachi etc.

6.1 DIGITAL LIBRARY SOFTWARE

The following are some other popular software for handling digital libraries:

(1) Active Paper Archive, Olive Software: "ActivePaper Archive revolutionizes the way you archive your newspaper, and the way you access your archive! This award-winning product uses groundbreaking technology to instantly convert print editions to rich

XML, which stores content, structure, and layout. The XML is interlaced with images of newspaper information objects, creating a perfect synergy between searchable database and hardcopy archive.

(2) **CONTENTdm:** "Fully web-compatible, the CONTENTdm Multimedia Software Suite provides easy loading, management and access to media archives. Tools are provided to assist with every phase of collection development. You can start small with a few items or CONTENTdm can handle databases with millions of objects. The CONTENTdm technology is based on years of university research and testing that have resulted in a proven set of programs."

(3) **DB TextWorks, Inmagic, Inc.:** DB/TextWorks(R), the core of all Inmagic(R) products, is a powerful, flexible combination of database and text retrieval software that enables you to quickly build networked and standalone textbases to manage diverse types of information including text, images, and multimedia. What's more DB/TextWorks is lightning fast—searching across 100,000 records in less than a second!

(4) **Dienst, Cornell Digital Library Research Group** (CDLRG): The word "Dienst" refers to a number of things. A conceptual architecture for distributed digital libraries. A protocol for service communication in that architecture. A software system that implements that protocol. Conceptually, Dienst is a system for configuring a set of individual services running on distributed servers to cooperate in providing the services of a digital library. The open architecture of the Dienst system - exposure of the functionality through an defined protocol - makes it possible to combine Dienst services in flexible ways and augment the existing services with other mediator services, which build on the functionality of the existing services.

(5) **Encompass, Endeavor, Inc.:** Created to manage today's digital library collections, Endeavor's ENCompass uses new technology standards like XML, EAD and Dublin Core to describe, index and search electronic resources. ENCompass provides seamless searching across diverse digital object collections while managing and controlling links for separate repositories. Despite the numerous and disparate descriptions applied to digital collections, ENCompass ties together descriptions, creating a more effective research base. With ENCompass, libraries truly have the ability to create, manage and navigate the digital collections that are becoming the standard in the library world.

(6) Insight Software Systems, Luna Imaging, Inc.

"Luna Imaging's new generation of Insight client/server software allows organizations to manage, access, and use image collections over the Internet. Find images using any of Insight's search modes, from simple point and click select lists, to keyword searching, to full Boolean searching. Unique to Luna software, Insight combines retrieval and display of text information with an equally rich set of tools for viewing, comparing, and organizing images. Zoom and pan images; create and save image groups. Insight offers a complete visual environment for working with images. Limit image access to local Intranets, or expand access to collections from almost anywhere."

(7) MetaLib, Ex Libris: MetaLib is the perfect platform for managing a hybrid library environment, including both the emerging electronic collection with its digital resources and the traditional library with its print resources. MetaLib serves as a gateway to local and remote databases.

(8) MrSID, Lizard Tech: LizardTech's MrSID Photo is a unique file format that enhances the workflow of imaging professionals. It encodes large, high-resolution images to a fraction of their original file size while maintaining the original image quality. Images become truly scalable and can be reduced, enlarged, zoomed, panned or printed without compromising integrity. And with MrSID's Selective Decompression and bandwidth-optimization capabilities, files move really fast!

(9) Past Perfect, Pastime Software Company, Inc.: PastPerfect provides full automation of accessions, loans, membership, cataloging, research, reports, and Digital Imaging. With Virtual Exhibit, the web interface module, you can produce HTML web pages for all or selected catalog records or create an entire website including themed exhibits with search capability.

Selecting software for the Digital Library by Chris Dobson and Digital/Electronic Library Softwares.

(10) CdsWare: CERN Document Server Software (CDSware) is the software developed by, maintained by, and used at, the CERN Document Server. It allows you to run your own electronic preprint server, your own online library catalogue or a document system on the web. It complies with the Open Archives Initiative metadata harvesting protocol (OAI-PMH) and uses MARC 21 as its underlying bibliographic standard. It is a free software issued under GNU-GPL license.

(11) Dienst: Dienst is a system for configuring a set of individual services running on distributed servers to cooperate in providing the services of a digital library. It has been written in PERL. It works more comfortably on Unix/Linux run webservers.

(12) DeSpace: DSpace is a digital library system to capture, store, index, preserve, and redistribute the intellectual output of a university's research faculty in digital formats. Dspace has been developed jointly by MIT Libraries and Hewlett-Packard (HP). It is now freely available to research institutions world-wide as an open source system.

(13) Eprints: Eprints is generic archive software under development by the University of Southampton. It is intended to create a highly configurable web-based archive. EPrints primary goal is to be set up as an open archive for research papers, but it could be easily used for other things such as images, research data, audio archives—anything that can be stored digitally by making changes in configuration. It works on Linux O/s and it needs MySQL, Perl modules and Apache Web server.

(14) Fedora: An Open-Source digital repository management system based on the Flexible Extensible Digital Object and Repository Architecture (Fedora). The Fedora repository system is open source software licensed under the Mozilla Public License. It requires Sun Java Software Development Kit, v1.4. Optionally one can use MySql or Oracale 9i to create relational databse. It works both on Windows and Unix versions of O/s.

(15) Greenstone: Greenstone is a suite of software for building and distributing digital library collections. It provides a new way of organizing information and publishing it on the Internet or on CD-ROM. It is available for both Windows and Linux O/S. It requires Perl software to build collections.

(16) Roads: The ROADS software is a suite of programs intended to aid in the setting up and day to day running of World Wide Web based catalogues of on-line resources. Although designed specifically to meet the requirements of the eLib subject gateways, ROADS software will be useful in a variety of other situations and for a variety of other purposes. It works on a variant of the Unix operating system with HTTP Apache webserver and Perl language.

(17) SiteSearch: The OCLC SiteSearch software provides a comprehensive solution for managing distributed library information resources in a World Wide Web environment. It offers tools that integrate electronic resources under one web interface, provide flexible access to resources, and build text and image databases locally. It works on both Digital Unix and Windows NT O/s.

(18) **Athenaeum Light:** Athenaeum Light can handle large cataloges and manage borrower lists easily. As per its developer it is maint for school libraries. As of now its limit of records is 50000. It is available for both Windows and Macintosh. It can even be used in the multiuser environment.

(19) **Avanthi Circulation System:** The Avanti circulation system is a simple, scalable, networkable, client/server circulation system that can be deployed in small to medium scale libraries. The end product will be modular: consisting of a system core, circulation module, minimal OPAC, and network and user interfaces. The entire system is written in Java. Avanthi 0.3.1 was developed using the Blackdown JDK v1.1.7 on a Linux system with the servlet classes being developed using the Apache Jserv servlet engine.

(20) **Java Book Cataloguing System:** This software is primarily to create a Book Catalog using barcode data from the freely avaliable cuecat(tm) bar code reader. It will use a rdbms backend database, and allow synchronization between different library branches.

(21) **Koha Open Source Library System:** Claimed as worlds first Free Open Source Library System. Made in New Zealand by the Horowhenua Library Trust and Katipo Communications Ltd the Koha system is a full catalogue, opac, circulation and acquisitions system, which in turn depends on the C CDK libraries http://www.vexus.ca/CDK.html

(22) **ITIL Library Management System:** Library Management System is a workflow application which aims at automating various library functions encompassing Member Registration, Requisitions for New Items & their Approval, Accession & Cataloging of library items, Reservation/ Issue / Return of library items, Order Generation and maintenance of details pertaining to Inter-Library Loans, Bills, Orders and Vendors. It requires Lotus Notes Domino server.

Emilda: Emilda is a complete Integrated Library System that features amongst others an OPAC, circulation and administration functions, Z39.50 capabilities and 100 per cent MARC compatibility. MARC compatibility is achieved using Zebra in conjunction with MySQL.

FireFly: FireFly is a Complete Public Library system. It is being written in Python, Perl, with all data being stored in XML. The driving force behind this project is to give public libraries a Free-Software set to run and maintain library systems.

Glibms: Glibms is Library Management software developed using PHP and PostgreSQL to automate the different activities carried out in the library.

Mylibrarian: This is a integrated suite of school library books management software targeted at secondary schools. The language used is totally Microsoft Visual BASIC 6.0 and Microsoft SQL 2000 server.

Mylibrary: MyLibrary is a user-driven customizable interface to collections of Internet resources—a portal. Primarily designed for libraries, the system's purpose is to reduce information overload by allowing patrons to select as little or as much information as desired for their personal pages.

OpenBiblio: OpenBiblio is a open source library software written in PHP containing OPAC, Circulation, Cataloguing and Staff administration. It is good for small school and public libraries. It requires MySql, PHP and works on both Linux and Windows platforms.

OpenBook Open Source Library System: OpenBook is full feature open source library system developed for use of small school and public libraries. It has been developed based the original Koha open source library system of New Zealand. At present it consists of three modules viz. OPAC, Cataloguing and Circulation. In future includes Acquisition module too. All the modules are web-interface based. It works on Linux O/S with Apache, Perl and MySql.

Open-ILS: It is Website of an open source Integrated Library System (ILS), named Evergreen. This software is being developed and maintained by the Georgia Public Library Service for use by the Georgia Library PINES Program, a consortium of 249 public libraries. This software can be downloaded for free, and anyone can contribute to development efforts.

PhpMyLibrary: PhpMyLibrary is a open source Web-based library software having cataloguing, circulation, webopac, file management modules, etc. It also supports import of data from ISIS databases. It requires PHP and MYSQL.

Python Simple Libary Circulation System: The software useful for carrying out circulation activities of the library.

Research Guide: This software helps librarians to build Web-based subject guides.

Seansoft Library Loan Management System 1.17B

A useful software for something like a school library. Each student can have a login, from which they loan books. Keeps track of weeks loaned and will alert a supervisor of overdue books. Has mail merge facilities to produce letters to individuals or a list for a noticeboard. Also features backup and basic repair facilities for the database. It requires VB5 Runtimes, DAO.

WebLis: WEBLIS is a free-of-charge Web-based Library Integrated System based on CDS/ISIS. Presently it has Cataloguing system , OPAC (search), LOAN module and Statistical modules.

6.2 OPTICAL CHARACTER READERS (OCR)

Unlike MICR, optical character recognition (OCR) techniques permit the direct reading of any printed character. No special ink is required. This OCR flexibility makes it possible for organizations to eliminate or reduce the input keying bottleneck. You've already seen how optical character readers can read bar codes and merchandise tags to enter data directly into a computer system. But readers designed to interpret handmade marks and machine-printed characters are also commonly used in off-line operations to prepare data for processing.

This is literally what it says. The card or form is divided up into boxes, in which a mark is made by pencil or pen. A character is represented by marking the correct combination of boxes in any one column, as opposed to displacing holes from a punched card. Forms and cards are pre-printed for special purposes so that a mark can be made in a certain position to represent a YES or NO, to answer a market survey question for example, or to signify a number, as on insurance forms, gas and electricity recording cards. The same can be very well used for sensing the correctness of an examination answer book of objective type.

In one form of detection, the conductivity of graphite marks is sensed. The method necessitates the use of a soft pencil, and non-graphite pen or printed marks are not acceptable. Another method uses equipment which reads marks optically. Quite simply a light source senses the presence of a mark. In this case, special pencils are not required to mark the cards or documents. A mark reader may be designed to be insensitive to certain colours. These colours can then be safely used in the pre-printing of the cards or documents without risk of being read when marks made later are sensed.

Mark reading is performed in two different ways. One method relies on the conductivity of graphite to determine the presence of a pencil mark. This reading process is sometimes referred to as mark sensing. The marks must be made only in pencil and a soft pencil is recommended since the graphite deposit is more pronounced. The second method of reading is based on the reflectance of light. The presence of a mark is indicated by a drop in light reflectance. Marks other than by pencil may be acceptable.

Data can be transferred directly to a computer but, because of the slow operating speeds of mark readers, typically 200 documents per minute or less, the information is more likely to be transferred to magnetic tape first. Some mark readers are designed to handle both punched cards and marked cards.

Unlike MICR, optical character recognition (OCR) techniques permit the direct reading of any printed character (not just 14). No special ink is required. This OCR flexibility makes it possible for organizations to eliminate or reduce the input keying bottleneck. You've already seen how optical character readers can read bar codes and merchandise tags to enter data directly into a computer system. But readers designed to interpret handmade marks and machine-printed characters are also commonly used in off-line operations to prepare data for processing.

OCR readers examine each character as if it were made up of a collection of minute spots. Once the whole character has been scanned, the pattern detected is matched against a set of patterns stored in the computer. Whichever pattern it matches, or nearly matches, is considered to be the character read. Patterns which cannot be identified are rejected. OCR readers can read at a rate of up to 2400 characters per second. They are generally designed to operate at slower speeds, typically 300–800 characters per second, at which they are more accurate and can handle characters which are not quite so perfectly formed. OCR readers are expensive devices and are only likely to be used when there are large quantities of data to process.

A wide range of fonts, using ordinary inks, can now be accepted by OCR readers, including many common typewriter fonts. The standard fonts used are OCR-A (American standard) and OCR-B (European standard). Some OCR readers can accept computer print-out and complete pages of typed text. It is possible that a computer could be programmed to accept some signatures, but it is unlikely that it could ever be programmed to accept every type of signature.

Even so, devices have been developed which can read neat hand printing (capital letters rather than lower case) in black ink, and with sufficient accuracy for this to become a viable form of input.

The readers are expensive compared with most other reading devices. However, many firms and organizations with a large and regular daily volume of data for processing by computer make use of OCR. The advantages gained from preparing documents, designed for specific tasks, using conventional typing methods, and from not having to key the information a second time to get it into a machine-acceptable form, are considerable. Banks, insurance companies, airlines, and numerous commercial and business concerns make use of OCR. Some retail outlets also use optical scanners connected to modified cash registers or terminal devices to record the sale of goods. Each item on sale is marked with a product code which can be read by the device.

The primary advantage of OCR is that it eliminates some of the duplication of human effort required to get data into the computer. This reduction in effort can improve data accuracy and can increase the timeliness of the information processed. However, difficulties in using OCR equipment may be encountered when documents to be read are poorly typed or have strike overs or erasures. Also, form design and ink specifications may become more critical than is the case when people key the data from the forms.

6.3 IMAGE EDITORS AND VECTOR ILLUSTRATION SOFTWARE

Image editors and illustration software are the two primary categories of graphics software. While there are potentially many subcategories of graphics software, most applications fall into one or the other of these two main categories. This is because there are two very different types of graphics: vector and bitmap graphics.

Image Editors (pixel/bitmap-based)

Also known as paint applications, this is the most popular type of graphics software. Image editors allow you to create bitmap-based images from scratch with a variety of painting tools intended to mimic natural painting tools. Almost all image editing software includes features for photo enhancement and retouching as well.

6.4 BITMAPPED IMAGE EDITING SOFTWARE

Images that are created and edited in this type of application are referred to as bitmaps because they are made up a series of pixels in a grid.

The advantage to bitmapped images are that you can get a great deal of detail, as much detail as a photograph. The downside is that you will always be confined by the pixel resolution of the image.

Image editors are used to edit digital photos and scanned images (which are always bitmap-based) and for creating continuous-tone and photo-realistic artwork.

These are some well-known image editing programs:

- Adobe Photoshop
- Corel Photo-Paint
- Corel Painter
- Microsoft Paint
- Jasc Paint Shop Pro
- Ulead PhotoImpact

6.5 ILLUSTRATION SOFTWARE (VECTOR-BASED)

Illustration software allows you to create drawings using lines and curves. These drawings are called vector illustrations. Vector illustrations can be easily edited by moving points, adjusting curves, and changing the colors of various objects.

Vector illustrations, because they consist of points and objects rather than pixels on a grid, are free from the confines of pixel resolution. Unlike bitmapped images, they will always print at the highest resolution possible, giving you printed output with smooth fills and crisp lines no matter what size you print it. However, vector graphics generally cannot contain the high level of detail as bitmapped images so they often have a cartoon-like appearance.

Illustration software is most often used to create stylized, cartoon-like images, logos, maps and other design elements that need to be output at any size.

These are some well-known illustration programs:

- Adobe Illustrator
- CorelDRAW
- Macromedia Freehand

- Serif DrawPlus
- Xara X

6.6 OBJECT-BASED GRAPHICS SOFTWARE

In recent years, more and more graphics software is blending the features of both bitmap-based image editing software and vector-based illustration software. Today's graphics software uses more of an object-based way of working. In most graphics software today, your documents can contain both vector-based and pixel-based elements, and the tools provided in the software allow you to work on a pixel-editing basis or on a vector-editing basis.

For example, Photoshop, which is categorized as an image editor, has tools that allow you to create vector-based paths, shapes, and text. CorelDRAW, which is categorized as illustration software, allows you to work with bitmap-based images and perform photo enhancement functions. So why is Photoshop categorized as an image editor, but CorelDRAW as an illustration program? Here's the difference:

- Image editors are intended for outputting bitmap-based graphics. When working with vector objects in a bitmap editing program, they only preserve their vector-like qualities as you work in that particular software, saving your files in the program's native format. Once you want to output your documents from an image editor to use in other software, you are generally limited to bitmap-based formats.
- Illustration software is intended for outputting vector-based graphics. When working with bitmap-based objects in a vector program, the bitmaps do not become vector objects. Although you can sometimes output bitmap elements as part of a vector-based file format, these hybrid graphics then become metafiles, containing both bitmap and vector elements. In a metafile, only the vector-based portions of the file are scalable and resolution independent. Unlike bitmaps and simple vector graphics, metafiles cannot easily be converted from one program's format to another. Related Links:

7.1 INPUT CAPTURE DEVICES

Image capture is defined here as digital image capture, or digitisation. Image capture is essentially the deployment of analogue to digital conversion technology to produce a digital version of an image. This involves combinations of hardware and software selected for optimal conversion of a given collection of images. There are considerations to make either side of this stage, which influence how this process is approached.

To put the process in context, the identifiable stages are:

- Planning and feasibility study:
- Assessment of the nature of the resource;
- Digitisation of what, for whom, how and why;
- Assessment of copyright implications;
- Assessment of desired level of quality and of cost implications;
- Selection of hardware and software combinations;
- Assessment of required/available human resources and skills;
- Assessment of networking requirements;
- Input: capture by scanning, digital camera, other input devices ;
- Intermediate processes manipulation, cleaning, sharpening contrast, colour adjustment.
- Output file formats, incorporation of metadata, storage, archiving and distribution, database management system, user interface, typical specification of user access points.

Planning: Assessing the nature of the images is essential to ensure that, in the context of a networked resource, there is some stated aim of purpose which includes reference to the potential of the images and any limitations (fragility, size, state of conservation) matched against intended user communities. The outcome of such assessment then determines specification of equipment and level of human skill required. Capturing very detailed color images at a high resolution for demanding and discerning experts, requires much more software expertise and high end equipment than batch scanning of simple black and white documents.

Copyright issues are central to the feasibility of networked access to images and after establishing rights and clearance, image capture should include cost estimates for watermarking and encryption.

Preparation of images (any conservation required, for example) needs to be included in the planning stage.

7.2 INPUT: CAPTURE

Image Capture Devices Include

Scanners: flatbed, drum, film, handheld, 3-D laser, data pens. Considerations when selecting model are: scan area, optical resolution, bit depth colour accuracy and scan time. Developments in film scanners with Advanced Photo System capability extending the area of film scanned, have resonance for hybrid library applications, as does 3-D scanning which is rapidly increasing in importance.

Digital cameras: apart from a few high end models, the capability of most digital cameras falls below requirements for high quality image capture for network archiving, due to shortfall in resolution and use of lossy compression.

Important benchmarks are:

(1) **Resolution:** optical and interpolated determined by the number of pixels in the scanner sensor array, and measured in dots per inch (dpi)

(2) **Bit depth:** 30-36 bit is the minimum for capture of range of tonal contrast

(3) **Scan area:** with attachments for eg. transparent media or batch feeding

(4) **Noise and calibration**

(5) **Colour accuracy:** some systems incorporate colour calibration

(6) **Intermediate Processes**

Unless done as a batch scan, most images will need some degree of optimisation after the capture stage. The most usual are sharpening, removal of noise, contrast, colour adjustment. In the case of colour calibrated systems colour adjustment is not strictly necessary.

It is at this and the preceding stage that calculations concerning human resources and available skills are most relevant.

Selections of interfaces between the image capture device and the computer, and the software and the operator/user are also critical for this stage.

7.3 IMPLICATIONS FOR HYBRID LIBRARIES

The role and importance of images both independently, and in conjunction with text and as hypermedia, are impossible to estimate. Images are central to the absorption of all kinds of information and concepts. A recent user study conducted for the Visual Arts Data Service (see below) placed images second only to text across a range of types of data accessed over the Internet. For this and other reasons, as far as hybrid libraries are concerned it is important to capture not just the content of the image but also the context. This may extend from capture of detail of surface on which the original image rests, to incorporation of vital accompanying information at metadata level. This kind of integral incorporation is vital for facilitating cross-disciplinary use within the humanities.

The resulting demands on display technologies and networks need to be balanced against evaluation of user requirements, and estimated typical technical specification of access equipment.

Recommendations

- That wherever possible, that digital capture is done directly from the object rather than via an intermediary stage such as microfilm
- That emerging imaging and networking technologies, both for access to large resources, and for enabling individual user access via self captured input, are kept under close review within the project.
- That the initiative between the university library and the multimedia research group at Southampton University, to link resources directly from shelf to web, be monitored and the outcomes dispersed throughout the project.

7.4 VIDEO HARDWARE

Video hardware conists of devices to display and capture video. Video display devices, often called graphics adapters, enable the computer to present information on montiors capable of displaying up to 16 million colors.

The VGA (video graphic adapter), most common for personal computers since the late 1980s, provides a 640 by 480 pixel display. A pixel is the smallest sized space in which unique information is presented on a display, usually as unique information is capable of displaying a range of colors from two to 16 million. Super VGA extends the display from 640 by 480 pixels to 800 by 600 and 1024 by 768 pixels. In all cases, note that the aspect ratio, the width versus the height of the display for VGA is 4:3.

Video capture devices enable the computer the computer to connect to video cameras and video cassette recorders (VCRs) to record analog video as digital video. These devices can handle various types of television signals including American NTSC and European /

Japanese PAL. Video capture devices are capable of converting annealing video directly into a raw digital video stream that can be stored on a computer hard drive in both raw and/or compressed formats that significantly reduce the size of the video data stream.

7.5 SCANNING

Scanning used to be a costly process limited to designers and graphics experts. Scanners were costly and delicate, difficult to connect and configure. And they were an absolute no—no for homes. But things have changed and changed rapidly. Prices have crashed, they are easier to use and certainly have found their way to homes and other more casual use. You can use a scanner for just about anything these days. Store the volumes of family album as digital images. Your kid can use it for school project or even you can scan images for the business presentation that you are preparing.

If you are new to scanning, here are a few tips that you need to remember.

7.5.1 Interfaces to Choose From

A scanner today can be connected to a computer in three ways—through a SCSI adapter, through the parallel port or through the USB port. While SCSI interface scanners are for high-end and professional use, the parallel port and USB port scanners are for the common mass. USB scanners are plug and play and are the easiest ones to install. But then again USB ports are available only on the more recent computers. Parallel port scanners follow the USB scanners in terms of ease of use and setup.

7.5.2 Software Required

The next obvious question is which software do I need? Well, the scanner usually ships with all the software that you might need to start off. But as you improve your graphics skills or if you are a serious designer, you will need advanced image editing software. Typically speaking, scanners need the TWAIN drivers and image management software. So what is TWAIN? TWAIN (short for Technology Without An Interesting Name) is the programming interface, which will allow image management/ editing software like Adobe Photoshop or Paint Shop Pro to interact with a image capturing device (say a scanner) and transfer the captured image into the computer.

7.5.3 Scanning Process

Though the actual process of scanning differs from scanner to scanner, they follow the same hierarchy of steps:

1. Make sure the required scanner hardware and software has been installed. Start the program that you the scanned image to be inserted (Adobe PhotoShop, Paint Shop Pro, etc). In most popular programs, the scanner menu is accessed from FILE > IMPORT > FROM TWAIN or FILE > ACQUIRE > FROM TWAIN or INSERT > PICTURE > FROM SCANNER. The scanner goes into action and does a prescan. A prescan is a preview scan of the complete scan table. The time taken to do a prescan depends upon your previous settings.
2. Now pick up the part of the prescan that you actually want. You will have to crop the prescan image to the edge of the picture by dragging in scan boundaries. Adjust image width and height, image type, pixel depth, and DPI to suit your purposes. Pay attention to the image size. Higher resolutions and bigger sizes mean bigger files.
3. Prepare for final scan. Again, this may take a while depending upon your settings. Make final settings match the prescan settings. Set file destination and name. Click Scan. Save the image to disk as a *.TIF or *.PSP file with a descriptive name.
4. Edit the scanned image. Depending upon the graphics program, you can resize the image, add text, shapes, make special effects, crop, adjust colors, etc. Set the file name and type according to its purpose.

7.6 FULL-MOTION VIDEO

For the first few years of its existence, the PC was a text-only device. As users' needs increased and the technology improved, graphics became commonplace on the computer screen. Graphics are now viewed as a prerequisite for virtually all PCs, and our insatiable appetite for PCs that can do more and more has led us into the age of so-called "multimedia". Most people use PCs today that can show not just graphics but sound, animation and full-motion video as well.

The term "multimedia" is, unfortunately, one of those buzzwords that means different things depending on who says it and what they are talking about. It generally refers to the use of software that incorporates elements of sound, graphics and moving video (animation or recorded). Graphics have been around for years and so has sound. What is really new on the PC, relatively speaking, is full-motion video. This section briefly takes a look at the topic, concentrating on the new technologies involved in delivering full motion video to the PC desktop.

7.6.1 Applications for Full-Motion Video

Full-motion video is being seen increasingly on PCs as they expand their horizons to fill roles they never have before. Some of the areas where full-motion video is increasingly being seen:

Games: Many games today use video to add realism to the gaming experience. In particular, many games use video to make them more "cinematic"; these games are often a cross between a movie and a classical computer game. They are sometimes called "interactive movies" in fact.

Video Playback: Some movie companies are now making versions of movies and putting them on CDs for playback on the computer desktop. This trend is new but is expected to continue.

Documentation and Training: Many software (and even hardware) packages now ship with one or more instructional video segments that show in much clearer detail how the software works, or how to perform certain operations.

TV on the PC: An interesting trend is hardware that allows you to watch TV in a window on the desktop. In fact, the television and the PC are becoming more and more intermixed, with TV showing up on the PC and PCs that use the TV for monitors also becoming popular, plus "PC appliances" like WebTV gained ground in the marketplace.

7.6.2 Video Performance Issues and the MPEG Format

The idea of playing a movie or TV program on the PC is a relatively new one, and these applications make demands on the performance of a PC system that few others do. The main problem is the sheer volume of data that is involved.

Let's suppose that we wanted to display a 2-hour movie in high-quality, high-resolution graphics. To keep things from getting out of hand, let's choose just 640 × 480 resolution in high color (16 bits per pixel) instead of going "all out" to say, 1024 × 768 in true color. Full-motion video typically means 30 frames per second. To store the data required for a full-motion, two-hour movie using these parameters, would require approximately 133 GB. (That's not a typo—133 billion bytes, and that doesn't include sound.) Also, the data to support this video stream would need to be pumped to the video card at the rate of about 150 million bits per second. In ten years these numbers won't seem like that big of a deal but of course, right now they sure are.

There is a solution to this dilemma: data compression. Due to the nature of the data, video can be easily compressed by a large factor. There are several reasons for this:

Every image that is displayed has large areas of redundancy that can be represented by a smaller amount of data using an encoding algorithm. For example, if the bottom half of the screen in a particular frame is black, that can be represented using a small number of bytes; we don't need to have 150,000 2-byte pixels all filled with zeroes.

The human eye is not sensitive to certain details in a video, which can be removed without appreciable loss to the perceived signal.

If you take any two consecutive frames, the changes from one to the next are usually rather small. For example, if you have a scene centered on someone's face, the background images are probably going to remain static for hundreds or thousands of consecutive frames. There are special algorithms that can describe a frame only by the changes that it represents from the one before, which dramatically cuts down both on the amount of storage required and the time to transmit each frame of data.

The most popular compressed video format is MPEG, which stands for the Motion Pictures Expert Group <http://www.mpeg.org>, the body that defined the standard. MPEG can typically compress as high as 100 to 1, bringing our 2-hour movie down to a much more manageable 1.33 GB. To use MPEG the video must be encoded using an MPEG encoder, and then viewed using an MPEG decoder. Not surprisingly, MPEG has its own problems. The chief one is that the process of encoding and decoding takes a lot of processing power, which is why dedicated MPEG encoding and decoding hardware is used for high-performance applications.

7.6.3 Video Decoding Hardware and Software

There are several ways that MPEG (or other video) files can be played on a PC. In every case, the process involves reading the file, decompressing the video data, and then feeding it to the video card for display on the monitor. Since the data is compressed to reduce storage, it must be uncompressed to restore it to a visible image. This takes a great deal of processing power. There are software programs that can be used to decode these files, allowing you to play an MPEG file on your computer. As with every other case where an application takes a lot of processing power, it is more efficient (but more costly) to let dedicated hardware do the function. This is the purpose of an MPEG Decoder.

Much as is the case with 3D accelerators, it is possible to get a dedicated card for MPEG decoding, or to get a regular video card that has an MPEG decoder built into it. Dedicated MPEG decoders come with different capabilities. They often use the video feature connector on the regular video card, to supply video information directly to the card without tying up the system bus with an enormous amount of video data.

Warning: Beware of some video cards that advertise that they "support MPEG decoding" when all they mean is that they will work with a software MPEG decoder. Many of these cards aren't doing hardware decoding at all. They still may provide acceptable performance, but make sure you know what you are buying.

The quality of MPEG decoders varies, with generally the more expensive ones providing better quality. A high-quality decoder will allow the display of full-screen, smooth video animation while leaving the system CPU free to do other tasks. A cheaper hardware decoding solution, or a software package, may have problems keeping the supply of data to the video card which can result in choppy performance or lost frames. In some cases, this is compensated for by running the video in a small window instead of full-screen, and in others by slowing the frame rate down to less than 30 frames per second. Others reduce the resolution of the screen.

Video encoding and decoding software and hardware is still evolving today and is likely to improve dramatically in coming years. By the end of the decade it is likely that a PC without smooth video capability will be as out of place as one without graphical abilities is today. However at present video on the PC still does not provide very good images compared to a regular VCR and TV combination, unless you are using high-end hardware.

7.7 Digital Movie Camera

Point and Shoot Digital Cameras

Most consumer cameras on the market today fall into the category of Point and Shoot digital cameras. These cameras usually provide resolutions of between 2 and 4 megapixels (million pixels). These cameras are the digital version of your typical 35mm consumer camera. A point and shoot digital camera is designed to do just that - take snapshots. They provide limited flexibility with options, although many newer models have some limited options for special effects.

The image quality from a point and shoot digital camera is generally good enough for common uses, such as auction photos, and even 4 × 6 prints. If you just want a standard, entry level digital camera for family snapshots and auction photos, then a point and shoot digital camera is a budget conscious choice.

7.7.1 Semi-Professional Digital Cameras

A semi-professional digital camera is a more advanced version of a point and shoot digital camera. Typically, these cameras provide resolutions between 5 and 6 megapixels (million pixels). They also often offer the user additional settings and greater control over the camera. The image quality of these cameras is good enough to produce clear prints up to about 8 × 10. A semi-professional digital camera may be the best option for you if you seek greater creative control than you can get with a point and shoot model. The increased flexibility does come with a price. Most semi-professional models are priced much higher than point and shoot models.

7.7.2 Digital SLR Cameras

For professional grade photography, a Digital SLR model is required. SLR stands for Single Lens Reflex. Digital SLR cameras are very similar in operation to their 35 mm SLR counterparts. Also, many camera companies (such as Cannon and Nikon) have built their digital SLR models to be compatible with 35 mm SLR lenses. This is a great benefit to professional photographers who can select a model that is compatible with their existing lenses. Digital SLR cameras provide an amazing amount of creative control over the camera and the resulting images. Also, these cameras provide resolutions over 6 Megapixels and can produce prints of poster size without noticable pixelazation. A Digital SLR camera (without lenses) is a pricey investment indeed. Currently, you can expect to spend $1500 or more on the camera alone. Accessories and lenses cost extra.

7.7.3 Movie Camera

The movie camera is a type of photographic camera which takes a rapid sequence of photographs on film; once developed this film can be projected as a motion picture. In contrast to a still camera which captures a single snapshot at a time, the movie camera takes a series of images, each called a "frame". This is accomplished through an intermittent mechanism. The frames are later played back in a movie projector at a specific speed, called the "frame rate" (number of frames per second) to give the illusion of motion. Our eyes and brain merge the separate pictures together to generate the illusion, a phenomenon called the "persistence of vision".

Fig. 7.5: Movie Camera

Most of the optical and mechanical elements of a movie camera are present in the movie projector. The camera will not have an illumination source and will maintain its film stock in a light-tight enclosure. A camera will also have exposure control via an iris aperture located on the lens. Otherwise, the requirements for film tensioning, take-up, intermittent motion, loops, and rack positioning are almost identical. See the movie projector article for these details and for the various film formats used.

Video cameras (not movie cameras) can broadly be classified into two categories:

- Analog and
- Digital.

Analog cameras are of the "tube" type; they are bulky and generally used in professional studios. Digital cameras are of the "CCD type"; they are light and portable, and generally used for home/office purposes.

The normal frame rate for commercial film is 24 frames per second (fps). The standard commercial, *i.e.* movie theater film width is 35 millimeters, while many other film formats exist. The standard aspect ratios are 1.66, 1.85, and 2.39 (anamorphic widescreen). NTSC video (common in the U.S. and Japan) plays at 29.97 fps; PAL (common in most other countries) plays at 25 fps. Both of these video formats also has different resolution and color encoding. Many of the technical difficulties involving film and video concern translation between the different formats. Video aspect ratios, always measured as whole numbers, are 4:3 for fullscreen and 16:9 for widescreen.

One of the continuing problems in film is synchronizing a sound recording with the film. No film cameras record sound; instead, the sound is captured by a precision audio device separately. The clapper board which typically starts a take is used as a reference point for the editor to sync the picture to the sound (provided the scene and take are also called out so that the editor knows which picture take goes with any given sound take).

It also permits scene and take numbers and other essential information to be seen on the film itself. Aaton cameras have a system called AatonCode that can "jam sync" with an timecode-based audio recorder and prints a digital timecode directly on the edge of the film itself. However, the most commonly used system at the moment is unique identifier numbers exposed on the edge of the film by the film stock manufacturer (KeyKode is the name for Kodak's system). These are then logged (usually by a computer editing system, but sometimes by hand) and recorded along with audio timecode during editing. In the case of no better alternative, a handclap can work if done clearly and properly, but oftentimes a quick tap on the mike (provided it's in frame for this gesture) is preferred.

Some cameras have low-accuracy ("non-sync" or MOS) film-advance systems. One of the most common uses of these cameras in commercial films are the spring-wound cameras used in hazardous special effects, known as "crash cams". Scenes shot with these have to be kept short, or resynchronized manually with the sound. MOS cameras are also often used for second-unit work or anything involving variable or non-standard speed filming. Due to their non-sync nature, some designs forgo traditional low-noise considerations for a studio camera and thus are quite noisy.

The most popular 35mm cameras in use today are Arriflex, Moviecam (now owned by the Arri Group), and Panavision models. For very high speed filming, PhotoSonics are used.

7.7.4 Using Digital Camera

Digital camera allows you to capture images in digital form and then transfer them to a computer so that you can save or manipulate them using image editing software.

(**Note:** The images used in this tutorial have been taken from the camera manual and modified.)

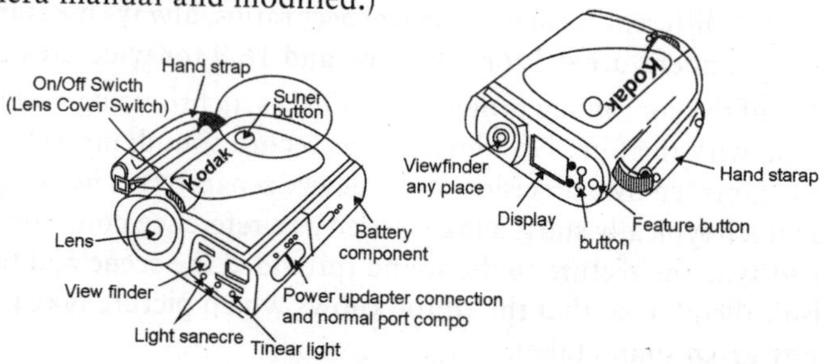

Fig. 7.6: Parts of Digital Camera

To turn on the camera, push the lens cover switch toward the ON symbol.

The lens will be uncovered, the camera will beep and the feature control icons will appear in the display.

To turn off the camera, push the lens cover switch away from the ON symbol.

The camera reverts to an inactive mode if no action is taken for 60 seconds. To return the camera to its active state, press and release the shutter button. Please make sure that your camera is active at the time of taking photograph. To find out whether camera is active, make sure that all icons are displayed on the display panel.

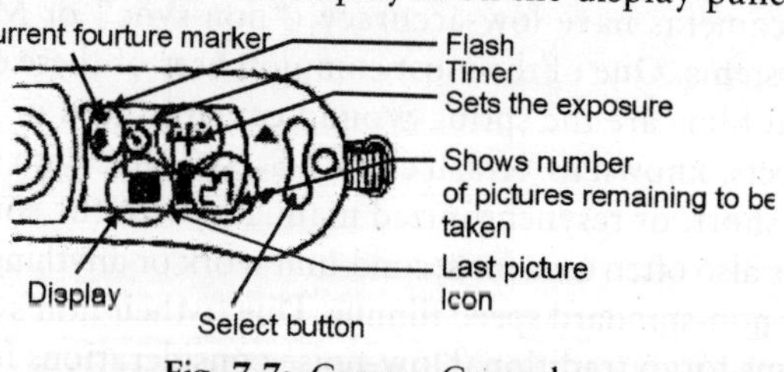

Fig. 7.7: Camera Controls

To access camera controls, use display, select button and feature button of the camera. As part of this exercise, you are not supposed to manipulate controls. However, you are free to do so if you are trying to be extremely creative and have a hang of the various features.

To take pictures, hold the camera so that the handstrap is on the right side and the top of the camera is facing up. Hold the camera up to your face and look through the viewfinder. Frame the subject in the view finder window and press the shutter button on top of the camera.

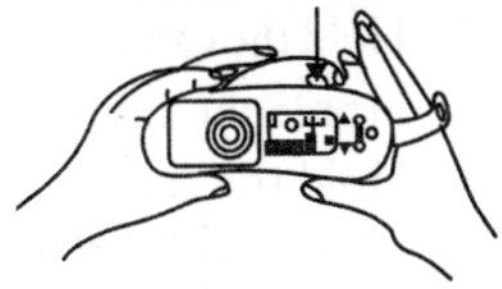

Fig. 7.8: Shutter Button

To connect camera to a Macintosh, turn off the camera by sliding the lens switch to the OFF position. To hook the camera to the Mac in the DURP computer lab near the scanner, use black Macintosh cable, one end of which is plugged in the modem port on the back of the computer.

Lift the rubber tab to uncover the camera's serial port and insert the other end of the cable (that has a label saying 'connect this end to your camera') into it.

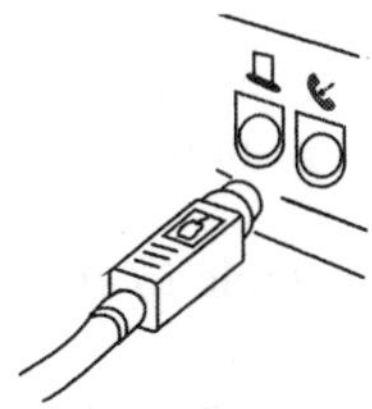
Fig. 7.9: Modem Port

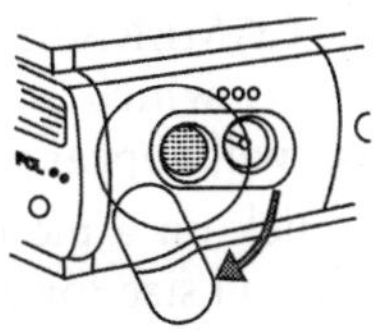
Fig. 7.10: Serial Port

To download pictures from the camera, use the same Mac to which you hooked up your camera. Launch Photo Enhancer from the hardrive. Choose CameraI View Slides in camera. . It will display all the pictures taken by the camera, on the monitor. Choose the one you want to use for your purposes by double-clicking on it. This enlarges the photo to 100 per cent of its size. To save this picture, click File/ Save as. Save it as TIFF file.

After the images are saved in a different location be sure to delete the images from the camera. Delete the images by selecting Camera Controls under the Camera menu. Then click on the Right arrow button until the triangle marker is directly above the black box with an "X" on it. Then click on the down arrow. PhotoEnhancer will ask if you want to delete the images in the camera, click yes. (Thanks to Norene Hough for this set of instructions.)

Taking Pictures with the DC-40

1. Hold the camera so that the handstrap is on the right side and the top of the camera is facing up.

2. Push the fingers of your right hand up through the handstrap, letting the camera rest on your thumb. Your index finger should rest on the shutter button.
3. Hold the camera up to your face and look through the viewfinder.
4. Frame the subject in the viewfinder window.
5. Hold the camera steady and press the shutter button. There is a brief delay between the time the shutter button is pressed and the time the picture is actually taken. Be sure to hold the camera still until you hear the click of the shutter. A red light will appear in the viewfinder. When the red light turns off and the camera beeps, it is ready to take another picture.
6. Take another picture.

Now the DC-40 has taken two pictures.

Downloading images from the DC-40 to a Computer

1. Start PhotoEnhance software.
2. From the CAMERA menu, choose "View Slides in Camera". A slide table, which is a graphic showing miniature camera pictures, will appear.
3. Double-click on a picture from the slide table to enlarge the picture to full size for your viewing.
4. Click the Close Box button in upper left corner of window.
5. Click on an image you want to save.
6. Create a directory (MITDC40).
7. Go to FILE FORMAT pop-up window and select "PICT".
8. Save the this image (as "myimg1" in the directory).
9. Repeat Steps 5 thru 8 for the other image 11. Save the second image (as "myimg2" in the directory)

Now the images have been downloaded from the DC-40 to your computer.

Downloading images from the DC-40

1. Create a directory (MITDC40) directory
2. Start PhotoEnhance software.
3. From the CAMERA menu, choose "View Slides in Camera". A slide table, which is a graphic showing miniature camera pictures, will appear.
4. Double-click on a picture from the slide table to enlarge the picture to full size for your viewing.

5. Click the Close Box button in upper left corner of window
6. Click on an image you want to save
7. Choose "Save As" from the FILE menu
8. Go to FILE FORMAT pop-up window and select "PICT".
9. Save the this image (as "myimg1" in the directory).
10. Repeat Steps 5 thru 8 for the other image 11. Save the second image (as "myimg2" in the directory).

7.7.5 Digital Cameras In Libraries

Librarians are becoming multimedia publishers as never before. Web documents or printed materials may contain staff photographs, book illustrations (and illustrations of books), building layouts and images of the many services and facilities available to the technology-hungry users. Presentations, to conferences, to colleagues or to senior management, are increasingly expected to include electronic illustrations or slide shows.

Whether in an academic, public or special library, the potential uses of digital images are limited (so the saying goes) only by your imagination:visual database of staff or users, photos for ID cards or administrative records a digital catalogue of images for quick reference, perhaps of your map or poster collection images to add to publicity, promotional and user support documentation, whether on the Web or on printed materials adding screen images to reports, to OHPs for training, and to Powerpoint presentations

Converting archive or Database Material to CD-Rom format

To meet this demand for images, there is a range of options:

- images bought in as clip art on disk;
- ordinary' photographs scanned by a bureau or inhouse scanning service;
- images 'grabbed' using video capture software;
- photographs converted to PhotoCD for mat, thus made available for electronic manipulation and use; and
- images taken by digital cameras.

Although all these options require financial and physical resources at some level, digital cameras offer an imaging solution that is comparatively cheap and simple.

7.5.1 Digital Photography

Digital cameras combine some features of traditional cameras with some of scanners. The lens, aperture and shutter focus the image, but on to a Charge Coupled Device (CCD) instead of light-sensitive silver halide film. Digital data is compressed and recorded either on to the integral memory chip or on to a removable PCMCIA memory card. This can then be downloaded to a PC running the relevant software, manipulated and output as needed, to screen or printer, or saved to disk, stored or sent as an e-mail attachment.

In practical terms, at this level at least, the cameras are 'autofocus', with distance and light calculated by the camera itself—no need for fiddling with controls. This automatic adjustment causes a slight time-delay between the pressing of the shutter button and the recording of the image, as the camera alters its settings. This has implications for photographing moving images.

Once taken, there is another few seconds delay as the image is converted to digital form and compressed, then written to the memory, again with implications for those wanting 'rapid-fire' sequences of images. Some high-end cameras can record images at the rate of over two per second - as ever it is a question of setting price against performance.

7.7.6 Formats

Digital cameras save images in a range of formats. The Casio reviewed here, for example, initially saves images in Cam format, but the supplied QV-Link software supports conversion to BMP, Tiff and JPeg. Using standard graphics software, such as Adobe Photoshop or Paint Shop Pro, these can be converted to any desired format.

Like many digital cameras, the Kodak DC20 is supplied with a Twain driver; the Twain standard allows images to be imported directly into Photoshop, Paint Shop Pro or similar compatible programs, allowing the camera to be used almost as though it were a scanner.

7.7.7 Quality Issues

Digital cameras produce images with lower resolution than those from scanners, meaning smaller file sizes, but at a cost in quality. If the images are for screen display, however, on Web pages or Powerpoint presentations for example, then the VGA standard of cameras like the Casio QV-100 is enough for most needs. High resolution images from the Casio were 45-55K of file space, with low resolution images about one-third of that.

Low-light conditions where a flash cannot be used may favour the higher-end digital camera over SLRs, due to the consistent performance of pixel resolution in all conditions, compared with the reaction of film to exposure times and settings. SLR cameras can compensate by using longer exposure times, but this then may need a tripod to avoid camera motion. Using a flash, where available, improves the overall colour and contrast of digital images, while post-transfer enhancement using the software allows for improvements to pictures taken without flash in surprisingly dark conditions.

Digital cameras have a clear advantage over other technologies when it comes to ease of use, portability, equipment cost, and the ability to capture real-life images. Incidentally, they are also fun to use, with the benefit of immediate feedback rather than waiting for the film to be developed.

Most camera manufacturers now have digital versions in their range. Prices vary from 69 (the monochrome Connectix QuickCam) to 24,950 and up (Kodak DCS-460). A growing number of companies supply digital cameras priced below 500, a semi-arbitrary point below which it may be worth buying equipment for experimental or even personal use. Suppliers in this price band include Canon, Agfa, Sanyo, Kodak, Casio and others.

8.1 DATA WAREHOUSING

Data warehousing is the enterprise-wide transformation of historical raw data into business intelligence. To compete and grow in today's dynamic business market, decision-makers need to have their hands directly on the pulse of the business. Business strategy is formed. Plans are put into action. Actions are measured. Analysts need flexible, immediate access to high quality, actionable information that can be used to measure today's performance and formulate tomorrow's winning strategies. Data warehousing provides the information critical to business survival.

The warehousing process requires standardization, transformation and integration of shared data extracted from heterogeneous operational systems. As we go through the data acquisition process, we gain an enormous amount of knowledge about the extent of integration, and level of quality of source system data. To gain full benefit from data warehousing initiatives, we need to leverage that knowledge to improve the architecture of the operational systems. We need to establish a "closed loop" IT architecture assessment process, where the information that the warehousing team gains in the data acquisition phase can be used as input to the operational systems assessment and improvement process. This requires that we need to think of the systems—operational and warehouse — as one enterprise information systems architecture.

All uses of enterprise information require high quality data—planning, analysis and control. Data has high quality when it serves its business purpose, satisfies quality characteristics, is shareable and its management is cost effective. The operational systems may, but do not necessarily, manage data quality or enforce all required business rules.

Data Warehousing is open to an almost limitless range of definitions. Simply put, Data Warehouses store an aggregation of a company's data.

Data Warehouses are an important asset for organizations to maintain efficiency, profitability and competitive advantages. Organizations collect data through many sources—Online, Call Center, Sales Leads, Inventory Management. The data collected have degrees of value and business relevance. As data is collected, it is passed through a 'conveyor belt', call the Data Life Cycle Management.

A data warehouse is, primarily, a record of an enterprise's past transactional and operational information, stored in a database designed to favour efficient data analysis and reporting (especially OLAP). Data warehousing is not meant for current "live" data.

Data warehouses often hold large amounts of information which are sometimes subdivided into smaller logical units called dependent data marts.

Usually, two basic ideas guide the creation of a data warehouse:

Integration of data from distributed and differently structured databases, which facilitates a global overview and comprehensive analysis in the data warehouse.

Separation of data used in daily operations from data used in the data warehouse for purposes of reporting, decision support, analysis and controlling.

Periodically, one imports data from enterprise resource planning (ERP) systems and other related business software systems into the data warehouse for further processing. It is common practice to "stage" data prior to merging it into a data warehouse. In this sense, to "stage data" means to queue it for preprocessing, usually with an ETL tool. The preprocessing program reads the staged data (often a business's primary OLTP databases), performs qualitative preprocessing or filtering (including denormalization, if deemed necessary), and writes it into the warehouse.

Scope of Warehousing

Data Warehousing is a Subject-oriented, integrated, time-variant, nonvolatile collection of data in support of management's decision-making process.

- **Subject-oriented:** Data organized by entity rather than by the application that uses it.
- **Integrated:** Data stored in place in the data warehouse, even though it originates from everywhere in the organization and from a variety of external sources.
- **Time-variant:** Data represents a snapshot at 1 point of time in the past, unlike an operational application whose data is accurate as of the moment.
- **Non-volatile:** Data is read-only; data is loaded into it periodically but users cannot update a data warehouse directly.
- Making better business decisions quickly is the key to succeeding in today's competitive marketplace. Organizations seeking to improve their decision-making ability can be overwhelmed by the sheer volume and complexity of data available from their varied operational and production systems. Making this data accessible is one of the most significant challenges for today's information technology professionals.
- In response to this challenge, many organizations choose to build a data warehouse to unlock the information in their operational systems. A data warehouse is an integrated store of information collected from other systems that becomes the foundation for decision support and data analysis. While there are many types of data warehouses, based on different design methodologies and philosophical approaches, they all have these common traits.

- Information is organized around the major subjects of the enterprise (for example, customers, products, sales, or vendors), reflecting a data-driven design.
- Raw data is gathered from nonintegrated operational and legacy applications, cleansed, and then summarized and presented in a way that makes sense to end users.
- Based on feedback from users and discoveries in the data warehouse, the data warehouse architecture will change over time, reflecting the iterative nature of the process.

The data warehousing process is inherently complex and, as a result, is costly and time-consuming. Over the past several years, Microsoft has been working within the software industry to create a data warehousing platform that consists of both component technology and leading products that can be used to lower the costs and improve the effectiveness of data warehouse creation, administration, and usage. Microsoft has also been developing a number of products and facilities, such as Microsoft SQL Server version 7.0, that are well suited to the data warehousing process. Coupled with third-party products that can be integrated using the Microsoft Data Warehousing Framework, customers have a large selection of interoperable, best-of-breed products from which to choose for their data warehousing needs.

A Data Warehouse is a complete architecture, and as such, requires a rigorous, yet iterative design approach and development methodology to ensure successful deployment. This presentation will focus on architecture considerations, design principles, and a development methodology for data warehouse construction and Discoverer 3.1 metadata build. Following this solid "Data Driven" methodology can ensure the success of your Oracle Data Warehousing project.

In contrast with the OLTP environment which is a volatile transactional data entry-oriented environment, data warehouses are much more historical in nature, tend to be measured in terabytes of information, are organized and queried by subject rather than process, integrated within the enterprise, and are accessed by all levels within the organization. The huge volume of data and data sources inherent in the data warehouse necessitates a solid requirements oriented approach to its construction and a much different development methodology than the typical life-cycle approach favored in building OLTP systems.

A fully functional data warehouse is an architected environment that typically summarizes data to the lowest common denominator or atomic level to provide the informational requirements of an organization. A data warehouse is an architecture comprised of many integrated enterprise components based on data and technology.

Data Mart

A data mart is a departmental instance of a data warehouse structured to meet the focused informational requirements of a distinct group within the organization. Organizations often impatiently embark on a data mart development effort as a low cost way to quickly get at their enterprise data. Data marts are often developed without considering an enterprise architecture approach. As a result, it is often difficult to draw many data marts together into a corporate data warehouse due to this lack of an enterprise strategy.

It is imperative that before embarking on either a departmental data mart or an enterprise data warehouse that the development team considers the proposed system and data architecture with an eye toward a fully corporate view of all data within the organization. The development methodology and standards to follow during design and development of either a data mart or data warehouse should be flexible enough to encompass the current or future needs of all departments in the enterprise.

8.2 DATA WAREHOUSING PROCESS

From the information technology perspective, data warehousing is aimed at the timely delivery of the right information to the right individuals in an organization. This is an ongoing process, not a one-time solution, and requires a different approach from that required in the development of transaction-oriented systems.

A data warehouse is a collection of data in support of management's decision-making process that is subject-oriented, integrated, time-variant, and non-volatile. The data warehouse is focused on the concept (for example, sales) rather than the process (for example, issuing invoices). It contains all the relevant information on a concept gathered from multiple processing systems. This information is collected and stored at regular intervals and is relatively stable.

A data warehouse integrates operational data by using consistent naming conventions, measurements, physical attributes, and semantics. The first step in the physical design of the data warehouse are determining what subject areas should be included and developing a set of agreed-upon definitions. This requires interviewing users, analysts, and executives in order to understand and document the scope of the information requirements. The issues must be thoroughly understood before the logical process can be translated into a physical data warehouse.

Following the physical design, operational systems are put in place to populate the data warehouse. Because the operational systems and the data warehouse contain different representations of the data, populating the data warehouse requires transformations of the data: summarizing, translating, decoding, eliminating invalid data, and so on. These processes need to be automated so that they can be performed on an ongoing basis: extracting, transforming, and moving the source data as often as needed to meet the business requirements of the data warehouse.

In the operational system, data is current-valued and accurate as of the moment of access. For example, an order entry application always shows the current value of inventory on hand for each product. This value could differ between two queries issued just moments apart. In the data warehouse, data represents information gathered over a long period, and is accurate as of a particular point in time. In effect, the data warehouse contains a long series of snapshots about the key subject areas of the business.

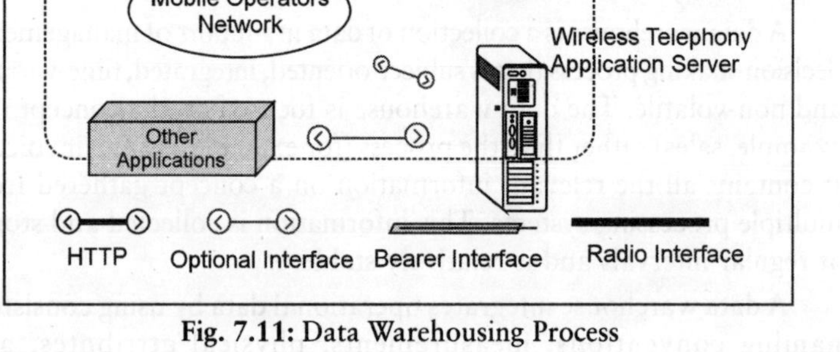

Fig. 7.11: Data Warehousing Process

Finally, information is made available for browsing, analyzing, and reporting. Many tools assist in analysis, from simple report writers to advanced data miners. Ultimately, analysis drives the final iterations of the data warehousing process, causing revisions in the design of the data warehouse to accommodate new information, improve system performance, or allow new types of analysis. With these changes, the process restarts and continues throughout the life of the data warehouse.

8.3 DATA WAREHOUSING ARCHITECTURE

Many methodologies have been proposed to simplify the information technology efforts required to support the data warehousing process on an ongoing basis. This has led to debates about the best architecture for delivering data warehouses in organizations.

Two basic types of data warehouse architecture exist:

- Enterprise data warehouses
- Data marts.

The enterprise data warehouse contains enterprise-wide information integrated from multiple operational data sources for consolidated data analysis. Typically, it is composed of several subject areas, such as customers, products, and sales, and is used for both tactical and strategic decision-making. The enterprise data warehouse contains both detailed point-in-time data and summarized information, and can range in size from 50 gigabytes (GB) to more than 1 terabyte. Enterprise data warehouses can be very expensive and time-consuming to build and manage. They are usually created from the top down by centralized information services organizations.

The data mart contains a subset of enterprise-wide data that is built for use by an individual department or division in an organization. Unlike the enterprise data warehouse, the data mart is usually built from the bottom up by departmental resources for a specific decision-support application or group of users. Data marts contain summarized and often detailed data about a subject area. The information in the data mart can be a subset of an enterprise data warehouse (dependent data mart) or can come directly from the operational data sources (independent data mart).

Enterprise data warehouses and data marts are constructed and maintained through the same iterative process described earlier. Furthermore, both approaches share a similar set of technological components.

8.4 Data Warehousing Components

A data warehouse always consists of a number of components, including:

- Operational data sources.
- Design/development tools.
- Data extraction and transformation tools.
- Database management system (DBMS).
- Data access and analysis tools.
- System management tools.

Several years ago, Microsoft recognized the need for a set of technologies that would integrate these components. This led to the creation of the Microsoft Data Warehousing Framework, a roadmap not only for the development of Microsoft products such as SQL Server 7.0, but also for the technologies necessary to integrate products from other vendors.

8.5 REASONS FOR WAREHOUSING

The following is a list of the basic reasons why organizations implement data warehousing. This list was put together because too much of the data warehousing literature confuses "next order" benefits with these basic reasons. For example, spend a little time reading data warehouse trade material and you will read about using a data warehouse to "convert data into business intelligence", "make management decision making based on facts not intuition", "get closer to the customers", and the seemingly ubiquitously used phrase "gain competitive advantage". In probably 99 per cent of the data warehousing implementations, data warehousing is only one step out of many in the long road toward the ultimate goal of accomplishing these highfalutin objectives.

The basic reasons organizations implement data warehouses are:

- To perform server/disk bound tasks associated with querying and reporting on servers/disks not used by transaction processing systems.
- Most firms want to set up transaction processing systems so there is a high probability that transactions will be completed in what is judged to be an acceptable amount of time. Reports and queries, which can require a much greater range of limited server/disk resources than transaction processing, run on the servers/disks

used by transaction processing systems can lower the probability that transactions complete in an acceptable amount of time. Or, running queries and reports, with their variable resource requirements, on the servers/disks used by transaction processing systems can make it quite complex to manage servers/disks so there is a high enough probability that acceptable response time can be achieved. Firms therefore may find that the least expensive and/or most organizationally expeditious way to obtain high probability of acceptable transaction processing response time is to implement a data warehousing architecture that uses separate servers/disks for some querying and reporting.

- To use data models and/or server technologies that speed up querying and reporting and that are not appropriate for transaction processing.
- There are ways of modeling data that usually speed up querying and reporting (*e.g.*, a star schema) and may not be appropriate for transaction processing because the modeling technique will slow down and complicate transaction processing. Also, there are server technologies that that may speed up query and reporting processing but may slow down transaction processing (*e.g.*, bit-mapped indexing) and server technologies that may speed up transaction processing but slow down query and report processing (*e.g.* technology for transaction recovery.)—Do note that whether and by how much a modeling technique or server technology is a help or hindrance to querying/reporting and transaction processing varies across vendors' products and according to the situation in which the technique or technology is used.
- To provide an environment where a relatively small amount of knowledge of the technical aspects of database technology is required to write and maintain queries and reports and/or to provide a means to speed up the writing and maintaining of queries and reports by technical personnel.
- Often a data warehouse can be set up so that simpler queries and reports can be written by less technically knowledgeable personnel. Nevertheless, less technically knowledgeable personnel often "hit a complexity wall" and need IS help. IS, however, may also be able to more quickly write and maintain queries and reports written against data warehouse data. It should be noted, however, that much of the improved IS productivity probably comes from the

lack of bureaucracy usually associated with establishing reports and queries in the data warehouse.

- To provide a repository of "cleaned up" transaction processing systems data that can be reported against and that does not necessarily require fixing the transaction processing systems.
- The data warehouse provides an opportunity to clean up the data without changing the transaction processing systems. Note, however, that some data warehousing implementations provide a means to capture corrections made to the data warehouse data and feed the corrections back into transaction processing systems. Sometimes it makes more sense to handle corrections this way than to apply changes directly to the transaction processing system.
- To make it easier, on a regular basis, to query and report data from multiple transaction processing systems and/or from external data sources an/or from data that must be stored for query/report purposes only.

  For a long time firms that need reports with data from multiple systems have been writing data extracts and then running sort/merge logic to combine the extracted data and then running reports against the sort/merged data. In many cases this is a perfectly adequate strategy. However, if a company has large amounts of data that need to be sort/merged frequently, if data purged from transaction processing systems needs to be reported upon, and most importantly, if the data need to be "cleaned", data warehousing may be appropriate.
- To provide a repository of transaction processing system data that contains data from a longer span of time than can efficiently be held in a transaction processing system and/or to be able to generate reports "as was" as of a previous point in time

  Older data are often purged from transaction processing systems so the expected response time can be better controlled. For querying and reporting, this purged data and the current data may be stored in the data warehouse where there presumably is less of a need to control expected response time or the expected response time is at a much higher level. —As for "as was" reporting, some times it is difficult, if not impossible, to generate a report based on some characteristic at a previous point in time. For example, if you want a report of the salaries of employees at grade Level 3 as of the beginning of each month in 1997, you may not be able to do this because you only have a record of current employee grade level. To be able to handle this type of reporting problem, firms may implement data warehouses that handle what is called the "slowly changing dimension" issue.

- To prevent persons who only need to query and report transaction processing system data from having any access whatsoever to transaction processing system databases and logic used to maintain those databases.

The concern here is security. For example, data warehousing may be interesting to firms that want to allow report and querying only over the Internet.

Some firms implement data warehousing for all the reasons cited. Some firm implement data warehousing for only one of the reasons cited.

If you examine the list you may be struck that need for data warehousing is mainly caused by the limitations of transaction processing systems. These limitations of transaction processing systems are not, however, inherent. That is, the limitations will not be in every implementation of a transaction processing system. Also, the limitations of transaction processing systems will vary in how crippling they are.

Finally, to repeat the point I made initially, a firm that expects to get business intelligence, better decision making, closeness to its customers, and competitive advantage simply by plopping down a data warehouse is in for a surprise. Obtaining these next order benefits requires firms to figure out, usually by trial and error, how to change business practices to best use the data warehouse and then to change their business practices. And that can be harder than implementing a data warehouse.

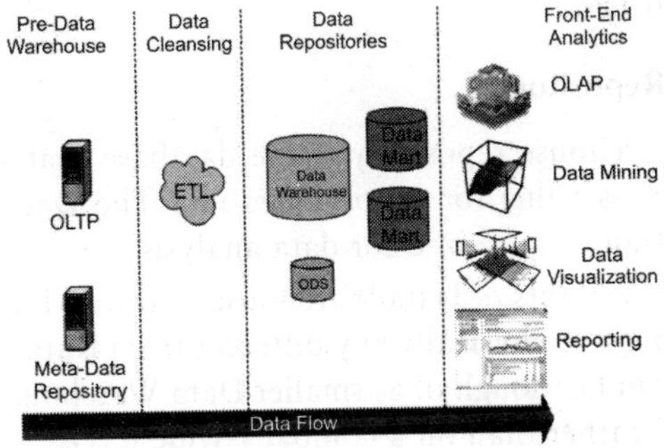

Fig.7.12: Overview of Data Warehousing Infrastructure

8.6 DATA LIFE CYCLE

An organization's data life cycle management's policy will dictate the data warehousing design and methodology.

The goal of Data Warehousing is to generate front-end analytics that will support business executives and operational managers.

8.6.1 Pre-Data Warehouse

The pre-Data Warehouse zone provides the data for data warehousing. Data Warehouse designers determine which data contains business value for insertion.

OLTP databases are where operational data are stored. OLTP databases can reside in transactional software applications such as Enterprise Resource Management (ERP), Supply Chain, Point of Sale, Customer Serving Software. OLTPs are design for transaction speed and accuracy.

Metadata ensures the sanctity and accuracy of data entering into the data lifecycle process. Meta-data ensures that data has the right format and relevancy. Organizations can take preventive action in reducing cost for the ETL stage by having a sound Metadata policy. The commonly used terminology to describe meta data is "data about data".

8.6.2 Data Cleansing

Before data enters the data warehouse, the extraction, transformation and cleaning (ETL) process ensures that the data passes the data quality threshold.

ETLs are also responsible for running scheduled tasks that extract data from OLTPs.

8.6.3 Data Repositories

The Data Warehouse repository is the database that stores active data of business value for an organization. The Data Warehouse modeling design is optimized for data analysis.

There are variants of Data Warehouses—Data Marts and ODS. Data Marts are not physically any different from Data Warehouses. Data Marts can be though of as smaller Data Warehouses built on a departmental rather than on a company-wide level.

Data Warehouses collects data and is the repository for historical data. Hence it is not always efficient for providing up-to-date analysis. This is where ODS, Operational Data Stores, come in. ODS are used to hold recent data before migration to the Data Warehouse.

ODS are used to hold data that have a deeper history that OLTPs. Keep large amounts of data in OLTPs can tie down computer resources and slow down processing - imagine waiting at the ATM for 10 minutes between the prompts for inputs. .

8.6.4 Front-End Analysis

The last and most critical potion of the Data Warehouse overview are the front-end applications that business users will use to interact with data stored in the repositories.

Data Mining is the discovery of useful patterns in data. Data Mining are used for prediction analysis and classification—*e.g.* what is the likelihood that a customer will migrate to a competitor.

OLAP, Online Analytical Processing is used to analyze historical data and slice the business information required. OLAPs are often used by marketing managers. Slices of data that are useful to marketing managers can be—How many customers between the ages 24–45, that live in New York state, buy over $ 2000 worth of groceries a month?

Reporting tools are used to provide reports on the data. Data are displayed to show relevancy to the business and keep track of key performance indicators (KPI).

Data visualization tools is used to display data from the data repository. Often data visualization is combined with Data Mining and OLAP tools. Data visualization can allow the user to manipulate data to show relevancy and patterns.

8.7 Introduction to Metadata

The primary rational for data warehousing is to provide businesses with analytics results from data mining, OLAP, scorecarding and reporting. The cost of obtaining front-end analytics are lowered if there is consistent data quality all along the pipeline from data source to analytical reporting.

Metadata is about controlling the quality of data entering the data stream.

Batch processes can be run to address data degradation or changes to data policy. Metadata policies are enhanced by using metadata repositories.

One of the projects we recently worked on was with a major insurance company in North America. The company had amalgamated over the years with acquisitions and also had developed external back-end data integrations to banks and reinsurance partners.

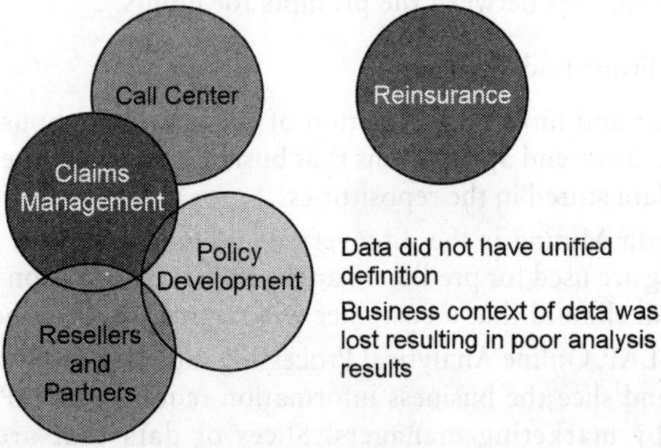

Fig. 7.13: Disparate Data Definition Policies in an Insurance Company

The client approached DWreview as they felt that they were not obtaining sufficient return-on-investments on their data warehouse. Prediction analysis, profit-loss ratio and OLAP reports were labor and time intensive to produce. The publicly listed insurance company was also in the process of implementing a financial scorecarding application to monitor compliance with the Sarbanes-Oxley act.

In consultation with the company's IT managers, we analyzed the potential quid-pro-quos of different design changes.

The first step in the process of realignment the data warehousing policies was the examination of the metadata policies and deriving a unified view that can work for all stakeholders. As the company was embarking on a new Scorecarding initiative it became feasible to bring the departments together and propose a new enterprise-wide metadata policy.

Departments had created their own data marts for generating quick access to reports as they had felt the central data warehouse was not responsive to their needs. This also created a bottleneck as data was not always replicated between the repositories.

With the IT manager's approval and buy-in of departmental managers, a gradual phase in of a company-wide metadata initiative was introduced. Big bang approaches rarely work—and the consequences are extremely high for competitive industries such as insurance.

The metaphor we used for the project was the quote from Julius Cesar by Shakespeare given at the start of the article. We felt that this was a potentially disruptive move but if the challenges were met positively, the rewards would be just.

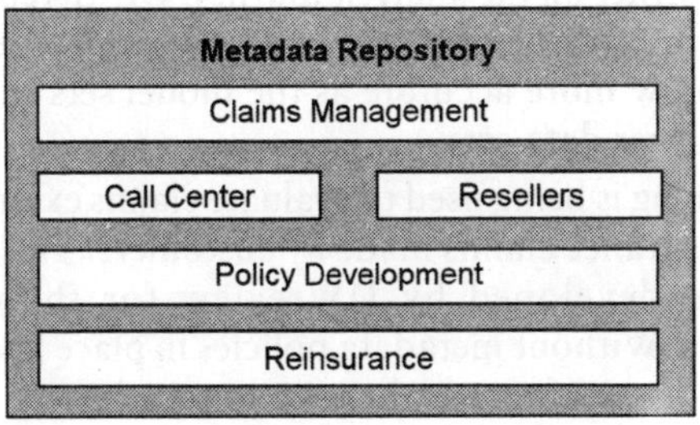

Fig. 7.14: Company-wide Metadata Policy

Industry metadata standards exists in industry verticals such as insurance, banks, manufacturing. OMG's Common Warehouse Metadata Initiative (CWMI) is a vendor back proposal to enable easy interchange of metadata between data warehousing tools and metadata repositories in distributed heterogeneous environments.

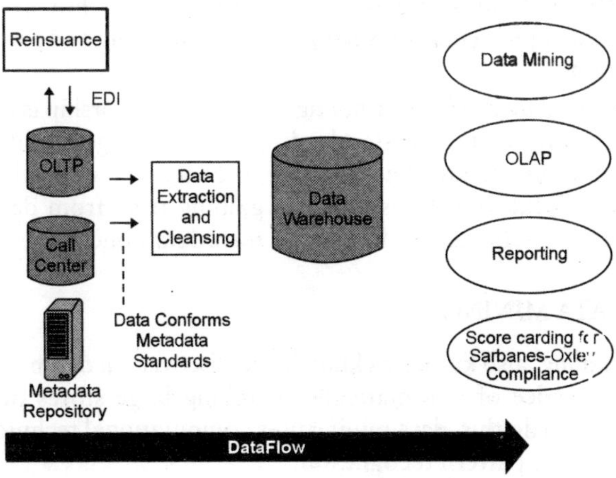

Fig. 7.15: Partial Schematic Overview of Data Flow after Company-wide Metadata Implementation

In the months since the implementation, the project has been moving along smoothly. There were training seminars given to keep staff abreast on the development and the responses were overwhelmingly positive.

The implementation of the Sarbanes-Oxley Scorecarding initiative was on time and relatively painless. Many of the challenges that would have been faced without a metadata policy were avoided.

With a unified data source and definition, the company is embarking further on the analysis journey. OLAP reporting is moving across stream with greater access to all employees. Data mining models are now more accurate as the model sets can be scored and trained on larger data sets.

Text mining is being used to evaluate claims examiners comments regarding insurance claims made by customers. The text mining tool was custom developed by DWreview for the client's unique requirements. Without metadata policies in place it would be next to impossible to perform coherent text mining. The metadata terminologies used in claims examination were developed in conjunction with insurance partners and brokers. Using the text mining application, the client can now monitor consistency in claims examination, identify trends for potential fraud analysis and provide feedback for insurance policy development.

Developing metadata policies for organizations falls into three project management spheres—generating project support, developing suitable guidelines and setting technical goals. For a successful metadata implementation strong executive backing and support must be obtained.

A tested method for gathering executive sponsorship is first setting departmental metadata standards and evaluating the difference in efficiency. As metadata is abstract in concept a visceral approach can be helpful. It will also help in gaining trust from departments that may be reluctant to hand over metadata policies.

8.8 DATA MINING

Data mining, also known as knowledge-discovery in databases (KDD), is the practice of automatically searching large stores of data for patterns. To do this, data mining uses computational techniques from statistics and pattern recognition.

Data mining has been defined as "The non-trivial extraction of implicit, previously unknown, and potentially useful information from data" and "The science of extracting useful information from large data sets or databases" Although it is usually used in relation to analysis of data, data mining, like artificial intelligence, is an umbrella term and is used with varied meaning in a wide range of contexts.

A simple example of data mining is its use in a retail sales department. If a store tracks the purchases of a customer and notices that a customer buys a lot of silk shirts, the data mining system will make a correlation between that customer and silk shirts. The sales department will look at that information and may begin direct mail marketing of silk shirts to that customer, or it may alternatively attempt to get the customer to buy a wider range of products. In this case, the data mining system used by the retail store discovered new information about the customer that was previously unknown to the company. Another widely used (though hypothetical) example is that of a very large North American chain of supermarkets. Through intensive analysis of the transactions and the goods bought over a period of time, analysts found that beers and diapers were often bought together. Though explaining this interrelation might be difficult, taking advantage of it, on the other hand, should not be hard (*e.g.* placing the high-profit diapers next to the high-profit beers). This technique is often referred to as "Market Basket Analysis".

8.9 DATA DREDGING

Used in the technical context of data warehousing and analysis, the term "data mining" is neutral. However, it sometimes has a more pejorative usage that implies imposing patterns (and particularly causal relationships) on data where none exist. This imposition of irrelevant, misleading or trivial attribute correlation is more properly criticized as "data dredging" in the statistical literature. Another term for this misuse of statistics is data fishing.

Used in this latter sense, data dredging implies scanning the data for any relationships, and then when one is found coming up with an interesting explanation. (This is also referred to as "overfitting the model".) The problem is that large data sets invariably happen to have some exciting relationships peculiar to that data. Therefore, any conclusions reached are likely to be highly suspect. In spite of this, some exploratory data work is always required in any applied statistical analysis to get a feel for the data, so sometimes the line between good statistical practice and data dredging is less than clear.

The common approach, in data mining, to overcoming the problem of overfitting is to separate the data into two or three separate data sets (called the training set, validation set, and testing set). The model is built using the training and validation set, and is then tested using the testing set; the procedure can be repeated many times by resembling the data sets, in order to be more certain that a real pattern has been found and that the model is not merely capitalizing on random chance (*i.e.* overfitting).

A more significant danger is finding correlations that do not really exist. Investment analysts appear to be particularly vulnerable to this. "There have always been a considerable number of pathetic people who busy themselves examining the last thousand numbers which have appeared on a roulette wheel, in search of some repeating pattern. Sadly enough, they have usually found it." However, when properly done, determining correlations in Investment analysis has proven to be very profitable for statistical arbitrage operations (such as pairs trading strategies), and furthermore correlation analysis has shown to be very useful in risk management. Indeed, finding correlations in the financial markets, when done properly, is not the same as finding false patterns in roulette wheels.

Most data mining efforts are focused on developing a finely-grained, highly detailed model of some large data set. Other researchers have described an alternate method that involves finding the minimal differences between elements in a data set, with the goal of developing simpler models that represent relevant data.

8.10 PRIVACY CONCERNS

There are also privacy concerns associated with data mining. For example, if an employer has access to medical records, they may screen out people who have diabetes or have had a heart attack. Screening out such employees will cut costs for insurance, but it creates ethical and legal problems.

Data mining government or commercial data sets for national security or law enforcement purposes has also raised privacy concerns.

There are many legitimate uses of data mining. For example, a database of prescription drugs taken by a group of people could be used to find combinations of drugs with an adverse reactions. Since the combination may occur in only 1 out of 1000 people, a single case may not be apparent. A project involving pharmacies could reduce the number of drug reactions and potentially save lives. Unfortunately, there is also a huge potential for abuse of such a database.

Essentially, data mining gives information that wouldn't be available otherwise. It must be properly interpreted to be useful. When the data collected involves individual people, there are many questions concerning privacy, legality, and ethics.

In statistical analyses in which there is no underlying theoretical model, data mining is often approximated via stepwise regression methods wherein the space of 2K possible relationships between a single outcome variable and k potential explanatory variables is smartly searched. With the advent of grid computing, it became possible (when k is less than approximately 40) to examine all 2K models. This procedure is called all subsets or exhaustive regression. Some of the first applications of exhaustive regression involved the study of clinical data.

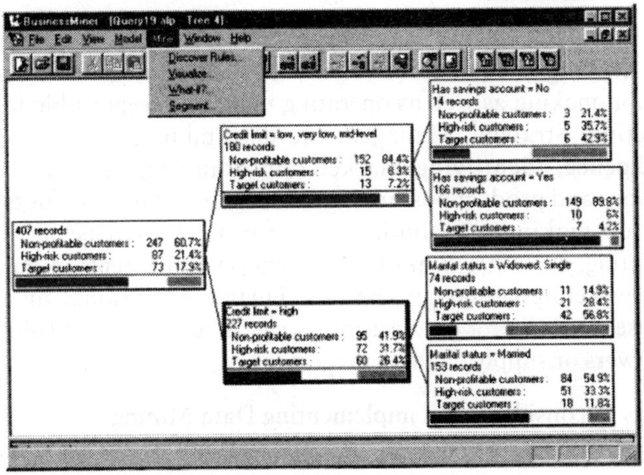

Fig. 7.16: Data Mining

Data Mining

Exploring and analyzing detailed business transactions. It implies "digging through tons of data" to uncover patterns and relationships contained within the business activity and history. Data mining can be done manually by slicing and dicing the data until a pattern becomes obvious. Or, it can be done with programs that analyze the data automatically. Data mining has become an important part of customer relationship management (CRM). In order to better understand customer behavior and preferences, businesses use data mining to wade through the huge amounts of information gathered via the Web.

What is Data Mining?

Data Mining is the process of analyzing large data sets in order to find patterns that can help to isolate key variables to build predictive models for management decision-making. In essence, data mining helps businesses to optimize their processes so that their customers receive the most relevant services and the costs of serving them are proportionate to the value of the profits earned from them, a company's exposure to risk is proportionate to premiums earned, *etc.* Data Mining enables companies to segment their customer base and to tailor products and services to the needs and purchasing power of individual groups of customers.

Who is Data Mining for?

Data Mining is for executives involved in strategic and tactical decision-making as well as operating managers responsible for cost reduction. Strategic Managers use data mining for competitive intelligence, identifying market opportunities, product launch decisions and product positioning. Managers responsible for tactical decision making use similar tools for sales forecasting, direct marketing, customer acquisition, retention and extension purposes and marketing campaign analysis. Finally, operational managers can use similar data for decisions such as the choice of sub-prime borrowers or supply chain management.

Things to consider when implementing Data Mining

- **Data Quality:** It's worthwhile remembering the adage "Garbage in and Garbage out" when implementing data mining. Poor quality of data can jeopardize any attempt to use data analysis for decision-making. This is especially true when data is purchased from external vendors.
- **Size of the Database:** Data mining comes in a variety shapes and forms depending on primarily the size of databases that have to

be analyzed. Smaller companies cannot afford packages, such as SAS, which are expensive to license. Excel, with its advanced data functions, is adequate for small companies and Access for data storage. Eventually, databases inevitably grow in size and companies have to plan for migration to more complex data mining tools.

Nature of Application

The choice of data mining methods depends a great deal on the kind of decision-making that a company wants to achieve. Sales related decision-making is best undertaken with OLAP tools such as those offered by Cognos or Siebel. Companies looking to do predictive modeling are better-off with SAS or SPSS. Similarly, data processing can be done with a variety of databases ranging from Access for small databases, MS SQL for medium size operations to more complex enterprises types like Oracle and Teradata for data coming from networks like credit card operations.

- **Human Resources:** Data mining requires a great deal of ingenuity that highly educated people can provide. Typically, the people hired for data mining have a graduate degree and bigger companies in the financial services companies hire Ph.D holders Data Mining requires a portfolio of skills in data management, statistical analysis and business decision-making which are hard to find.

  Data mining is an aide to strategic, tactical and operational decision-making in situations where numerous variables, affecting costs or benefits, impinge on the eventual outcome of the course of action that a company might decide to take. The modeling that accompanies data mining assimilates the information on costs and benefits of alternative courses of action as visualized in the form of familiar decision trees. Companies use such information to find new opportunities for growth, choose more effective means to achieve their business goals and streamline business processes to lower their costs.

  Customer churn is one of the several examples of complex business problems that data mining addresses. A great deal of marketing activity involves customer acquisition, retention and extension and companies invest considerable resources to achieve this. When all the competing companies in an industry exert themselves to woo customers, churn rates increase and higher costs have to be incurred to attract customers and to keep them. The benefits of attracting customers in the telecommunications industry, for example, depend on the length of time they stay with a company, the number and the duration of their calls. The costs, on the other hand, depend on network costs which are higher if interconnection charges are incurred or more outbound calls take place, time of day for calls, *i.e.*, costs are higher at peak time, customer service support and the duration of calls (shorter calls mean higher costs). None of these variables can be readily estimated at the outset.

Data mining uses statistical techniques, such as survival analysis, to determine the length of time for which a customer can be expected to stay with a company. Based on the profile of groups of customers, as indicated by demographics, psychographic, price sensitivity and knowledge of alternative vendors, the length of their expected stay with a company can be estimated. Similarly, the calling behavior of customers and attendant costs of servicing can be estimated from their profession, personality (introverted or extroverted), social networking behavior. Based on such data, companies can decide on the amount of money they can afford to spend to satisfy customers in order to acquire and retain their customers.

Profiling or segmentation of the customer base is the linchpin of much of the benefits that flow from data mining. The task of profiling consists of identifying homogenous groups of customers who exhibit similar patterns of behavior. Armed with this information, companies learn to target specific customers rather than randomly promote products to all of them. Customers are happier because junk mail otherwise inundates them and is aggravating. Companies don't have to drain their resources by sending mailers to customers who are often so indifferent that they toss an offer into a waste paper basket.

Customer segmentation is also a source of innovation such as in the design of insurance plans in the health industry. In the past, customers had access to only two broad types of insurance plans; the HMO and the PPO. The prices paid by customers have been uniform regardless of their actual use of health services. Consequently, customers are less responsible about their life-styles choices such as smoking, weight, diet and exercise. Worse, responsible customers, who invest in preventive care, by spending on yoga, alternative care, are not rewarded. As a result, health care costs have been rising rapidly without a commensurate improvement in quality. In recent years, however, companies such as WellPoint Health Networks Inc., PacifiCare Health Systems Inc. and Blue Shield of California consumer driven health plans have started to offer health insurance plans tailored to the risk profile of customers. Consumers are allowed to choose their price points and the benefits they receive and are provided information on the web to make a choice of their doctors. This is expected to lower waste in the health industry and encourage preventive care and patients would have a choice to spend on alternative care.

In general, profiling and segmentation of customers helps companies to efficiently align resources with the specific needs of customers. They can set prices, choose channels and design communication strategies based on the character of specific segments of customers.

Index